Edible Oil Structuring

Concepts, Methods and Applications

Food Chemistry, Function and Analysis

Series editors:
Gary Williamson, *University of Leeds, UK*
Alejandro G. Marangoni, *University of Guelph, Canada*
Juliet A. Gerrard, *University of Auckland, New Zealand*

Titles in the series:
1: Food Biosensors
2: Sensing Techniques for Food Safety and Quality Control
3: Edible Oil Structuring: Concepts, Methods and Applications

How to obtain future titles on publication:
A standing order plan is available for this series. A standing order will bring delivery of each new volume immediately on publication.

For further information please contact:
Book Sales Department, Royal Society of Chemistry, Thomas Graham House, Science Park, Milton Road, Cambridge, CB4 0WF, UK
Telephone: +44 (0)1223 420066, Fax: +44 (0)1223 420247
Email: booksales@rsc.org
Visit our website at www.rsc.org/books

Edible Oil Structuring
Concepts, Methods and Applications

Edited by

Ashok R. Patel
Sci Five Consulting Services, Belgium
Email: ashok2510@gmail.com

THE QUEEN'S AWARDS
FOR ENTERPRISE:
INTERNATIONAL TRADE
2013

Food Chemistry, Function and Analysis No. 3

Print ISBN: 978-1-78262-829-3
PDF eISBN: 978-1-78801-018-4
EPUB eISBN: 978-1-78801-222-5
ISSN: 2398-0656

A catalogue record for this book is available from the British Library

The Royal Society of Chemistry is a charity, registered in England and Wales, Number 207890, and a company incorporated in England by Royal Charter (Registered No. RC000524), registered office: Burlington House, Piccadilly, London W1J 0BA, UK, Telephone: +44 (0) 207 4378 6556.

Visit our website at www.rsc.org/books

Printed in the United Kingdom by CPI Group (UK) Ltd, Croydon, CR0 4YY, UK

Foreword

Organogelation refers to the formation of 'gel-like' solids consisting predominantly of a hydrophobic liquid. This is in stark contrast to the vast majority of gels, which are aqueous. In modern parlance, however, "organogelation" has come to encompass many other types of physical systems that may behave rheologically as a solid, such as foams, emulsions and polymers, but are not really gels. The field should be more appropriately called "hydrophobic soft matter"; however, the name "organogelation" has stuck and so it is that we resign ourselves to its use. An organogel purist from the early days of the field (pre-2000) would object to the use of "organogel" as an appellation to describe materials that are fundamentally not gels or even gel-like. The purist's grievance would also likely extend to the use of high-MW components as organogelators. To such a purist, an organogel is a gel formed by a low concentration (arbitrarily, less than 5%) of a low molecular weight organogelator, which has at most, a molecular weight of 1 kDa. However, it appears that the lack of a formal and agreed-upon definition of what constitutes a gel is partly to blame for the inclusion "creep" of other non-gel soft matter under the banner of "organogels".

Studies in the field of organogelation have grown at an astonishingly rapid rate over the last two decades, driven in part by (1) the vast promise of exciting new applications (such as tissue engineering and nanofabrication), (2) the search for solutions to existing, if mundane, problems (such as used frying oil disposal, oil-spill clean-up and art conservation), (3) the scientific allure of elucidating fundamental principles describing the formation of nano- and micro-scale structures from the molecular structures of the structural constituents, and (4) the ready access to the field owing to its interdisciplinary nature. Indeed, the study of organogelation is at the cross-roads of fields such as soft matter science, surface chemistry, colloid

Food Chemistry, Function and Analysis No. 3
Edible Oil Structuring: Concepts, Methods and Applications
Edited by Ashok R. Patel
© The Royal Society of Chemistry 2018
Published by the Royal Society of Chemistry, www.rsc.org

chemistry, organic chemistry, polymer chemistry and nanoscience, among many others. Many contributions to the study of organogelation come from a diverse range of fields and it is not uncommon to find mid- and end-career researchers switching their attention to this field from other disparate areas of research.

Despite the current excitement surrounding organogelation research, it must be noted that the observation that certain substances can thicken hydrophobic liquids is several millennia old. Naphtha, an antiquated catch-all term for a thickened material that we today may call crude oil thickened by pitch, has been in use as a fuel throughout the millennia and even merited mention in the Bible. Naphtha has also been used as an incendiary weapon, first as an ingredient in Byzantine Greek Fire and then later perfected as Louis Fieser's World War II invention: napalm (a portmanteau of napthenic and palmitic acids). Destructive applications aside, organogelation was also considered a nuisance, particularly when crude oil being transported *via* pipelines or fuel being injected into automobile engines gelled at low temperatures, obstructing flow.

Today, organogelation has moved away from its destructive and obstructionist origins. In the field of food science, which is the focus of this book, the investigation into the use of organogelators to structure liquid oil can be traced back to the year 2004 with the publication of Gandolfo, Bot and Floter's work on edible oil structuring by fatty acids and fatty alcohols. Organogelation has attracted considerable research attention in the field of food science for one major reason: as a method of structuring hydrophobic edible oils to minimize or replace saturated and *trans* fatty acids. The desire to reduce saturates and *trans* fatty acids in foods has led to a search for strategies to impart solidity to food oils while reducing the content of these label-unfriendly components. Organogelation offers a strategy to efficiently structure edible liquid oils with a minimum of saturated fatty acids and virtually no *trans* fatty acids.

As the editor of a 2011 volume on edible oil organogels, I can visibly see, by way of contrast, the expansion of organogel research to include new materials and physical systems in Ashok Patel's book. In my 2011 volume, the vast majority of organogelators studied were of the low molecular weight kind (no doubt, to a purist's delight) and the systems were invariably gels structured by dispersions of crystalline or liquid-crystalline particles in edible oils. Two systems reviewed that did not fit this bill (ethycellulose oleogels and dehydrated, crosslinked-protein-stabilized oil-in-water emulsion gels) were only beginning to be studied and were relegated to the end of the volume due to their brevity and recency.

Ashok Patel's current book is a good example of how far research into organogelation in foods has progressed and diversified. It contains plenty (to the point of predominance) of exciting new cutting-edge work on materials and systems that are traditionally not considered organogels. Worth mentioning are Davidovich, Mattice and Marangoni's ethylcellulose oleogels, Scholten's protein-based oleogels, Liua and Yang's cereal protein emulsion gels, Fameau's foam oleogels, John's bio-based molecular structuring agents and Wijaya, van der Meeren and Patel's emulsion oleogels stabilized by protein–polysaccharide complexes. Ashok Patel also reviews even-more-novel dispersion strategies, such as O/W/O double emulsions, jammed oil foams stabilized by crystalline materials and composite fat materials stabilized by hydrophilic (yes, hydrophilic) polymers.

For many who wax nostalgic for the simpler days when an organogel was indeed a gel (whatever that may have been), Patel's book also includes many of the oldies-but-goodies. In prime place is Bot and Floter's work on fatty acid/fatty alcohol organogels, which can be considered a progenitor of the field. Other work included in the volume that merits mention is Rogers' work on ceramide/fatty acid/phytosterol organogels, Toro-Vazquez, Charó-Alonso and Alvarez-Mitre's work on organogels formed from salts of aminated hydroxylated fatty acids and the work on wax organogels by the duos of Mattice and Marangoni and Yilmaz and Ok.

Research into edible oil organogelation continues to hold great promise. Scientists from various fields continue to make contributions and inroads into the field. For those seeking to be swept up by this current and participate in work on edible oil organogels, a work that summarizes the state-of-the-art is indispensable. This book is one such work.

Alejandro G. Marangoni
Edmund D. Co
Department of Food Science
University of Guelph
Guelph, Ontario, Canada

Preface

Driven both by real industrial needs and academic curiosity for fundamental research, edible oil structuring has emerged as a subject of growing interest in the last couple of years.

In this book, we aim to give a comprehensive and concise overview of the field of oil structuring with special emphasis on the updates from the last 5 years. This up-to-date information on the state-of-the-art of the field of oil structuring will be of significant interest to industrial scientists and academic researchers alike.

New insights into the mechanism of gelation in mono- and multi-component gels are discussed for several categories of previously known structuring agents (such as ethyl cellulose, waxes, fatty acids/fatty alcohols, ceramides and hydroxylated fatty acids) along with the potential food applications of some of these systems. In addition, use of alternative methods (such as solvent-exchange and use of emulsion templates) to explore structuring properties of hydrophilic biopolymers are presented with illustrative examples. Some new concepts, such as bio-based synthesis of supergelators, foamed oleogels and use of innovative dispersion techniques, are also introduced to give a broader picture of the current research in the field of edible oil structuring.

The first chapter is of special character. It deals with the concepts of oil structuring with a focus on understanding gelation in mono-component, multi-component and polymer gels through a colloidal gel perspective along with a bird's eye view of the reported literature. It concludes with a few pointers that highlight the expected future trends in this field.

Chapter 2 comprises bio-based synthesis of gelators using sugar esters as the starting material. These supergelators display a very potent gelation behavior owing to intermolecular non-covalent interactions.

Food Chemistry, Function and Analysis No. 3
Edible Oil Structuring: Concepts, Methods and Applications
Edited by Ashok R. Patel
© The Royal Society of Chemistry 2018
Published by the Royal Society of Chemistry, www.rsc.org

Chapter 3 discusses an interesting bio-mimicry approach for creating a multicomponent gel using ceramide, fatty acid and plant sterol as structuring agents.

From Chapters 4 to 6, new insights into the structuring mechanism of previously known small molecular weight gelators are discussed. Structuring agents such as natural waxes, fatty acids/fatty alcohols and hydroxylated fatty acids that form building blocks such as crystalline units and self-assembled structures are discussed.

Chapters 7 to 10 deal with polymeric gels prepared by direct dispersion of a polymer, such as ethylcellulose, or through indirect routes, such as whey protein gels by solvent exchange technique, protein–polysaccharide and cereal protein gels by emulsion-templated approach.

Recently explored edible applications of gels prepared from waxes and ethylcellulose are discussed in Chapters 11 to 12.

The book concludes with two chapters focusing on functional colloids based on oleogels (foamed oleogels) and a collection of three distinct innovative dispersion technologies that could be used for structuring purposes in food product formulations.

This book will be of interest to students, academics and scientists involved in the field of edible oil structuring. I further believe that this book will be an important reference as it provides up-to-date information, which is made possible by high-quality contributions from different research groups active in the field of oil structuring.

Ashok R. Patel
Braga (Portugal)

With the support of the Marie Curie Alumni Association

"You know, this gel sample has more structure than I do in my life these days"

Contents

Section I: Introduction

**Chapter 1   Oil Structuring: Concepts, Overview and Future
Perspectives**   3
Ashok R. Patel

1.1 Introduction   3
1.2 Oleogelation: Concepts   4
    1.2.1 Oleogelation from a Colloidal Gel Perspective   5
1.3 Oleogelation: Overview   12
1.4 Oleogelation: Future Perspectives   15
1.5 Conclusions   17
References   18

Section II: Structuring Units

Chapter 2   Biobased Molecular Structuring Agents   25
Sai Sateesh Sagiri, Malick Samateh and George John

2.1 Introduction   25
2.2 Vegetable Oil Structuring: Chemical Methods   28
    2.2.1 Hydrogenation   28
    2.2.2 Interesterification   29
    2.2.3 Fractionation and Fat Blending   30

Food Chemistry, Function and Analysis No. 3
Edible Oil Structuring: Concepts, Methods and Applications
Edited by Ashok R. Patel
© The Royal Society of Chemistry 2018
Published by the Royal Society of Chemistry, www.rsc.org

2.3 Vegetable Oil Structuring: Biobased Methods   33
       2.3.1 Molecular Gelators or Low Molecular
                 Weight Gelators (LMWGs)   33
       2.3.2 Polymeric Gelators (Cellulose Derivatives)   43
2.4 Multifunctional Molecular Gelators as
       Next-generation Oil Structuring Agents:
       Design, Synthesis and Self-assembly   44
2.5 Conclusions   48
Acknowledgements   48
References   48

Chapter 3   Biomimicry: An Approach for Oil Structuring   53
                 Michael A. Rogers

3.1 Introduction   53
3.2 The Stratum Corneum   54
3.3 Ceramides   55
       3.3.1 Health Aspects of Ceramides   56
       3.3.2 Ceramide Oleogels   57
3.4 Mimicking the Stratum Corneum Lipid Domains   61
3.5 Conclusions   65
Acknowledgement   65
References   65

Section III: Structuring Units: Crystalline Particles and Self-assembled Structures

Chapter 4   New Insights into Wax Crystal Networks in Oleogels   71
                 K. D. Mattice and A. G. Marangoni

4.1 Introduction   71
4.2 Natural Waxes   72
4.3 The Gelation of Oil by Waxes   74
4.4 Wax Crystal Network Microstructure   79
4.5 Types of Natural Wax Gelators   82
       4.5.1 Rice Bran Wax (RBX)   82
       4.5.2 Sunflower Wax (SFX)   84
       4.5.3 Candelilla Wax (CLX)   84
       4.5.4 Carnauba Wax (CRX)   85
       4.5.5 Other Natural Wax Gelators   85
4.6 Oil Binding Capacity of Wax Crystal Networks   86
4.7 Rheological Profiling of Wax Crystal Networks   88
4.8 The Effect of Cooling Rate on the Properties
       of Wax Crystal Networks   90

4.9 The Effect of Shear on the Properties of
     Wax Crystal Networks   91
4.10 Conclusions   92
References   92

Chapter 5 **Structuring Edible Oil Phases with Fatty Acids
             and Alcohols**   **95**
             Arjen Bot and Eckhard Flöter

5.1 Introduction   95
5.2 Fatty Acids (FA)   97
5.3 Fatty Alcohols   98
5.4 Fatty Acids + Fatty Alcohols   98
5.5 Potential Applications   101
5.6 Conclusions   104
References   104

Chapter 6 **Gelation Properties of Gelator Molecules
             Derived from 12-Hydroxystearic Acid**   **106**
             *J. F. Toro-Vazquez, M. A. Charó-Alonso and
             F. M. Alvarez-Mitre*

6.1 Introduction   106
6.2 Molecular Structure and Mechanism for
     Self-assembly of HSA   110
6.3 Shear Rate and Cooling Rate Effect on the
     Microstructure, Self-assembly, and Rheological
     Properties of Organogels   112
     6.3.1 Independent Effect of Shearing and
            Cooling Rate on the Formation of
            Organogels with Vegetable Oil   112
     6.3.2 Combined Effect of Shearing and Cooling
            Rate on the Formation of Organogels
            Developed with HSA, HOA, and OHOA in
            Vegetable Oil   116
6.4 Organogels Developed by "Polar" Gelator
     Molecules Derived from HSA   123
     6.4.1 The Self-assembly Mechanism of
            Ammonium Chloride Salt Derivatives
            of HSA   124
     6.4.2 Rheological Behavior of Organogels
            Developed with Ammonium Chloride
            Salt Derivatives of HSA   124
6.5 Conclusions   129
References   129

Section IV: Structuring Units: Polymeric Strands and Network

Chapter 7 Thermo-gelation of Ethyl-cellulose Oleogels   135
Maya Davidovich-Pinhas

7.1 Introduction   135
7.2 Ethyl-cellulose Characteristics   136
7.3 Thermo-gelation of Ethyl-cellulose Oleogels   138
7.4 Ethyl-cellulose Gelation Mechanism   140
7.5 Ethyl-cellulose Gel Properties   141
    7.5.1 The Effect of Polymer Concentration and
        Molecular Weight on the Gel Properties   142
    7.5.2 The Effect of Oil Type on the Gel Properties   142
    7.5.3 The Effect of Surface-active Molecule
        Addition on Gel Properties   144
    7.5.4 The Effect of Thermal Treatment on
        Gel Properties   145
    7.5.5 Ethyl-cellulose Oleogel Fractionation   146
7.6 Summary and Conclusion   146
References   147

Chapter 8 Proteins as Building Blocks for Oil Structuring   150
E. Scholten and A. de Vries

8.1 Introduction   150
8.2 Solvent Exchange Route   152
8.3 Protein Oleogels Prepared from Protein
    Hydrogels   153
8.4 Protein Oleogels Prepared from Protein
    Aggregates   156
8.5 Effect of Oil Type   160
8.6 Suitable Protein Building Blocks   164
8.7 Role of Capillary Interactions   165
8.8 Potential Applications: First Trials   169
8.9 Conclusion   171
References   172

**Chapter 9 Oleogels from Emulsion (HIPE) Templates Stabilized
by Protein–Polysaccharide Complexes**   175
Wahyu Wijaya, Paul Van der Meeren and Ashok R. Patel

9.1 Introduction   175
9.2 Formation of Biopolymer Complexes   178
    9.2.1 Preparation of Complexes: Effect of
        pH and Concentration   178

9.2.2 Properties of Complexes 179
9.3 HIPEs Stabilized by Biopolymer Complexes 181
9.3.1 Preparation and Microstructure of HIPE 181
9.3.2 Properties of HIPEs 182
9.4 HIPE-templated Oleogels 186
9.4.1 Preparation of Oleogels 186
9.4.2 Properties of Oleogels 188
9.5 Potential Food Applications of
Biopolymer-based Oleogels: Where
Can We Use Them? 191
9.6 Conclusion 193
Acknowledgements 194
References 194

Chapter 10 **Cereal Protein-based Emulsion Gels for Edible
Oil Structuring** 198
Xiao Liu and Xiao-Quan Yang

10.1 Introduction 198
10.2 Zein 200
10.3 Kafirin 205
10.4 Gliadin 206
10.5 Wheat Gluten 208
10.6 Conclusions and Outlook 210
Acknowledgements 211
References 211

Section V: Edible Applications

Chapter 11 **Edible Applications of Wax-based Oleogels** 217
E. Yilmaz and S. Ok

11.1 Introduction 217
11.2 Wax Oleogels in Margarine and Spread
Production 218
11.3 Wax Oleogels in Bakery Products 225
11.4 Wax Oleogels in Chocolate and
Confectionery Products 233
11.5 Wax Oleogels in Dairy Products 237
11.6 Wax Oleogels in Comminuted Meat
Products 242
11.7 Wax Oleogels in Other Food
Applications 244
11.8 Conclusions and Recommendations 246
References 247

Chapter 12   Edible Applications of Ethylcellulose Oleogels   250
             K. D. Mattice and A. G. Marangoni

             12.1 Introduction   250
             12.2 Ethylcellulose Oleogels   251
             12.3 Physical Properties of Ethylcellulose Oleogels   252
                   12.3.1 Techniques for the Analysis of
                           Oleogel Physical Properties   253
                   12.3.2 Parameters Affecting the Physical
                           Properties of Oleogels   254
             12.4 Health Implications of Ethylcellulose Oleogel
                   Consumption   255
                   12.4.1 *In Vitro* and *In Vivo* Digestion of
                           Ethylcellulose Oleogels   255
                   12.4.2 Ethylcellulose Oleogels for the
                           Controlled Release of Bioactive
                           Molecules   258
                   12.4.3 *In Vitro* Bioaccessibility of β-carotene in
                           Ethylcellulose Oleogels   259
             12.5 Considerations and Practicality of Ethylcellulose
                   Oleogels in Food Systems   260
             12.6 Edible Applications of Ethylcellulose
                   Oleogels   263
                   12.6.1 Cream Cheese   263
                   12.6.2 Frankfurters or Comminuted Meats   264
                   12.6.3 Sausages   265
                   12.6.4 Laminating Shortenings   267
                   12.6.5 Ethylcellulose for the Reduction of Oil
                           Migration   268
                   12.6.6 Heat-resistant Chocolate   270
             12.7 Conclusion   271
             References   272

Section VI: Functional Colloids from Structured Oils

Chapter 13   Non-aqueous Foams Based on Edible Oils   277
             Anne-Laure Fameau

             13.1 Introduction   277
             13.2 Aqueous Foam   278
                   13.2.1 Formation of Aqueous Foam   278
                   13.2.2 Classification of Aqueous Foams   278
                   13.2.3 Aqueous Foam: A Multiscale System   279
                   13.2.4 Methods of Aqueous Foam Production   279

13.2.5 Mechanisms of Aqueous Foam
Destabilization 280
13.2.6 Mechanisms of Aqueous Foam
Stabilization 280
13.2.7 Importance of the Surface Tension
and Viscoelastic Properties 281
13.2.8 Main Differences Between Aqueous
and Non-aqueous Foams 282
13.3 Non-aqueous Foams Based on Surfactants 283
13.3.1 Non-aqueous Foams Based on
Hydrocarbon-type Surfactants 283
13.3.2 Non-aqueous Foams Based on
Polymethylsiloxane-type Surfactants 284
13.3.3 Non-aqueous Foams Based on
Fluoroalkyl-type Surfactants 284
13.3.4 Non-aqueous Foams Based on
Asphaltenes and Resins 285
13.4 Non-aqueous Foams Based on Solid
Particles 285
13.4.1 Wettability of Solid Particles 285
13.4.2 Formation and Properties of
Non-aqueous Foams Obtained from
Solid Particles 287
13.4.3 Modification of the Contact Angle by
the Non-aqueous Liquid Surface
Tension 289
13.4.4 Modification of the Contact Angle
by the Surface Chemistry 291
13.5 Non-aqueous Foams Based on Oleogels 293
13.5.1 Formation of Oleogel Systems to
Produce Non-aqueous Foams 293
13.5.2 Production of Non-aqueous Foams
Based on Oleogels 294
13.5.3 Properties of Non-aqueous Foams
Based on Oleogels 297
13.5.4 Foamability and Solubility Boundary
of Oleogel Systems 298
13.5.5 Foam Stability and Solubility Boundary
of Oleogel Systems 299
13.5.6 Rheological Properties of Non-aqueous
Foams Based on Oleogels 301
13.5.7 Responsive Non-aqueous Foams Based
on Oleogels 301
13.6 Conclusion and Outlook 303
References 305

Chapter 14 **Innovative Dispersion Strategies for Creating Structured Oil Systems** 308
Ashok R. Patel

14.1 Introduction 308
14.2 Structured O/W/O Double Emulsions 309
14.3 'Arrested' Oil Foams 316
14.4 Polymer-coated Crystalline Microcapsules 321
14.5 Conclusion 326
References 327

Subject Index 331

Section I

Introduction

CHAPTER 1

Oil Structuring: Concepts, Overview and Future Perspectives

ASHOK R. PATEL*

Sci Five Consulting Services, Coupure 164F, 9000 Gent, Belgium
*E-mail: ashok2510@gmail.com

1.1 Introduction

Emerging evidence related to the negative cardiovascular effects of increased fat consumption has resulted in increased regulation of *trans* fats in food products by regulatory authorities the world over. Starting from the mandatory labelling of the amount of *trans* fats in food products in the mid-2000s, to concrete steps taken towards complete removal in recent times, there has been a consistent decline in the use of *trans* fats in food products over the last few years.[1,2] Accordingly, the food manufacturing industry has been under pressure to innovate and find alternative solutions to formulate products with the complete absence of *trans* fats. The solution currently used by the food industry is replacing *trans* fats with saturated fats from natural sources, such as palm oil.[3] Although quite effective in terms of replicating the

Food Chemistry, Function and Analysis No. 3
Edible Oil Structuring: Concepts, Methods and Applications
Edited by Ashok R. Patel
© The Royal Society of Chemistry 2018
Published by the Royal Society of Chemistry, www.rsc.org

functionality of *trans* fats, the use of saturated fats is criticized for two main reasons: (a) negative image of palm oil owing to the ecological damage that is often linked to palm plantations, and (b) possible negative cardiovascular effects of the long-term consumption of saturated fats.[4] The latter point has seen its share of debate in recent times where some authors have argued the scientific validity of such claims based on empirical evidence.[5] However, it is widely accepted that consumption of polyunsaturated fats over saturated fats is beneficial for cardiovascular health.[6] The current dietary guideline with respect to saturated fat consumption is to restrict the daily consumption to less than 10% of total calories and this is unlikely to change in the near future. As the food industry is currently phasing out *trans* fats in formulations after the ban imposed by the US FDA in 2015, the demand for palm oil has increased. In order to shed the negative image linked to palm oil, many food manufacturers have publicly shared information about their engagement with sustainable palm oil product suppliers (Roundtable on Sustainable Palm Oil or RSPO) to minimize environmental damage done by palm plantations. However, recent scientific opinion[7] from the EFSA panel on Contaminants in the Food Chain (CONTAM) regarding the negative health effects of 3- and 2-monochloropropanediol, as well as their fatty acid esters and glycidyl fatty acid esters, has somewhat tarnished the image of palm oil in the eyes of consumers. Hence, there is a great deal of interest in finding ways to formulate products with a better nutritional profile (*i.e. trans* fat-free and high in unsaturated fats), and preferably without the use of palm oil.

In recent years, oleogelation has gained popularity as an approach for formulating food products where the functionality provided by saturated fats (*i.e.* texturing, oil binding, rheological characteristics, organoleptics and stabilizing properties) can be replicated by non-fat components used as structuring agents. The possibility of gelling ≥90 wt% of liquid oil at a relatively lower mass fraction of gelator molecules makes oleogelation a very efficient approach for oil structuring. This efficient structuring in oleogels is usually achieved by supramolecular assemblies (building blocks) of gelator molecules that organize into a three-dimensional network that can physically imprison a large volume of mobile liquid oil into an 'arrested' gel-like structure. In the following sections, a general concept and an overview of the field of oleogelation is presented with special emphasis on the research done in the recent years.

1.2   Oleogelation: Concepts

By definition, oleogels are a sub-class of a broader class of colloidal structures called organogels and are defined as a gel system where an oil continuous liquid phase is immobilized in a network of self-assembled molecules of an oleogelator or a combination of gelators. In the last few years, oleogelation has changed from a relatively unknown term into a heavily investigated research domain. Academic research in this area has progressed rapidly as researchers from a wide range of backgrounds (such as applied chemistry, colloid science, material science and process engineering) have taken up the challenge of identifying novel ingredients and innovative processing methods to create oleogels. At the same time, industrial scientists have shown increased interest in investigating edible applications of these oleogels in a range of food products, including chocolates, baking fats and meat fat replacements. Fundamentally, it is fascinating to fabricate and characterize this new class of edible soft matter systems as there is still a lot to be learnt about their structuring mechanisms and the 'tunability' of their bulk properties through microstructural alterations. From an application point-of-view, the oleogelation approach has the potential to cater to a number of generalized and 'niche' applications. Such applications include: (a) reduction of saturated fat levels in food products, (b) stabilization of surfactant-free emulsions, (c) decreasing oil mobility and migration in chocolate products (controlling fat bloom in filled chocolates), (d) improving temperature stability in certain food products (heat-resistant chocolates), (e) developing controlled-release systems for delivery of low molecular weight bioactives, and (f) transforming liquid oils into a new range of edible materials, such as foams, films and soft pliable solids.

1.2.1 Oleogelation from a Colloidal Gel Perspective

1.2.1.1 Oleogelators and Monocomponent Gels

One of the main bottlenecks in the field of oleogelation is finding structuring agents (oleogelators) that are effective at low concentrations, cheap, readily available and, most importantly, have the required regulatory approval for use in edible products.[4,6] Mechanistically, the most important requirement for a material to act as an oleogelator is to display a suitable balance between its affinity for edible oils (*i.e.* weak interactions with unsaturated triacylglycerols) and sufficient insolubility in these solvents in order to first trigger a 'bottom-up' phenomenon leading to the formation of primary particles as a function of super saturation and external factors, such as temperature, that can alter the solute–solvent interactions. These primary particles then need to undergo molecular self-assembly (a process by which individual molecules form defined aggregates) and subsequent self-organization (a process by which the aggregates create higher-ordered structures) to create supramolecular structures depending on the delicate balance between solute–solute and solute–solvent interactions. These supramolecular structures include: crystal lattice, liquid crystals, micelles, bilayers, fibrils, and agglomerates, which may form a three-dimensional network to physically trap the liquid oil in a gel-like system. These supramolecular structures or molecular assemblies or building blocks are usually stabilized by non-covalent interactions, such as H-bonding, van der Waals attractive interactions (London dispersion forces) and π–π stacking. The growth of molecular clusters or assemblies into a continuous network is further governed by the anisotropy of the specific surface free energy (at active sites) as well as the anisotropy of the mobility of the diffusing molecules (or the kinetic coefficient).[8]

So far, the mechanisms responsible for the formation of the structuring units in edible oleogels have not been studied in detail. Most of the knowledge related to self-assembly and self-organization of edible oleogelators has been obtained mainly from in-depth studies done using 12-hydroxystearic acid and its derivatives as gelators in a range of organic solvents.[9,10] In rare cases, the correlation of self-assembled structure formation with solvent properties (Hansen solubility parameters) has also been explored to better understand the mechanism involved in the formation of supramolecular gels.[11-13] As reported in the literature, a weak solvent–gelator interaction results in dominant gelator–gelator interactions, which may lead to the formation of a continuous network. However, a much stronger gelator–gelator interaction will eventually lead to the precipitation of crystalline or amorphous molecule clusters resulting in phase separation. Therefore, a suitable balance between solvent–gelator and gelator–gelator interactions ultimately governs the formation of a continuous (percolated) network of self-assembled gelator molecules that provides the structural framework for gelation of the solvent. As seen with the colloidal gels,[14] arrested phase separation may be considered as one of the mechanisms that drives 'out-of-equilibrium' gelation in oleogels. Basically, the micro- to macro-phase separation is interrupted by a dynamical arrest, which leads to the immobilization of the solvent (gelation). In the majority of cases, equilibrium gelation may follow the conventional route, which involves primary particles forming transient clusters, which in turn form a transient network that further transform into a percolation of long-lived clusters that cause gelation of the solvent. Hence, oleogelation can be considered to be a rather complex process as it involves supramolecular interactions at the primary, secondary and even tertiary levels.

Based on the above discussed mechanisms of gelation, some of the important conditions that need to be fulfilled for gelation of organic solvents include: (i) directional intermolecular interactions of gelator molecules that promote unidirectional growth of supramolecular aggregates; (ii) formation of internetwork secondary interactions leading to intertwined aggregates, and (iii) prevention of neat crystallization of gelator molecules.[15] Accordingly, based on this understanding, molecular and crystal engineering concepts can be utilized to design and identify new gelator molecules. For instance, molecular features such as having moieties with hydrogen bonding functionalities and long alkyl chains that are capable of self-assembling (*via* London dispersion forces) could provide an early indication of gelation properties of components based on their molecular structures. On the other hand, molecules displaying crystallization properties such as unidirectional crystal growth (and suppression of lateral growth) and/or minimal post-crystallization events (such as crystal aggregation) may result in a more 'spread-out' crystalline mass in the continuous phase, forming a space-spanning network of crystals at a very low volume fraction. Additionally, information about thermodynamic dissolution parameters (enthalpies and entropies) of structuring agents in the chosen solvents could offer new insights into the gelation behaviour of gelators as well as the critical role played by the solvent in gelation.[16]

This information could also pave the way to creating novel 'solvent-mediated' gels where the poor solubilizing capacity of solvents and/or the difference in the crystallizing/aggregating/self-assembling properties of the solute in different solvents could be exploited to generate new types of gels.[17] One of the most common examples in this category is lecithin-based oleogels, which are typically formed at certain critical ratios of vegetable oil and water as solvents.[18] A similar concept could even be exploited to create water-free oleogels by investigating the crystallization/self-assembly behaviour of gelators in mixed solvent systems consisting of different types of oils (long chain triglycerides, medium chain triglycerides, oils rich in diglycerides and essential oils). For instance, natural waxes have been known to display different gelation behaviours (critical gelling concentration, sol–gel transformation temperature *etc.*) in vegetable oils with differing fatty acid profiles, and this is attributed to the difference in the gelator–solvent interactions that affects their crystallization properties (kinetics, molecular packing and crystal aggregation).[19,20] A thorough investigation focusing on different gelator–solvent combinations may provide information on creating multi-solvent gels where the formation of molecular assemblies and their subsequent organization into a continuous network could be controlled.

Another molecular feature that may play a vital role in the molecular ordering of gelators is the chirality of the molecules. As demonstrated by Kim and co-workers,[21] the presence of a stereogenic centre in the molecular structure strongly influences the molecular ordering, resulting in preferential one-directional growth of crystalline units, which in turn causes a decrease in the minimum gelation concentration of the structuring agent. Close to the edible field, the concept of chirality can be explored to probe the gelation properties of monoacylglycerols (MAGs). MAGs are esters of glycerol in which only one of the hydroxyl groups is esterified with a fatty acid, and they can exist in three different structural forms: sn-1, sn-2 and sn-3 isomers, depending on the location of the fatty acids on the glycerol backbone. Of these, sn-1 and sn-3 are not distinguished from each other and are termed 'α-MAGs', while the sn-2 isomers are called 'β-MAGs'. The second carbon atom on the glycerol backbone is a stereogenic centre, while β-MAGs are devoid of such a stereogenic centre. Commercial MAGs used in the food industry have more than 90% α-MAGs with β-MAGs at less than 5%. Owing to their broad functionalities, commercial MAGs are among the most common food emulsifiers used in a range of edible products. They have also been explored for their oil structuring properties both in monocomponent[22,23] and mixed component gels.[24–26] However, one of the drawbacks associated with MAGs includes polymorphic transition to gritty β-crystals on aging. This drawback could be solved by using β-MAGs instead of α-MAGs as the non-chiral nature of the former may help with a better crystallization profile and reduce the unwanted polymorphic transitions that cause stability issues in gels. Moreover, β-MAGs are considered to have higher surface activity compared to α-MAGs (which is reflected in their anti-bacterial effects), which may further enhance their self-assembly in oil medium. It is important to note, however, that the acyl migration phenomenon may pose an issue in obtaining a high yield of β-MAGs synthesized through conventional processes.[27]

1.2.1.2   Oleogelators and Multi-component Gels

Multi-component supramolecular gels have not yet been explored to their full potential. As the name indicates, these gels are composed of multiple solute components that directly or indirectly assist in gelling the solvents. The simplest of these gels are two-component gels where synergistic interactions of two solutes are exploited to alter the formation of microstructure as well as the gel-supporting structural framework. What makes this category of gels particularly interesting is the possibility of tuning their properties by simple alteration of the proportions of the components. The two-component gels can be categorized into three general classes (Figure 1.1) including: (a) two component gel-phase materials where both components are required for gelation as the individual components either cannot form structured materials (higher ordered self-organized structures) or gel the solvents on their own; (b) two gelator component gels where both components are themselves gelators (having properties of forming structured materials) and when used in combination they are capable of organizing into assembled structures either together (co-assembly) or independently of each other (self-sorting), and (c) gelator plus additive component gels, which are formed by a combination of a gelator and a non-gelling additive. The additive is required to either impact the self-assembling properties of the gelator or to promote an effective spatial distribution of building blocks formed by the gelator or to strengthen the network linkages among formed building blocks.[28]

Lecithin–sorbitan tristearate oleogels are good examples of two-component-gel phase systems as both lecithin and sorbitan tristearate (STS)

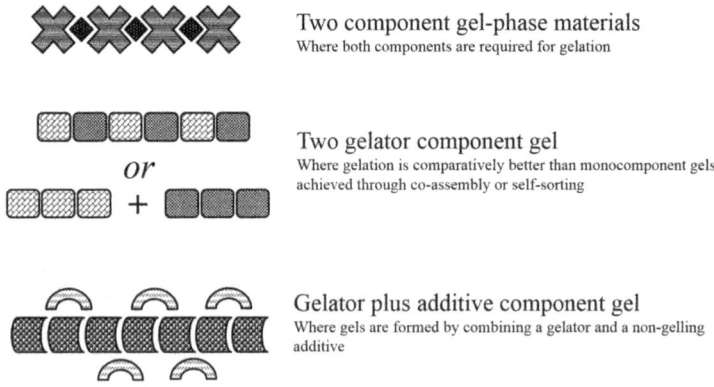

Figure 1.1   Schematic representation of categories of multicomponent gels. Reprinted from *Curr. Opin. Food Sci.*, Patel A. R., A colloidal gel perspective for understanding oleogelation, DOI: 10.1016/j.cofs.2017.02.013. Copyright 2017 with permission from Elsevier.

Figure 1.2 Photographs of (a) monocomponent oil systems containing lecithin and sucrose oligoester at 10 wt%, respectively, and (b) mixed component gel containing lecithin:sucrose oligoester at a 2:8 ratio (total concentration 10 wt%). Microscopy images of (c) monocomponent oil system of sucrose oligoester and (d) mixed component gel. Notice how the presence of lecithin results in a finer crystalline network (microscopy image of monocomponent oil system of lecithin did not show any discernible structures and is hence not included in the figure). Reprinted from *Curr. Opin. Food Sci.*, Patel A. R., A colloidal gel perspective for understanding oleogelation, DOI: 10.1016/j.cofs.2017.02.013. Copyright 2017 with permission from Elsevier.

are incapable of forming gels on their own but, when mixed at certain ratios, undergo synergistic association to form gels.[29] The crystalline units formed in these systems are based on STS, while the surfactant (lecithin) plays an important role in influencing the morphology of the crystalline units and strengthening the network junctions among the formed units. Similarly, this kind of synergistic association is also observed when lecithin is combined with sucrose oligoesters (Figure 1.2). As seen from the photographs in the figure, both components are incapable of gelling oil on their own but they form a gel when mixed together at a specific ratio. Microstructure studies suggest that the crystalline units in the gels are based on sucrose oligoesters and the inclusion of lecithin influences the crystallization and aggregation behaviour of the units, leading to a finer network that is capable of physically immobilizing the liquid oil. Some other examples explored for creating this class of gels are oleic acid + sodium oleate and lecithin + α-tocopherols[30,31] In most cases, the main component is responsible for providing structured materials (crystalline particles, lamellar phases) in the bulk oil phase, while the presence of the other component is necessary for modifying these basic structured materials into units that can form a space-filling network.

As far as co-assembled and self-sorted systems are concerned, they have been rarely studied in the edible field. Some examples that can be loosely fitted to this category are fatty acids + fatty alcohols (co-assembly),[32] sterol + sterol ester (co-assembly),[33] fatty acid + sorbitan ester (co-assembly)[34] and sterols + monoglycerides (self-sorting).[26] Fatty acids + fatty alcohols systems (stearic acid/stearyl alcohol, in particular) were among the first mixed component oleogels studied for potential use in edible fields.[35] The synergistic effect in these systems was explained based on mixed crystal formation (co-crystallization) with finer crystal sizes as well as an altered crystal morphology (needle-like for combination *versus* platelet-like for monocomponents).[35,36] It has also recently been reported that the synergistic effect is rather linked to an increase in the crystalline mass as well as the spatial distribution of the mass.[37]

Mixed component oleogels of β-sitosterol and γ-oryzanol are undeniably one of the most extensively studied oleogel systems. On their own, both components form crystalline particles in vegetable oils, but when mixed in certain proportions they co-assemble to form nanoscale tubular structures. Depending on the concentration, these nanoscale tubules aggregate to form a 3-D network that can physically immobilize liquid oil through capillary forces to form a viscoelastic gel.[38-40]

Recently, it was observed that the combination of phytosterols + saturated monoglycerides displayed synergistic gelation behaviour when used at certain ratios. Based on the data from several characterization techniques, it was concluded that the synergism was not based on co-crystallization owing to a large difference in the molecular structures of the two components.[26] Based on the microstructure of the mixed component gel (Figure 1.3), self-sorting behaviour was noticed as both components crystallize independently of each other to form characteristic structuring units (fibrous and sperulitic crystals) that contribute to a synergistic enhancement of the rheological properties of the gel.

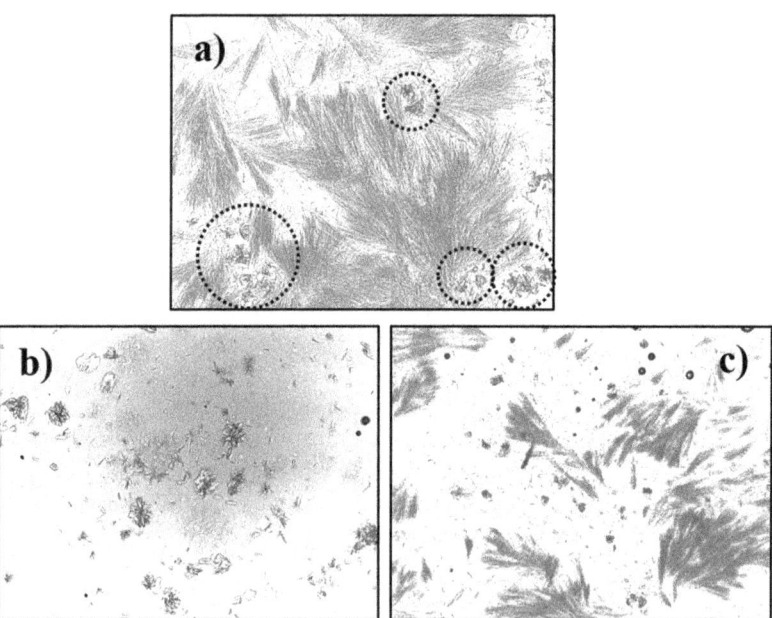

Figure 1.3   Microscopy images of mixed component system (a) and monocomponent gels of monoglycerides and phytosterols (b and c, respectively). Characteristic fibrous and spherulitic crystalline units (marked in dashed circles) are responsible for structuring the mixed component gel. Reprinted from *Curr. Opin. Food Sci.*, Patel A. R., A colloidal gel perspective for understanding oleogelation, DOI: 10.1016/j.cofs.2017.02.013. Copyright 2017 with permission from Elsevier.

In addition, the presence of different structuring units prevents aggregation of crystalline mass (which is an issue in the monocomponent gels), leading to a better spatial distribution of mass in the bulk liquid oil phase.[26]

The third and the last category of multi-component gels is created using a combination of gelator and a non-gelling additive, such as a surfactant, filler or polymer.[28] While the gelator is capable of forming a gel independently, the presence of an additive may be required for stabilizing the gel (suppressing post-crystallization events such as aggregation and contraction) and for tailoring their thermo/mechanical properties (melting behaviour, thixotropic recovery and shear stability). Some examples that can be categorized as gelator/additive gels include gels created from crystalline particle-forming gelators, such as sorbitan monostearate, sitosterol and stearic acid, in the presence of non-crystallizing surfactants, such as lecithin and Tween 20 (polyoxyethylene 20 sorbitan monolaureate).[41-43] Some functionalities provided by these surfactants include changes in the crystallization kinetics, crystal habit modification, a decrease in the mess size of the network and prevention of crystal aggregation.

It is also important to note that although a lot of examples discussed above show that lecithin contributes to gelation either by directly accessing the gel formation or by acting as an adjuvant non-gelling additive, it has also been known to cause disruption of gelation in mixed component gels. For instance, when mixed with 12 HSA, lecithin causes remarkable reduction in gelation functionality by affecting the basic structural units.[44] Similarly, lecithin also shows strong antagonistic effects when combined with a range of natural waxes (unpublished results), including waxes predominantly rich in wax esters (sunflower wax) and hydrocarbons (carnauba wax).

1.2.1.3 Oleogelators and Polymer Gels

The concepts of gelation mentioned in the previous section relate mostly to low molecular weight structuring agents (mol. wt. < 3000), such as natural waxes, crystallizing (partial glycerides, sorbitan esters, sucrose esters *etc.*) and non-crystallizing (lecithin, polysorbates *etc.*) emulsifiers, hydroxylated fatty acids (12-HSA) and other miscellaneous compounds, such as fatty acids/alcohols, tocopherols, and sterols/sterol esters. Beside these gels, there are other structured systems where long-chain polymers are used to produce molecular assemblies (polymeric strands) from a 'bottom-up' approach or create a pre-formed structural framework using colloids (such as hydrogels, emulsions) as templates.

Like the gels described in the previous section, polymer gels are prepared by dispersing the polymer in liquid oil above the glass transition temperature of the polymer followed by cooling. On cooling, the dispersed polymer shows a nanophase separation and a consequent formation of a polymeric network through different orders of random crosslinking based on the self-complementary supramolecular interactions (often mediated through hydrogen bonding or ionic interactions).[45]

Edible polymer gels of ethylcellulose (EC) (a food-grade hydrophobic cellulose derivative) have been extensively investigated both for fundamental exploration (to understand the mechanism of gelation)[46-48] as well as for practical applications in food systems.[49,50] EC oleogelation follows the typical polymer gel formation steps, starting from dispersion of the polymer at high temperature and ending with network formation based on self-complimentary supramolecular interactions. Factors such as the molecular weight of the polymer, the polarity and fatty acid profile of oil medium, the type of surfactant (if used) and the polymer:surfactant ratio are all known to have a strong effect on the properties of the formed oleogels.[51] The molecular weight (viscosity) of the polymer has a significant effect on the mechanical strength of the gels (higher molecular weight = increased firmness), irrespective of the oil type used.[48] The fatty acid profile of the oil influences the density of the oil, which in turn has an effect on the packing of oil droplets in EC scaffolding, while polar entities present in the oil phase interact with EC to strengthen the network (*i.e.* high polarity = higher gel strength).[47] Surfactants play two important roles: they have a plasticizing effect and they provide crystalline mass; together this leads to an increased gel strength by improving the rigidity of the polymer network and providing an adjuvant crystalline network, respectively.[52]

The second category of polymer gels is created using hydrophilic polymers (with surface active properties) that do not have the right solubility characteristics to be dispersed in liquid oil. The approach used for creating these gels includes the use of biphasic colloid templates (water continuous emulsions or foams) where the hydrophilic polymers are dispersed and hydrated in water.[6,53,54] Owing to their surface active properties, they accumulate at the air–water or oil–water interfaces. Subsequent removal of water from these templates results in the formation of a dense structural framework where oil droplets are packed within the matrix of the polymer (in case of emulsion templates) or a porous architecture comprising an open reticulated network of polymer (in case of foam templates).[55-57] In addition to using molecularly dispersed polymers, emulsion templates are sometimes stabilized using surface active colloidal particles (Pickering stabilization).[58,59] These kinds of polymer gels can be hybridized[58,60] to improve their mechanical properties and when used as a substitute for solid fat in food products (such as cakes, cookie cream and peanut butter) they have shown very promising results.[61-63]

Polymer gels can also be prepared from natural proteins *via* solvent-exchange method where the aqueous continuous medium of hydrogel (structured with protein aggregates) is replaced with liquid oil in a stepwise fashion.[64,65]

1.3   Oleogelation: Overview

An overview of the field of edible oil structuring has already been comprehensively covered in a recent review article. The field has seen a tremendous progress in the last decade, as confirmed by the significant increase in

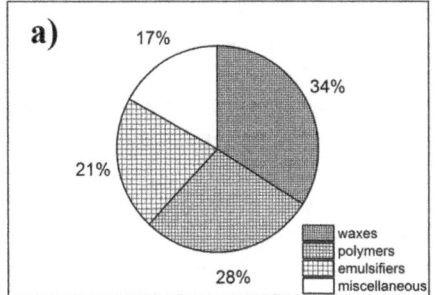 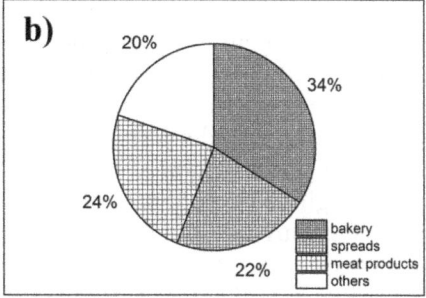

Figure 1.4 Pie charts categorizing literature (from Web of Science™) in the field of edible oil structuring based on: (a) type of structurant and (b) type of food application.

the number of papers published in this area.[6] A closer look at the numbers reveals details about the specific categories of structurants explored for oleogelation as well as the food product categories that could benefit from it. As per Figure 1.4a, natural waxes have been the most investigated structuring agents (34%), closely followed by polymers (28%). The rest of the structuring agents explored so far can be categorized into emulsifiers (21%) and miscellaneous agents (17%), such as sterol/sterol esters, hydroxylated fatty acids and inorganic particles. The popularity of natural waxes as preferred structuring agents is understandable from the fact that a wide range of natural waxes are capable of gelling liquid oil (at very low concentration) through a simple heating–cooling protocol. Moreover, depending on the type of waxes used (differing in melting behaviour, chemical constituents and sources), gels with different thermo/mechanical properties can be prepared.[66–68] In addition, depending on the concentration used for oleogel preparation, fat-mimicking behaviour is easy to obtain both in the presence and absence of water.[19]

With respect to polymers, a major bulk of the publications are focused on EC oleogels. EC has been investigated extensively as it is the only food-grade polymer that can be dispersed directly into liquid oil. Other polymers that have been tried for oil structuring (through indirect approaches) include: hydrophilic cellulose derivatives (hydroxyl propyl methylcellulose and methylcellulose); hydrophilic proteins (gelatin and whey protein); and hydrophobic proteins such as zein, wheat proteins and cellulose (nanocrystals).

Emulsifiers offer themselves as suitable candidates for oil structuring and, accordingly, they are the next class of materials explored as structurants. Some of the properties that make them attractive as structurants include: (a) amphiphilic nature (providing scope for self-assembly); (b) crystalline mass provider (for crystallizing surfactants); (c) good dispersibility in oil; (d) fat-like crystallization properties (partial glycerides); (e) providing Pickering stabilization (in the case of oleogel-based emulsions), and (f) favourable regulatory status. Emulsifiers are usually known to show synergistic interactions and are thus used in combination with other structurants in mixed component gels. They play diverse roles in these gels, as crystal initiators (through templating effect), crystal modifiers (affecting crystal habit and polymorphic transitions), involvement in heterogeneous supramolecular interactions, and minimizing post-crystallization events such as aggregation and agglomeration.[6]

Although most publications have strongly focused on the fundamental understanding of the mechanism involved in the formation and properties of oleogels, some studies related to the practical applications in food products have recently emerged. Reports involving applications of oleogels (or structured oil) account for only 28% of the total literature published in this field. On further breaking them down into different categories of products (Figure 1.4b), it can be seen that bakery applications (cakes, cookies, pastries *etc.*) enjoy a mega share (34%) of the research. These products are formulated using at least 25–40% fat phase, which almost entirely comprises solid fat and so the use of structured oil as a solid fat substitute could really benefit this category of food products. Systems such as wax-based gels,[69–71] polymer gels,[72] emulsifier-based gels[73] and structured emulsions[74,75] are some examples that have been explored for potential applications in bakery products. In most cases, the end products with structured oil are shown to have comparable results to commercial references. Meat products (such as sausages,[76] meat batters,[77] patties[78] and frankfurters[79,80]) are the second most investigated category of food products (24%), where structured oil systems like oleogels, structured emulsions and oil bulking systems have been incorporated to achieve both total fat reduction and an improvement in the fatty acid profile (high in poly and mono unsaturated fatty acids and low in saturated fatty acids).[81] In addition, structuring systems have also been used for imparting physical stability to products (such as sauces and suspensions) without significantly altering the fatty acid profile or fat content.[82,83]

For most structured systems, the incorporation of water results in a structure collapse. This is one of the main drawbacks that limits the use of structured oils in products such as table spreads and margarines. Among the structured systems studied so far, wax-based oleogels have been found to be the most tolerant to water incorporation. In fact, crystallization of waxes in presence of water–oil interfaces even promotes Pickering stabilization of emulsions, enabling the formation of surfactant-free emulsions.[60] Accordingly, most research published on applications in spreads focuses on wax-based systems.[24,84] In some cases, oil-based spreads have also been formulated without water incorporation.[85–87]

In addition to the above discussed categories of food product, other applications of structured oil systems include: (a) replacement of fat in water continuous emulsion systems, such as ice-creams[88] and cream cheese products;[89] (b) stability enhancement in chocolate products (heat resistant chocolate bars,[90] chocolate spreads,[91] pralines[92] and confectionary fillings[93]); (c) creation of novel colloidal systems, such as non-aqueous foams;[94] (d) as a matrix for controlled delivery of bioactives.[95]

1.4 Oleogelation: Future Perspectives

The field of oleogelation has seen tremendous progress in the last decade with respect to identification of different structuring agents. However, lipidic components (such as waxes, partial glycerides, emulsifiers and sterols) are still considered as the ideal structuring agents as they can closely mimic the crystallization properties of high melting TAGs. Space-spanning networks formed through surface interactions among crystalline particles play an important role in providing the main characteristic properties of structured oil (*i.e.* plasticity or malleability). In engineering, plasticity is defined as the ability of a material to be deformed repeatedly without rupture by the action of a force, and to remain deformed after the removal of force. For example, clays, where the plasticity results because of the cluster of grains rolling over each other under applied force. The cohesive forces among clusters of grains and the solid–solvent surface interactions determine this workable behaviour of clays. In the case of the oil system structured with crystalline TAGs, the plasticity arises from weak surface interactions (among clusters of anisometric fat crystal aggregates) that can be easily overhauled by small mechanical stress to cause a spatial rearrangement of clusters that immediately form weak surface interactions in their new fixed positions. On comparing the mechanical properties of different oleogel systems, plastic behaviour similar to that of fat structured systems is only seen with fumed silica-based gels. As seen from Figure 1.5, the structural organization of fat[96] leading to final network formation is quite comparable to that of fumed silica.[97] Primary particles of fat (nano platelets) and fumed silica (nanoscale particles) undergo a secondary arrangement to form clusters and agglomerates (of aggregates of primary particles), respectively. The clusters and agglomerates

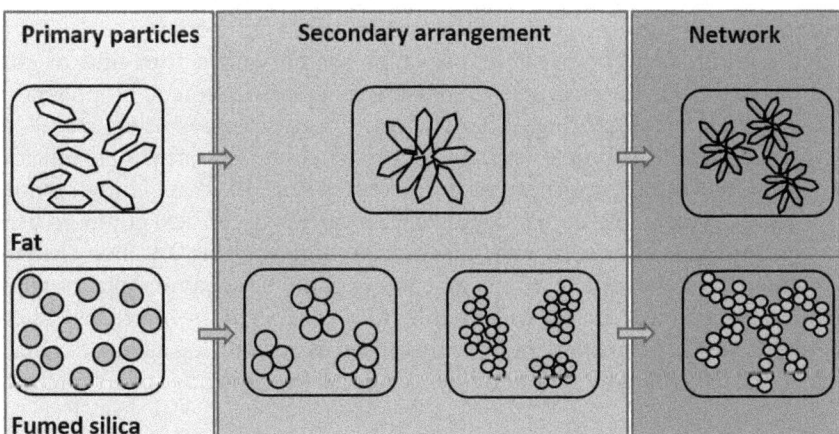

Figure 1.5   Different structural levels in fat (nanoplatelets → clusters → network) and fumed silica (nanoscale particles → aggregates → agglomerates → network).

interact through weak surface forces (dispersive forces and hydrogen bonding, respectively) to form a space-spanning network. On application of a small external force, the clusters and agglomerates can transiently reorganize without permanent disruption of the overall network, resulting in a plastic/malleable system.

From the above discussion, it is clear that creating an alternative plastic structured system will require the application of particle and surface engineering principles. Fat crystals will still be an important element in such alternative systems but there may be a possibility of replacing a major part of it with components such as biopolymers. Two scenarios where this could be made possible are:

(1) Creating a biopolymeric framework in the bulk oil phase such that it strengthens and provides additional linkages in the crystalline network formed at a relatively lower mass fraction than conventional fat structured systems. As most biopolymers are hydrophilic in nature, the challenge lies in exploring a possible route for introducing them in the bulk oil phase. Depositing a layer of hydrated biopolymers on the surface of molten fat droplets could lead to the formation of polymer-coated fat crystal capsules. These capsules could then allow the incorporation of polymer sheets/strands in oil and strengthen the crystalline network of fat. The first step involves emulsification of molten fat in an aqueous solution of polymer; the amphiphilic nature of the polymer leads to an accumulation at the oil–water interfaces. By controlling the droplet size of the emulsion, it is possible to separate polymer-coated fat capsules through a programmed time–temperature cycle. The polymer-coated fat capsules can then serve as templates to create a fat crystalline network substantiated by polymer strands. More details on the preliminary results from this concept are discussed in the Chapter 14, Innovative Dispersion Strategies for Creating Structured Oil Systems.

(2) Creating biopolymeric microcapsules coated with a thin film of condensed crystalline particles and using these crystal-coated biopolymeric capsules as structuring units. The concept is explained in Figure 1.6. In contrast to the above mentioned approach, here we start with an oil continuous emulsion with water droplets structured with a gelling biopolymer. Crystallizing emulsifiers (such as partial glycerides) could be used to promote interfacial crystallization of high melting TAGs on the surface of water droplets, which on cooling could be separated out as gelled microcapsules coated with a thin film of crystalline particles. These coated microcapsules could be then used as structuring agents to obtain a plastic system similar to the ones created by aggregates of fat crystals.

The above mentioned approaches are representative of some of the innovative dispersive techniques that could be utilized to obtain plastic fats with lower levels of solid crystalline phase. Such techniques could also pave the way to exploiting structuring properties of other hydrophilic food polymers in oil continuous phases.

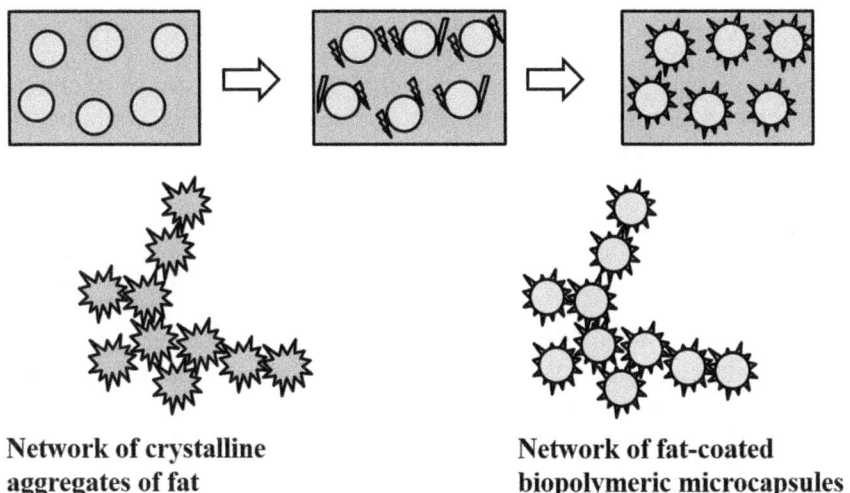

**Network of crystalline
aggregates of fat**

**Network of fat-coated
biopolymeric microcapsules**

Figure 1.6 (Above) Step-by-step presentation of approach for creating fat-coated biopolymeric microcapsules. (Below) Schematic representation of network structure created by fat aggregates in conventionally structured system and network structure created by fat-coated biopolymeric capsules. Based on this hypothesis, similar plasticity can be obtained at a relatively lower mass fraction of crystalline phase, resulting in lowering of saturated fats.

1.5 Conclusions

Given the current trends in the field of edible oleogelation, the next phase of research in this field should be focused on:

(i) Understanding the concepts of supramolecular (colloidal) gelation, solvent–solute interactions, self-assembly and self-organization in order to gain new in-sights into the gelation of organic solvents. This could involve an investigation of the link between structural molecular features (such as H-bonding functional groups, linearity of the molecular structure, polarity and stereo specificity/region specificity) and the gelation properties of known gelators, generating a database of dissolution parameters (such as entropy and enthalpies) of known gelators in solvents or mixtures of solvents, and understanding the synergistic interactions among known gelators to create multicomponent gels.

(ii) Exploring further potential applications of oleogel systems in different food product formats. If complete replacement of solid fats with oleogels is not possible, hybrid mixtures of oleogels with solid fats could be tested.

(iii) Structured edible oils fortified with polyunsaturated fatty acids (PUFAs) with the goal of both texturizing these oils to mimic solid fats and to improve the stability of PUFAs.

   (iv) Using oleogels as matrices for controlled delivery of lipid-soluble bioactives.

   (v) Fabricating oil-based novel colloidal systems (oleocolloids), such as foams, films and capsules, from structured oil, which could be used as new product formats.

References

1. D. R. Kodali, *Trans Fats Replacement Solutions*, AOCS Press, Urbana, IL, USA, 2014.
2. S. Stender, A. Astrup and J. Dyerberg, *BMJ Open*, 2012, **2**, e000859.
3. A. G. Marangoni and N. Garti, *Edible Oleogels: Structure and Health Implications*, AOCS Press, Urbana, Illinois, 2011.
4. A. R. Patel, *Alternative Routes to Oil Structuring*, Springer International Publishing, 2015.
5. R. Chowdhury and S. Warnakula, *et al.*, *Ann. Intern. Med.*, 2014, **160**, 398–406.
6. A. R. Patel and K. Dewettinck, *Food Funct.*, 2016, **7**, 20–29.
7. EFSA Panel on Contaminants in the Food Chain (CONTAM), *EFSA J.*, 2016, **14**, e04426.
8. J. S. Wettlaufer, M. Jackson and M. Elbaum, *J. Phys. A: Math. Gen.*, 1994, **27**, 5957.
9. P. Terech and R. G. Weiss, *Chem. Rev.*, 1997, **97**, 3133–3160.
10. Y. Lan and M. A. Rogers, *CrystEngComm*, 2015, **17**, 8031–8038.
11. M. Raynal and L. Bouteiller, *Chem. Commun.*, 2011, **47**, 8271–8273.
12. J. Gao, S. Wu and M. A. Rogers, *J. Mater. Chem.*, 2012, **22**, 12651–12658.
13. K. K. Diehn, H. Oh, R. Hashemipour, R. G. Weiss and S. R. Raghavan, *Soft Matter*, 2014, **10**, 2632–2640.
14. Z. Emanuela, *J. Phys.: Condens. Matter*, 2007, **19**, 323101.
15. P. Dastidar, *Chem. Soc. Rev.*, 2008, **37**, 2699–2715.
16. M. L. Muro-Small, J. Chen and A. J. McNeil, *Langmuir*, 2011, **27**, 13248–13253.
17. M.-M. Su, H.-K. Yang, L.-J. Ren, P. Zheng and W. Wang, *Soft Matter*, 2015, **11**, 741–748.
18. M. Bodennec, Q. Guo and D. Rousseau, *RSC Adv.*, 2016, **6**, 47373–47381.
19. A. R. Patel, *Alternative Routes to Oil Structuring*, Springer International Publishing, 2015, pp. 15–27.
20. A. J. Martins, M. A. Cerqueira, L. H. Fasolin, R. L. Cunha and A. A. Vicente, *Food Res. Int.*, 2016, **84**, 170–179.
21. J.-U. Kim, D. Schollmeyer, M. Brehmer and R. Zentel, *J. Colloid Interface Sci.*, 2011, **357**, 428–433.
22. N. K. O. Ojijo, I. Neeman, S. Eger and E. Shimoni, *J. Sci. Food Agric.*, 2004, **84**, 1585–1593.
23. S. Da Pieve, S. Calligaris, E. Co, M. C. Nicoli and A. G. Marangoni, *Food Biophys.*, 2010, **5**, 211–217.

24. M. Öğütcü, N. Arifoğlu and E. Yılmaz, *J. Am. Oil Chem. Soc.*, 2015, **92**, 459–471.
25. J. F. Toro-Vazquez, R. Mauricio-Pérez, M. M. González-Chávez, M. Sánchez-Becerril, J. d. J. Ornelas-Paz and J. D. Pérez-Martínez, *Food Res. Int.*, 2013, **54**, 1360–1368.
26. M. D. Bin Sintang, T. Rimaux, D. Van de Walle, K. Dewettinck and A. R. Patel, *Eur. J. Lipid Sci. Technol.*, 2017, **119**, 1500517.
27. A. Rodríguez, L. Esteban, L. Martín, M. J. Jiménez, E. Hita, B. Castillo, P. A. González and A. Robles, *Enzyme Microb. Technol.*, 2012, **51**, 148–155.
28. L. E. Buerkle and S. J. Rowan, *Chem. Soc. Rev.*, 2012, **41**, 6089–6102.
29. M. Pernetti, K. van Malssen, D. Kalnin and E. Flöter, *Food Hydrocolloids*, 2007, **21**, 855–861.
30. C. V. Nikiforidis, E. P. Gilbert and E. Scholten, *RSC Adv.*, 2015, **5**, 47466–47475.
31. C. V. Nikiforidis and E. Scholten, *RSC Adv.*, 2014, **4**, 2466–2473.
32. F. Gandolfo, A. Bot and E. Flöter, *J. Am. Oil Chem. Soc.*, 2004, **81**, 1–6.
33. A. Bot and W. G. M. Agterof, *J. Am. Oil Chem. Soc.*, 2006, **83**, 513–521.
34. K. Uvanesh, S. S. Sagiri, K. Senthilguru, K. Pramanik, I. Banerjee, A. Anis, S. M. Al-Zahrani and K. Pal, *J. Food Sci.*, 2016, **81**, E380–E387.
35. F. G. Gandolfo, A. Bot and E. Flöter, *J. Am. Oil Chem. Soc.*, 2004, **81**, 1–6.
36. H. M. Schaink, K. F. van Malssen, S. Morgado-Alves, D. Kalnin and E. van der Linden, *Food Res. Int.*, 2007, **40**, 1185–1193.
37. C. Blach, A. J. Gravelle, F. Peyronel, J. Weiss, S. Barbut and A. G. Marangoni, *RSC Adv.*, 2016, **6**, 81151–81163.
38. F. M. AlHasawi and M. A. Rogers, *J. Am. Oil Chem. Soc.*, 2013, **90**, 1533–1540.
39. H. Sawalha, G. Margry, R. den Adel, P. Venema, A. Bot, E. Flöter and E. van der Linden, *Eur. J. Lipid Sci. Technol.*, 2013, **115**, 295–300.
40. H. Sawalha, P. Venema, A. Bot, E. Flöter, R. d. Adel and E. van der Linden, *J. Am. Oil Chem. Soc.*, 2015, **92**, 1651–1659.
41. S. Murdan, G. Gregoriadis and A. T. Florence, *J. Pharm. Sci.*, 1999, **88**, 608–614.
42. K. Uvanesh, S. S. Sagiri, I. Banerjee, H. Shaikh, K. Pramanik, A. Anis and K. Pal, *J. Am. Oil Chem. Soc.*, 2016, **93**, 711–719.
43. L.-J. Han, L. Li, L. Zhao, B. Li, G.-Q. Liu, X.-Q. Liu and X.-D. Wang, *Food Res. Int.*, 2013, **53**, 42–48.
44. T. Tamura and M. Ichikawa, *J. Am. Oil Chem. Soc.*, 1997, **74**, 491–495.
45. A. Noro, M. Hayashi and Y. Matsushita, *Soft Matter*, 2012, **8**, 6416–6429.
46. A. K. Zetzl, A. J. Gravelle, M. Kurylowicz, J. Dutcher, S. Barbut and A. G. Marangoni, *Food Struct.*, 2014, **2**, 27–40.
47. A. J. Gravelle, M. Davidovich-Pinhas, A. K. Zetzl, S. Barbut and A. G. Marangoni, *Carbohydr. Polym.*, 2016, **135**, 169–179.
48. M. Davidovich-Pinhas, S. Barbut and A. G. Marangoni, *Carbohydr. Polym.*, 2015, **117**, 869–878.

49. M. Davidovich-Pinhas, S. Barbut and A. G. Marangoni, *Annu. Rev. Food Sci. Technol.*, 2016, **7**, 65–91.
50. A. K. Zetzl, A. G. Marangoni and S. Barbut, *Food Funct.*, 2012, **3**, 327–337.
51. A. J. Gravelle, S. Barbut, M. Quinton and A. G. Marangoni, *J. Food Eng.*, 2014, **143**, 114–122.
52. M. Davidovich-Pinhas, S. Barbut and A. G. Marangoni, *Carbohydr. Polym.*, 2015, **127**, 355–362.
53. A. R. Patel and K. Dewettinck, *Eur. J. Lipid Sci. Technol.*, 2015, **117**, 1772–1781.
54. R. Mezzenga and S. Ulrich, *Langmuir*, 2010, **26**, 16658–16661.
55. A. R. Patel, N. Cludts, M. D. Bin Sintang, B. Lewille, A. Lesaffer and K. Dewettinck, *ChemPhysChem*, 2014, **15**, 3435–3439.
56. A. R. Patel, P. S. Rajarethinem, N. Cludts, B. Lewille, W. H. De Vos, A. Lesaffer and K. Dewettinck, *Langmuir*, 2015, **31**, 2065–2073.
57. A. R. Patel, D. Schatteman, A. Lesaffer and K. Dewettinck, *RSC Adv.*, 2013, **3**, 22900–22903.
58. Z.-M. Gao, X.-Q. Yang, N.-N. Wu, L.-J. Wang, J.-M. Wang, J. Guo and S.-W. Yin, *J. Agric. Food Chem.*, 2014, **62**, 2672–2678.
59. Z. Hu, H. S. Marway, H. Kasem, R. Pelton and E. D. Cranston, *ACS Macro Lett.*, 2016, **5**, 185–189.
60. A. R. Patel and K. Dewettinck, *Eur. J. Lipid Sci. Technol.*, 2015, **117**, 1772–1781.
61. A. R. Patel, N. Cludts, M. D. B. Sintang, A. Lesaffer and K. Dewettinck, *Food Funct.*, 2014, **5**, 2833–2841.
62. R. Tanti, S. Barbut and A. G. Marangoni, *Food Hydrocolloids*, 2016, **61**, 329–337.
63. R. Tanti, S. Barbut and A. G. Marangoni, *Food Hydrocolloids*, 2016, **61**, 399–408.
64. A. de Vries, J. Hendriks, E. van der Linden and E. Scholten, *Langmuir*, 2015, **31**, 13850–13859.
65. A. de Vries, A. Wesseling, E. van der Linden and E. Scholten, *J. Colloid Interface Sci.*, 2017, **486**, 75–83.
66. C. Doan, D. Van de Walle, K. Dewettinck and A. Patel, *J. Am. Oil Chem. Soc.*, 2015, **92**, 801–811.
67. A. R. Patel, M. Babaahmadi, A. Lesaffer and K. Dewettinck, *J. Agric. Food Chem.*, 2015, **63**, 4862–4869.
68. C. D. Doan, C. M. To, M. De Vrieze, F. Lynen, S. Danthine, A. Brown, K. Dewettinck and A. R. Patel, *Food Chem.*, 2017, **214**, 717–725.
69. B. Mert and I. Demirkesen, *LWT–Food Sci. Technol.*, 2016, **68**, 477–484.
70. B. Mert and I. Demirkesen, *Food Chem.*, 2016, **199**, 809–816.
71. H.-S. Hwang, M. Singh and S. Lee, *J. Food Sci.*, 2016, **81**, C1045–C1054.
72. X.-W. Chen, S.-Y. Fu, J.-J. Hou, J. Guo, J.-M. Wang and X.-Q. Yang, *Food Chem.*, 2016, **211**, 836–844.
73. I. Heertje, E. C. Roijers and H. A. C. M. Hendrickx, *LWT–Food Sci. Technol.*, 1998, **31**, 387–396.

74. F. C. Wang, A. J. Gravelle, A. I. Blake and A. G. Marangoni, *Curr. Opin. Food Sci.*, 2016, 7, 27–34.
75. F. C. Wang and A. G. Marangoni, *J. Colloid Interface Sci.*, 2016, **483**, 394–403.
76. S. Barbut, J. Wood and A. Marangoni, *Meat Sci.*, 2016, **122**, 84–89.
77. C. Ruiz-Capillas, P. Carmona, F. Jiménez-Colmenero and A. M. Herrero, *Food Chem.*, 2013, **141**, 3688–3694.
78. L. Salcedo-Sandoval, S. Cofrades, C. Ruiz-Capillas, J. Carballo and F. Jiménez-Colmenero, *Meat Sci.*, 2015, **101**, 95–102.
79. S. Barbut, J. Wood and A. Marangoni, *Meat Sci.*, 2016, **122**, 155–162.
80. S. Barbut, J. Wood and A. G. Marangoni, *J. Food Sci.*, 2016, **81**, C2183–C2188.
81. F. Jimenez-Colmenero, L. Salcedo-Sandoval, R. Bou, S. Cofrades, A. M. Herrero and C. Ruiz-Capillas, *Trends Food Sci. Technol.*, 2015, **44**, 177–188.
82. F. R. Lupi, D. Gabriele, N. Baldino, L. Seta, B. de Cindio and C. De Rose, *Eur. J. Lipid Sci. Technol.*, 2012, **114**, 1381–1389.
83. F. R. Lupi, D. Gabriele, L. Seta, N. Baldino and B. de Cindio, *Eur. J. Lipid Sci. Technol.*, 2014, **116**, 1734–1744.
84. H.-S. Hwang, M. Singh, E. Bakota, J. Winkler-Moser, S. Kim and S. Liu, *J. Am. Oil Chem. Soc.*, 2013, **90**, 1705–1712.
85. E. Yılmaz and M. Öğütcü, *J. Am. Oil Chem. Soc.*, 2014, **91**, 1007–1017.
86. E. Yılmaz and M. Öğütcü, *J. Food Sci.*, 2014, **79**, E1732–E1738.
87. E. Ylmaz and M. Ogutcu, *RSC Adv.*, 2015, **5**, 50259–50267.
88. D. C. Zulim Botega, A. G. Marangoni, A. K. Smith and H. D. Goff, *J. Food Sci.*, 2013, **78**, C1334–C1339.
89. H. L. Bemer, M. Limbaugh, E. D. Cramer, W. J. Harper and F. Maleky, *Food Res. Int.*, 2016, **85**, 67–75.
90. T. A. Stortz and A. G. Marangoni, *Food Res. Int.*, 2013, **51**, 797–803.
91. A. R. Patel, P. S. Rajarethinem, A. Gredowska, O. Turhan, A. Lesaffer, W. H. De Vos, D. Van de Walle and K. Dewettinck, *Food Funct.*, 2014, **5**, 645–652.
92. H. Si, L.-Z. Cheong, J. Huang, X. Wang and H. Zhang, *J. Am. Oil Chem. Soc.*, 2016, **93**, 1075–1084.
93. C. D. Doan, A. R. Patel, I. Tavernier, N. De Clercq, K. Van Raemdonck, D. Van de Walle, C. Delbaere and K. Dewettinck, *Eur. J. Lipid Sci. Technol.*, 2016, **118**, 1903–1914.
94. A.-L. Fameau, S. Lam, A. Arnould, C. Gaillard, O. D. Velev and A. Saint-Jalmes, *Langmuir*, 2015, **31**, 13501–13510.
95. C. M. O'Sullivan, S. Barbut and A. G. Marangoni, *Trends Food Sci. Technol.*, 2016, **57**(pt A), 59–73.
96. D. Tang and A. G. Marangoni, *J. Am. Oil Chem. Soc.*, 2006, **83**, 377–388.
97. A. R. Patel, B. Mankoc, M. D. Bin Sintang, A. Lesaffer and K. Dewettinck, *RSC Adv.*, 2015, **5**, 9703–9708.

Section II

Structuring Units

CHAPTER 2

Biobased Molecular Structuring Agents

SAI SATEESH SAGIRI[a], MALICK SAMATEH[a,b] AND GEORGE JOHN*[a,b]

[a]Department of Chemistry and Biochemistry & Center for Discovery and Innovation (CDI), The City College of New York, New York, NY 10031, USA;
[b]Ph.D. Program in Chemistry, The Graduate Center of the City University of New York, New York, NY 10016, USA
*E-mail: gjohn@ccny.cuny.edu

2.1   Introduction

Nature offers abundant biomass (as opportunities) in the form of carbohydrates, amino acids, proteins, fatty acids, triglycerides, waxes and nucleotides for designing structural and functional materials. Over the years, these functional molecules have been used in food, pharmaceutical, medicinal and cosmetic products. However, their use has hitherto been restricted to adjunct or supplementary roles in commercial products. In the contemporary market, few products are available with bioderived functional molecules as major constituents. For instance, food products such as margarine, shortening, spreads, chocolates, bakery and meat products possess either plant- or animal-derived fats as one of their major ingredients. These fats are largely responsible for differences in textural properties (consistency, spreadability, hardness, firmness, brittleness, *etc.*) and organoleptic properties (sensory attributes: appearance, smell, taste and touch). Since fat is a key

Food Chemistry, Function and Analysis No. 3
Edible Oil Structuring: Concepts, Methods and Applications
Edited by Ashok R. Patel
© The Royal Society of Chemistry 2018
Published by the Royal Society of Chemistry, www.rsc.org

component, food industries tend to use a wide range of fat sources for different applications. The major animal-derived solid fats currently being used in food products are butter, beef fat (tallow), pork fat (lard) and chicken fat. Besides animal sources, plant-derived vegetable oils containing saturated fatty acids, *viz.* coconut oil, palm kernel oil and palm oil, are also being used in many food products. Though the oils with saturated fatty acids are largely used, vegetable oils with lower saturated fat content such as canola oil, corn oil, cotton seed oil, olive oil, soybean oil and sunflower oil are also employed for some specific applications.

Saturated fatty acids contribute to the solid fraction of the oil. Therefore, the use of vegetable oils containing a high proportion of saturated fatty acids has become a common practice at the industrial level. The solid fraction of the total fat in an oil/fat source is improved by adding saturated fats to the liquid oils or modifying the unsaturated fatty acids present in the liquid oils into saturated or *trans* fats. The phenomenon of converting liquid oils to semi-solid fats is called oil structuring. In general, structuring is mainly affected by the solid component present in vegetable oils, which is commonly referred to as the structuring agent. At the industrial level, structuring of oils has been achieved mainly by either complete or partial hydrogenation methods. Despite the culinary and economic allure of these chemical treatments, food industries face several challenges in developing market-acceptable food products. These challenges include (a) achieving functional characteristics (texture, snap, appearance and stability), (b) minimizing costs, (c) avoiding adverse health implications (hydrogenation of oils yield *trans* fats, which are reported to be responsible for cardiovascular diseases) and (d) the ability to modify physical properties. Thus, there is a need for unremitting effort to develop new and improved functional and healthy fat materials for food applications. The solidification of vegetable oils using chemical modification (partial hydrogenation) yields undesirable *trans* fats. Thus, scientists have been trying to negate this problem by investigating methods such as mixing liquid oils with one or more external agents. These external agents or structuring agents can transform liquid oils to self-standing solid gels. Structured oils prepared *via* gelation are referred to as organogels, oleogels or lipid gels. Over the past two decades, various structuring agents have either been discovered or designed for vegetable oil structuring. These molecular gelators offer a potential alternative to the commercial/traditional edible oil structuring processes *viz.* complete/partial hydrogenation.

Molecular gelation resembles the mechanistic entrapment of liquid molecules within a fat crystal network as observed in butter, lard, margarines and shortenings. The direct replacement of solid fats with structured fats is technically challenging because the mouthfeel and textural properties of food products are mainly due to the saturated fats. Structured vegetable oils (*e.g.* oleogels, emulsions) possess the functionality of fats with the nutritional profile of vegetable oils (high unsaturated fatty acids, low saturated fatty acids and free of *trans* fats). As a result, structured oils can act as solid fat mimicking agents (fat replacers, fat alternatives) in shortenings, margarines, ice creams, chocolates, cookies, bakery products and meat products. In general, oil leakage and textural problems are the major concerns with the use of unsaturated fats or liquid oils in food products. These concerns can be answered by using structured vegetable oils as they have shown significant reduction in the oil leakage or oiling out and improvement in the textural properties of food products.[1] In the recent past, formulations containing structured oils have also been explored as delivery vehicles for drugs, nutraceuticals and other active ingredients.[2,3] The structuring strategies for vegetable oils and the possible applications of the resulting structured vegetable oils are schematically represented in Scheme 2.1. The development of bioderived structuring agents has enhanced the scope of molecular gels, emulsions and related formulations in food, pharmaceutical, cosmetics and petrochemical industries towards green and sustainable value added products.

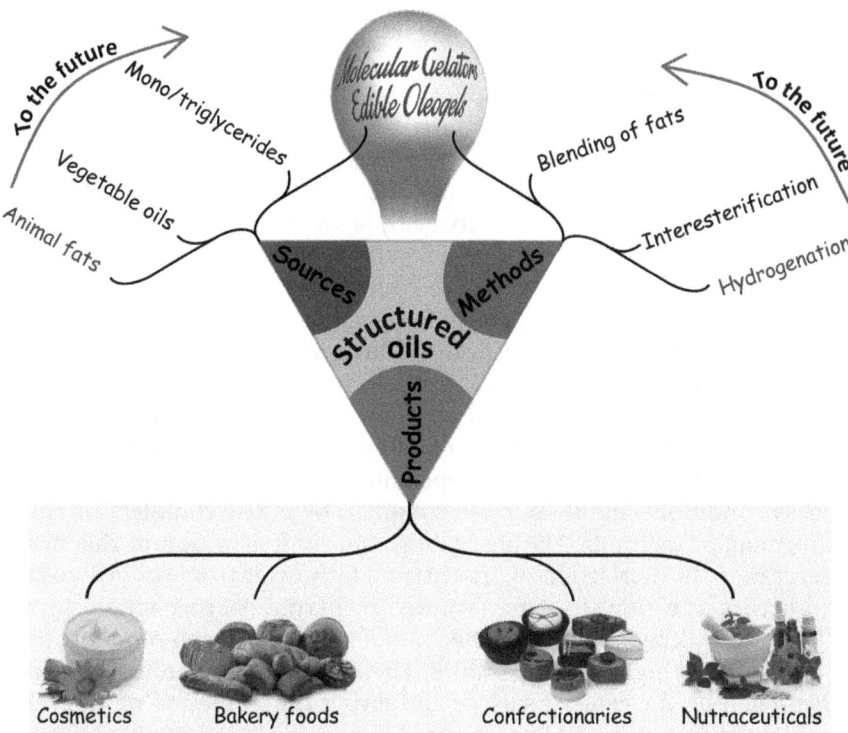

Scheme 2.1 Structuring strategies of vegetable/edible oils and the possible applications of structured oils.

2.2   Vegetable Oil Structuring: Chemical Methods

2.2.1   Hydrogenation

The successful creation of innumerable confectioneries, chocolates, bakery and dairy-based food products can be attributed to a simple reaction: hydrogenation of edible oils, invented in 1902.[4] This technology was patented in 1906 by the British company Joseph Crossfield and Sons and established the protocol for hydrogenation of edible oils.[5] This technology was introduced in the USA by Procter and Gamble (Cincinnati, OH) in 1911.[5] Then it was recognized that animal fat (butter) could be substituted with vegetable oils by formulating margarine. However, the technology for hydrogenation of edible oils was not extensively used until the introduction of margarine to the market by food industries in the 1950s. Margarine was found to have a longer shelf-life, higher melting point, lower cost and better organoleptic properties compared to butter and lard. Since then, various kinds of margarines, shortenings, vegetable spreads, creams and pastes have been synthesized using edible oils. The main objective of hydrogenation is to improve the oxidative stability of liquid oils by reducing the concentration of unsaturated fatty acids, which are prone to oxidation at room temperature. Hydrogenation changes the physical properties of liquid oils by converting unsaturated fatty acids to saturated fatty acids, which are stable and solid at room temperature, thus extending their application.[6] Hence, the addition of hydrogen results in the reduction of unsaturated fatty acids to saturated fatty acids, which enhances the van der Waals interactions between the aliphatic chains to form solidified oils. The conversion of oleic acid (monounsaturated fatty acid) to stearic acid (saturated fatty acid) is schematically shown in Figure 2.1. The chemical process for hydrogenation of vegetable oils involves the addition of hydrogen atoms to the unsaturated carbons of fatty acids in the presence of a catalyst, namely, nickel, copper, platinum, or palladium. Nickel has been used most extensively at the industrial level owing to its good catalytic potential, selectivity, reusability, less influence on unsaturated fatty acid oxidation and cost effectiveness compared to platinum and palladium. Hydrogenation mainly depends on oil temperature, hydrogen pressure, rate of mixing, reaction time, catalyst type and concentration. According to the process conditions, edible oils are solidified by either complete or partial hydrogenation methods. During hydrogenation, in addition to the change in saturation, isomerisation of unsaturated fatty acids from *cis* form to *trans* form occurs as a side reaction, favoured by partial hydrogenation. In general, partial hydrogenation can be achieved at low hydrogen pressure, minimal mixing and high oil temperature. These conditions lead to the scarcity of hydrogen on the catalyst surface and favour the formation of *trans* fats.[6] Complete hydrogenation of unsaturated fatty acids in liquid oils lead to the formation of hard or solid fat, while partial hydrogenation leads to soft or semi-solid fat. Thus, the textural and organoleptic properties of the food products are being fine-tuned by the extent of hydrogenation (from 0% to

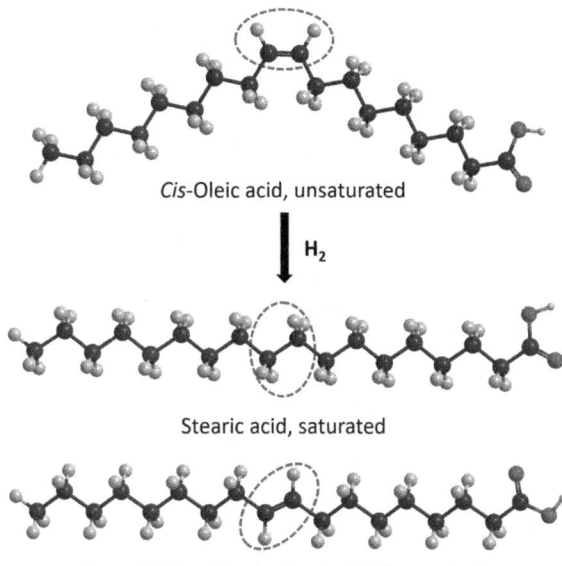

Cis-Oleic acid, unsaturated

H₂

Stearic acid, saturated

Trans-Oleic acid, unsaturated (side product)

Figure 2.1 Conversion of oleic acid to stearic acid, showing the formation of *trans* oleic acid during partial hydrogenation.

100%) in vegetable oils. Regardless of the advantages, hydrogenation yields *trans* fats, which are undesirable.[7] Hence, the food industry and academic researchers are on a quest to find sustainable vegetable oil structuring strategies that provide the benefits of healthy fats whilst reducing the production of unhealthier forms of fat as by-products. The challenge then is to control *cis/trans* isomerisation, since *trans* fats are suspected to increase the incidence of cardiovascular diseases.

2.2.2 Interesterification

Interesterification of oils was proposed and patented for the first time in 1920 by Wilhelm Normann, who was also the inventor of vegetable oil hydrogenation.[8] However, this was not fully commercialized until the 1960s as an alternative to hydrogenation. While hydrogenation yields *trans* fatty acids in the process of generating fats with desired melting points, interesterification produces customized fats with a wide range of melting points and zero *trans* fatty acids, making it ideal for commercial food applications. It is an acyl rearrangement reaction that improves or alters the functionality and physical properties of fats and oils. Through interesterification, the molecular composition of triglycerides can be varied without affecting their fatty acid composition. In this process, fatty acids are exchanged within or between triglycerides either chemically or enzymatically in a random or controlled manner. During chemical interesterification, the fatty acid esters are changed into fats when heated to 80 °C in the presence of sodium methoxide as a catalyst. It is carried out first by the hydrolysis of triacylglycerides (TAGs) to yield fatty acids, which are then subsequently re-esterified on the glycerol backbone in a random fashion. Since chemical interesterification yields random fatty acid esters and involves a substantial loss of oil (approximately 30%) owing to the formation of fatty acid methyl esters, microbial lipases were investigated as alternative catalysts. Enzymatic interesterification using stereoselective lipases (*e.g.* 1,3-lipases) is preferred over chemical interesterification as the former produces specific triglycerides. However, enzymatic catalysis also involves some drawbacks, *viz.* longer reaction times (hours to days), sensitivity to reaction parameters (pH, temperature) and high cost.

In general, interesterification is done by blending oils with a high proportion of saturated fatty acids (palm oil, coconut oil and palm kernel oil) or hydrogenated solid fats with unsaturated, liquid edible oils so as to achieve fats with intermediate characteristics. Complete hydrogenation yields virtually no unsaturated fatty acids—and therefore no *trans* fats—generating hardstock, which is blended with unsaturated oils prior to interesterification. For example, completely hydrogenated soybean oil is interesterified with natural soybean oil to provide a suitable oil blend for the synthesis of margarine. Some of the commercial oil blends or fats are given in Table 2.1.

2.2.3 Fractionation and Fat Blending

Vegetable oils with high proportions of saturated fatty acids (50%), and thus high melting points, have been considered as suitable candidates to replace partially hydrogenated oils in bakery and confectionary food products. However, most researchers and much of the general public have the

Table 2.1 Some commercial oil blends with traditionally structured fats.[a]

Product	Key ingredients	Applications
Crisco® all vegetable shortenings[b]	Soybean oil, fully hydrogenated palm oil, partially hydrogenated palm and soybean oils, mono- and diglycerides	Multiple uses
Elite Vreamay RighT[c]	Partially hydrogenated shortening from soybean and cottonseed oils	Cakes and icings
BUNGE ULTRA 219[c]	Blend of enzymatic interesterified soybean oil and hydrogenated soybean oil	Very cold pie doughs
BUNGE ULTRA 327, 368[c]	Soybean oil, interesterified soybean oil, hydrogenated cottonseed oil, with mono- and diglycerides and polysorbate 60	Icing shortenings
Regal™ Donut shortening[d]	Palm oil, soybean oil, and hydrogenated cotton seed oil	Donuts
Regal™ Puff Pastry shortening[d]	Soybean oil and hydrogenated soybean oil	Puffs and pastries

Regal™ all-purpose shortening[d]	Soybean oil and hydrogenated soybean oil	Multiple uses
Regal™ Cake & icing shortening[d]	Soybean oil, hydrogenated soybean oil, mono and diglycerides, and polysorbate 60	Cakes and icings
Caprenin[e]	Chemical interesterified coconut oil, palm kernel oil, and rapeseed oil, mono- and diacylglycerides	Confectionaries and soft candies
Benefat or salatrim[f]	Base catalyzed interesterified fully hydrogenated short chain tri-glycerides (acetic/propionic/butyric acids) with long chain triglycerides (stearic acid)	Chocolates, bakery and dairy products
Captex®[g]	Interesterified medium chain tri-glycerides (caprylic acids/capric) from coconut and palm kernel oils	Personal care products (lipstick, foundations)
NEOBEE®[h]	Interesterified medium chain tri-glycerides (caprylic acids/capric) from coconut and palm kernel oils	Confectionaries, baked foods, mixes, seasonings
BUNGE ULTRA 133, 157[c]	Enzymatic interesterified soybean oil	All-purpose shortenings
BUNGE ULTRA 415[c]	Enzymatic interesterified soybean oil	Donut frying shortening
BUNGE ULTRA 587[c]	Enzymatic interesterified soybean oil and monoglycerides	Bakers' margarine
Regal™ Icing Shortening NH[c]	Palm oil, canola oil, mono and diglycerides, polysorbate 60	Cakes
BUNGE NH 100 VREAM[c]	Palm oil	Very firm all-purpose shortenings
BUNGE NH 112[c]	Canola oil, palm oil, and palm kernel oil	Firm all-purpose shortenings
BUNGE NH 150[c]	Soybean oil and palm oil	Soft all-purpose shortenings
BUNGE NH 112[c]	Canola oil, palm oil, and palm kernel oil	Very soft all-purpose shortenings
BUNGE NH 200, 204[c]	Palm oil and soybean oil, with mono- and di-glycerides	Industrial pie manufacture
BUNGE NH 300 series[c]	Different ratios of palm oil, soybean oil, mono- and di-glycerides with varied saturated fats/total fats	Cake, bread, pastry, and icing shortenings

[a]NH: Non-hydrogenated fat blends.
[b]Smucker company, Orrville, Ohio.
[c]Bunge Oils Inc., St Louis, Missouri.
[d]Cargill Inc, Minneapolis, Minnesota.
[e]Proctor & Gamble Co. Cincinnati, Ohio.
[f]Cultor Food Science, NY, New York.
[g] ABITEC Corp., Columbus, Ohio.
[h]Stepan company, Maywood, New Jersey.

misconception that saturated fats are dangerous to health. Since the 1970s, diet regulations in the USA and other countries have been based on the hypothesis that saturated fats raise bad or LDL (low-density lipoprotein) cholesterol in blood, which blocks arteries and eventually causes heart disease. The initial Dietary Goals for Americans (1977) recommended higher carbohydrate intake and lower saturated fat, cholesterol and salt to combat obesity, diabetes and cardiovascular diseases.[9] Though the underlying hypothesis for the mentioned dietary rules remains unproven, the dietary shift is already taking place and the concerns over the diseases are still growing. Interestingly, several recent studies have found no link between saturated fat consumption and heart diseases. These studies have also quoted that the dietary regulations are based on misconceptions associated with saturated fats.[10-12] Furthermore, some studies have found that higher carbohydrate intake may be associated with a rise in some of the aforementioned diseases. For example, Schwab *et al.* and Siri-Tarino *et al.* reported that substituting saturated fats with refined carbohydrates may increase the risk of cardiovascular diseases.[13,14] Over the years, the role of saturated fats in body metabolism has been analysed. The fats are found to play a crucial role in strengthening the immune system and providing energy and structural integrity to the cells.[15,16] More importantly, saturated fats have the potential to lower cholesterol.[14] Moreover, short and medium chain saturated fatty acids possess antimicrobial properties as well.[15,17] These dietary and health benefits associated with saturated fats have motivated members of the food industry to search for vegetable oils that contain a high proportion of saturated fatty acids as alternative structuring agents in their food products.

At the industrial level, the most commonly used oils are palm oil, palm kernel oil and coconut oil. Palm fruit and kernel oils are rich in long chain saturated fatty acids such as palmitic acid (C_{16}) and stearic acid (C_{18}). On the other hand, coconut oil possesses a higher proportion of medium chain saturated fatty acids such as lauric acid (C_{12}) and myristic acid (C_{14}).[18] Depending on the type of food application, these tropical oils are used as whole or fractionated oils. Simple physical methods like fractionation, followed by isolation of fatty acids, facilitate the use of these oils for suitable applications. Palm oil is an excellent natural oil for fractionation and fat blending as it contains high melting triglycerides, which can be fractionated by either dry fractionation or solvent extraction.[19] Typical palm oil fractions are palm olein (soft fat), which is used as the main ingredient in cooking and salad oils, and palm stearin (hard fat), in margarines, shortenings and frying fats. In the process of achieving different grades of palm stearin with diverse melting points, partial hydrogenation is employed. Thus, it seems impossible to totally replace complete/partially hydrogenated oils with tropical oils. It is in this context that gelators have emerged as the paramount way to structure edible oils without adding saturated and *trans* fats. The evolution and progress of edible oil structuring is represented in Scheme 2.2 with a few examples of commercial food products.

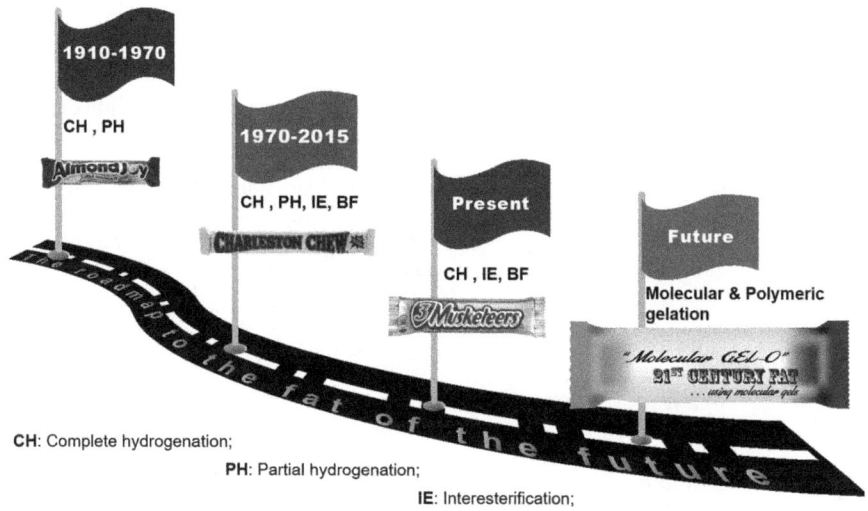

Scheme 2.2 The evolution of edible oil structuring procedures.

2.3 Vegetable Oil Structuring: Biobased Methods

As mentioned earlier, the main objective of structuring is to solidify the liquid vegetable oil. To achieve this, biobased strategies follow a simple method where the liquid oils are mixed with external structuring agents. The added molecules provide the necessary crystallinity and other desired textural properties to vegetable oils, and subsequently to the end-formulations. However, formulating "solid fat free" food products using structuring agents, without compromising the food texture and mouthfeel, is challenging. Structuring of vegetable oils using gelators is a relatively new area of research in lipid chemistry, with several gelators being identified every year. The most familiar biobased structuring agents are mono-, di-, and tri-glycerides, fatty acids, gums, waxes, phytochemicals, polymers, carbohydrate derivatives and proteins. For convenience, bioderived gelators have been divided into two categories, namely low molecular weight gelators (LMWGs) and polymeric gelators. These bioderived gelators have shown potential for structuring a wide range of edible oils (Table 2.2). Addition of gelators must preserve or improve the structural and palatable characteristics (*e.g.* mouth feel, plasticity and flavour) of the liquid oils. In this regard, edible oils are structured by either small molecules such as LMWGs or macromolecules such as polymers or proteins.

2.3.1 Molecular Gelators or Low Molecular Weight Gelators (LMWGs)

Molecular structuring agents or molecular gelators are a novel class of amphiphiles that have the potential to convert liquid organic solvents into semi-solid organogels (soft materials) at very low concentrations.

Table 2.2 Edible oil structuring bioderived gelators.

Category	Type of gelators	Edible oils
Low molecular weight gelators (LMWGs)	Mono-, di-, and tri-acyl glycerides	Avocado oil,[20,21] canola oil,[22] cod liver oil,[23] hazelnut oil,[24,25] olive oil,[26] safflower oil[27]
	Fatty acids	Castor oil,[28] lavender oil,[28] peanut oil,[28] rapeseed oil,[28,29] safflower oil,[30] sesame oil,[31] soybean oil,[29,31] sunflower oil[28,29]
	Fatty alcohols	Castor oil,[28] lavender oil,[28] olive oil,[32] peanut oil,[28] rapeseed oil,[28,29] soybean oil,[29] sunflower oil[28,29]
	Fatty acids + fatty alcohols	Rapeseed oil,[29] soybean oil,[29] sunflower oil[29,33]
	γ-Oryzanol + phytosterols	Sunflower oil[34–36]
	Ceramides	Canola oil[37,38]
	Cocoa butter	Olive oil[39,40]
	Sorbitan derivatives *e.g.* Span 60, Span 40	Almond oil,[41] avocado oil,[21] castor oil,[42] cottonseed oil,[43] corn oil,[43] olive oil,[43,44] sesame oil,[43] soybean oil,[43] sunflower oil[45,46]
	Lecithin	Sunflower oil[45]
	Carbohydrate derivatives	Canola oil,[47,48] coconut oil,[47] grape seed oil,[47,48] olive oil,[47,48] soybean oil,[47,48] sunflower oil,[47] jojoba oil[49]
	Beeswax	Avocado oil,[21] canola oil,[50] cod liver oil,[51] corn oil,[50] hazelnut oil,[24,52,53] olive oil,[50,53,54] safflower oil,[50] soybean oil,[50] sunflower oil[50]
	Candelilla wax	Canola oil,[55,56] safflower oil,[57–60] soybean oil,[61,62] sunflower oil[63]
	Carnauba wax	Canola oil,[56] cod liver oil,[51] olive oil,[64] sunflower oil[63]
	Rice bran wax	Canola oil,[56] soybean oil,[62] sunflower oil[63,65]
	Sunflower wax	Canola oil,[56] hazelnut oil,[53] olive oil,[52,53] soybean oil[62]
Polymeric gelators	Proteins *e.g.* β-lactoglobulin, gelatin	Olive oil,[66] sunflower oil[67]
	Gelatin+ xanthan gum	Sunflower oil[67]
	Resins *e.g.* Shellac	Rapeseed oil[68]
	Ethyl cellulose	Canola oil,[69–71] flaxseed oil,[70,71] soybean oil[70,71]
	Cellulose derivatives + xanthan gum	Sunflower oil[72,73]

Analogous to hydrogels (components: solid hydrogelator and water), organogels are also constituted by two major components: solid organogelator and organic solvent. During gelation, the solid organogelator immobilizes the solvent by forming self-assembled, three-dimensional networks.[2] The liquid gets trapped by capillary forces within the gelator network, thus forming a viscoelastic, semi-solid material—a gel (from the Latin *gelatus* meaning "frozen, immobile"). The same principle is adapted to synthesize oleogels in which vegetable oil is used instead of the liquid organic solvent. In short, oleogels are organogels with vegetable oil as the liquid component. Molecular gelators are small molecules (molecular weight < 3.0 kDa) that, when dispersed in a solvent, either undergo self-assembly or crystallization to form three-dimensional architecture.[2] Non-covalent intermolecular forces/attractions like London forces, van der Waals forces and hydrogen bonding are involved in the formation of the three-dimensional network during gelation. Since non-covalent interactions play a major role in the self-assembly of gelators, the network can be destroyed by either shearing or heating. Oleogels undergo deformation under high shear or above the gel-to-sol transition temperature (T_g) and can be reconstituted by lowering the shear or by temperature. Thus, thixotropy and thermoreversibility are two common phenomena associated with oleogels. These features make oleogels very intriguing and attractive for food, pharmaceutical, cosmetic and medical applications.

In general, molecular gelators undergo two kinds of gelation behaviour, namely, self-assembly and formation of crystal particles, to create network architecture in edible oils. In the context of self-assembly, the gelator molecules undergo molecular-level self-organization and form hierarchical three-dimensional structures in the oil phase. In the other system, gelators undergo stochastic nucleation, which results in the formation of crystals that immobilize the oil phase by forming network structures. A list of LMWGs with either of these phenomena is given in Table 2.3.

2.3.1.1 Oleogelators with Self-assembly Mechanism

Amphiphiles self-assemble *via* highly specific interactions in oleogels, promoting one-dimensional growth towards supramolecular hierarchical structures, *viz*. micelles and liquid crystals. These in turn organize themselves as self-assembled fibrillar networks (SAFiNs) such as tubules, fibers and helices (Scheme 2.3). In the late 1980s, lecithin was reported as an organogelator

Table 2.3 Classification of molecular gelators.

Gelation behaviour	Low molecular weight gelators (LMWGs)
Self-assembly	Monoacyl glycerols
	Ceramides
	Phytosterols + oryzanol
	Sorbitan derivatives
	Carbohydrate derivatives
Crystal particles	Fatty acids, fatty alcohols
	Fatty acids + fatty alcohols
	Hydroxylated fatty acids
	Waxes, wax esters
	Triacyl glycerols

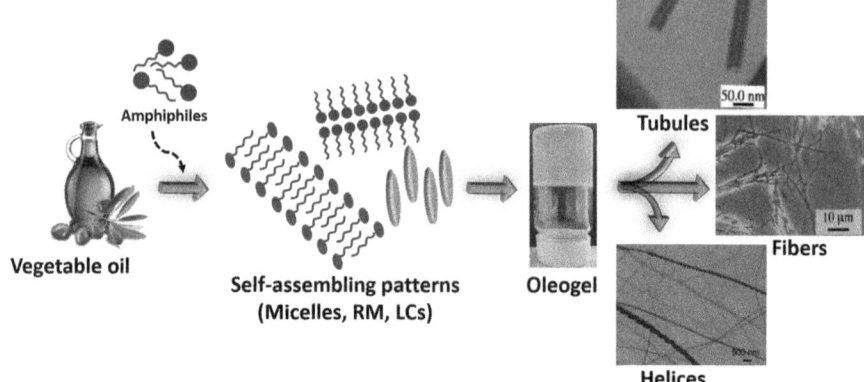

Vegetable oil → Amphiphiles → Self-assembling patterns (Micelles, RM, LCs) → Oleogel → Tubules / Fibers / Helices

Scheme 2.3   Self-assembling process of amphiphilic molecular gelators (RM: reverse micelles, LCs: liquid crystals) (SEM images are reprinted from ref. 73 with permission from the Royal Society of Chemistry and from ref. 72, *Advanced Materials*, G. John *et al.*, Nanotube formation from renewable resources *via* coiled nanofibers, Copyright © 2001 WILEY-VCH Verlag GmbH & Co. KGaA, Weinheim).

for the first time as it formed a jelly-like material in organic solvents. When a small amount of water is added, lecithin self-assembles to form entangled reverse worm-like micelles (2.0–2.5 nm in the radius and hundreds to thousands of micrometers in length) and immobilizes vegetable oils. In order to form these micelles, soy or egg lecithin must contain more than 95% phosphotidylcholine. Similar to lecithin, an entangled reverse micellar network is formed by sorbitan derivatives (non-ionic surfactants or amphiphiles). In vegetable oils, bilayered reverse micelles of sorbitan derivatives (sorbitan monostearate or Span 60, sorbitan monopalmitate or Span 40) self-assemble as toroidal vesicles, which further elongate or develop into rod-shaped tubules.[41] Addition of Tween (Tween 20) modifies the rod-shaped tubules and imparts stability to vegetable oil/Span 60-based oleogels.

In pure oils or a non-aqueous environment, monoacylglycerols (MAGs) orient themselves as inverse micelles, which act as precursors for the formation of reverse bilayered lamellar structures. Above a certain gelator concentration, the reverse bilayers acquire gel-like behaviour by forming an entangled network in liquid oil. Ojijo *et al.* characterized the gelation behaviour of a monoglyceride mixture of monopalmitoylglycerol (46.44%) and monostearoylglycerol (53.56%) in olive oil.[74] In aqueous phase, MAGs exhibit numerous mesomorphic phases, *viz.* cubic, micellar, lamellar and hexagonal structures owing to their amphiphilic nature. Hence, MAGs have the potential to structure lipid and aqueous phases and are thus being used commercially as an alternative to TAGs in spreads, shortenings and other fat-containing food products. During emulsification upon heating, MAGs form an L_α liquid-crystalline lamellar layer around the water droplet. When cooled below the Kraft point, the metastable α-phase thermodynamically transforms to the more stable β-phase plate-like crystals. These crystals can embed large amounts of water and impart a similar mouthfeel sensation to fat crystals in spreads and margarine.

Akin to MAGs, sterol derivatives such as plant sterols and sterol esters can also be used as lipid structurants or LMWGs. Phytosterols (cholesterol, dihydrocholesterol and stigmasterol) in combination with sterol ester (γ-oryzanol) have shown gelation ability in edible oils. These structuring agents have a history of human consumption; γ-oryzanol is naturally available in rice bran oil while phytosterols are commonly found in trace amounts in many vegetable oils. The hypocholesterolemic effect (lowering the serum levels of LDL cholesterol, which causes cardiovascular diseases) of phytosterols is evident and they have been used in margarines as a nutraceutical.[75] A γ-oryzanol and β-sitosterol mixture is capable of gelling sunflower oil at low concentrations, between 2% and 4%.[33] Phytosterol-based gels are transparent and form firm gels at lower concentrations as they self-assemble to nanoscale tubules while forming secondary structures. These tubules aggregate and interact to form a supramolecular space-filled network that immobilizes vegetable oil. The structural diversity of phytosterols plays an important role in their gelation ability. The hydroxyl group enhances the gelation potential, whereas the opposite effect was observed by means of the double bond in the chemical structure of phytosterols (Figure 2.2).[33] Hence, cholestane and ergosterol are incapable of forming gels as the former lacks an –OH group and the latter possesses two double bonds (Figure 2.2). The hydroxyl group at C-3 of the sterol ring is important for sterols and sterol ester assembly in oleogels. Hydrogen bonding between the phytosterol's hydroxyl group and the sterol ester's carbonyl oxygen allow them to stack in a wedge-shape instead of parallel to each other. Further stacking of the phytosterol-γ-oryzanol pair results in curved supramolecular helical tubules/ribbons, which are formed by stacking out the ferulic acid moiety of γ-oryzanol. These supramolecular structures grow along a single dimension and form tubular microstructures similar to a spiral staircase. The liquid oil phase is present both within and outside the tubules. The double bonds in the B-ring of sterols have a negative effect on the gelation kinetics. The increase in the number of double bonds increases the gelation time and decreases the tubule diameter. When sheared, the tubular structures of the sterol–sterol ester mixture aggregate to form a fibrillar network as tertiary structures in the oleogels. Based on the observations, the sterol–sterol ester mixture follows a two-step gelation process. In the first step, the gelator mixture forms tubular nanostructures, controlled by hydrogen bonding. In the second step, the tubular nanostructures aggregate to form a gel network under the influence of shear. The rate of collisions between nanostructures increases the rate of formation of the gel network and enhances the mechanical properties of oleogels.

Ceramides, a class of polar lipids, were the first food-grade vegetable oil structured gelators discovered. Ceramides differ from each other in terms of their fatty acid carbon chain length, degree of saturation and chemical substitution. Ceramides immobilize vegetable oils at 2 wt% and form different self-assembled structures depending on their purity. Pure ceramides

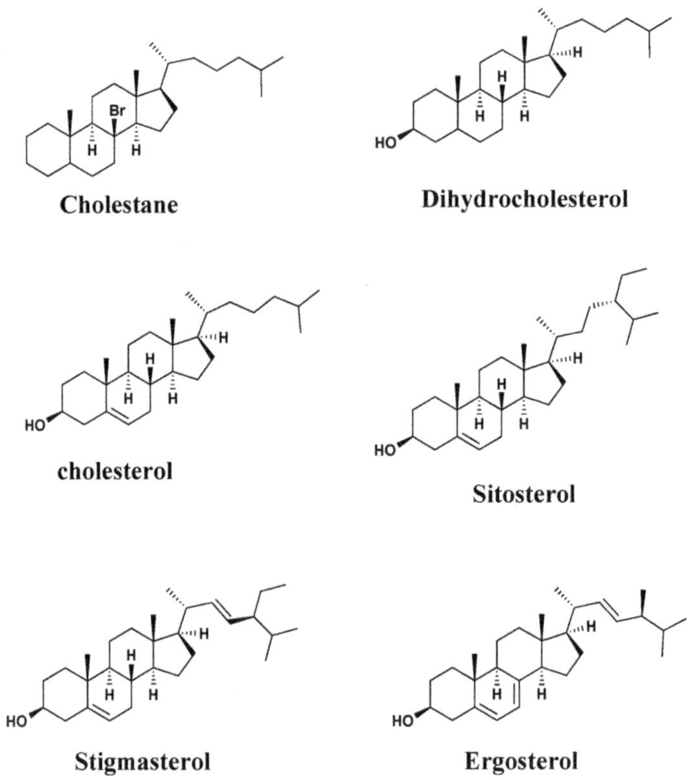

Figure 2.2 The chemical structures of phytosterols.

(ceramides with a short carbon chain, C_2) form long thin fibers, contrary to the small crystals formed by mixed ceramide systems (ceramides with long carbon chains, C_{16}–C_{24}). The positive health effects associated with ceramides were demonstrated by feeding them to rats along with sphinoglipids. The results showed a decrease in total serum cholesterol of 30% and an improvement in the chemical composition of serum lipoproteins.

In general, MAGs, TAGs, lecithin, phytosterols and ceramides are complex natural materials with the capability to gelate vegetable oils. Taking inspiration from the above, our group has been working to design new vegetable oil structuring gelators based on biomimicry since 2006.[76,77] The developed molecules are molecularly defined and simple, and have the ability to structure a wide range of vegetable oils for targeted applications. Similar to the natural gelators, the bioderived molecular gelators have the oleogelation ability by self-assembly process. Fatty acid diesters of trehalose (a disaccharide), sorbitol, mannitol (six-carbon sugar alcohol) and raspberry ketone glucoside were all synthesized by following a simple one-step biocatalytic pathway. The synthesis of oleogelators and their gelation behaviour *via* SAFiNs are explained in detail in the forthcoming section on the design, synthesis and self-assembly of molecular gelators as next-generation oil structuring agents.

2.3.1.2 Oleogelators with Crystal Particles System

There have been numerous oleogelators that can form a continuous network of crystals to immobilize vegetable oils. As mentioned before, this system mainly involves nucleation followed by the growth of crystals from nucleation sites. Fatty acids, fatty alcohols and waxes are important oleogelators under this category.

It is well known that molecular gelators are capable of gelling large volume of edible oil at low concentrations. This kind of behaviour of molecular gelators is one of the most interesting features that attracted scientists and motivated them to explore the potential of these materials in the first place. The gelation potential or gelling efficiency of molecular gelators depends on the network formed during gelation of edible oils. The crystals formed by the gelator molecules tend to aggregate to form SAFiNs during gelation and possess better gelation efficiency compared to the other class of molecular gelators. In liquid oils, SAFiNs are formed by long, thin, needle-like fibers spanning hundreds of micrometers in length. The high aspect ratio (length:-diameter) enhances gelator–solvent interactions by capillary forces and other physical forces, which in turn increases the gelation efficiency of molecular gelators. The network architecture of molecular gels involves the hierarchical arrangement of self-assembled gelator molecules from nanoscale primary structures to macroscale tertiary supramolecular structures. The architecture of a typical molecular gel includes primary, secondary and tertiary structures.[78,79] Primary and secondary structures can be modulated by designing the gelator and solvent molecules. On the other hand, the tertiary structure affects the macromolecular properties of the molecular gel. Aggregation or self-assembly of gelator molecules constitute the primary structure (*e.g.* lamellae). Unidirectional assembly of primary structures forms fibrous or secondary structures, such as rods, tubules and sheets. Further, the interconnection of secondary structures leads to the formation of a tertiary supramolecular network structure *i.e.* SAFiN. Physical interactions are in general responsible for the organization of secondary structures (fibers) into supramolecular structures. The fibrous network is held together by either transient or permanent junctions/interactions (Scheme 2.4). In the gelator network, transient junctions are seen as fiber entanglements, which are supported by dipolar and/or van der Waals forces. During branching, permanent junctions are formed at the tips or side walls of fibers by following either tip-branching or side-branching, respectively. The formation of SAFiNs from secondary

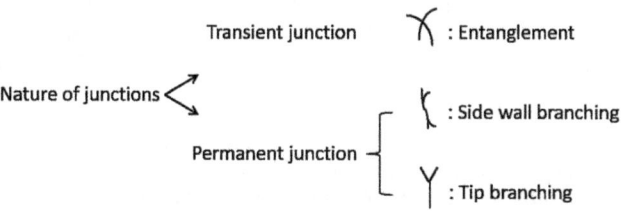

Scheme 2.4 Nature of junctions and types of branching in gelator networks.

structures can be understood by two mechanisms: (i) fibrous growth model and (ii) spherulitic growth model.

In the fibrous growth model, the fibers grow unidirectionally after nucleation and form a network mainly through entanglement of fibers. Here, gelation is triggered by instantaneous nucleation followed by one-dimensional growth of fibers so as to attain SAFiNs.[80,81] The spherulitic model involves branching of fibers that occurs to form tertiary supramolecular SAFiNs subsequent to the primary nucleation and one-dimensional growth of fibers. At the time of gelation or sol-to-gel transition, the solubility of the gelator decreases and it separates out as solid nuclei. From the nucleation sites, the nuclei grow as crystalline particles and form spherulites. These spherulites emanate radially from the stochastic nucleation sites by alternating fiber growth and branching until they impinge upon the neighbouring spherulites. The non-covalent interactions between the spherulites establish the gelator network in molecular gels.

After nucleation in both models, crystallographic mismatch branching (CMB) may occur when new layers of fibrils undergo a certain degree of structural mismatch during the fiber growth. At the site of mismatch, branching occurs either at the tips or side walls which lead to the formation of new fibers *i.e.* secondary nucleation. Thus, heterogeneous nucleation is considered to be associated with CMB, also called non-crystallographic branching. CMB can be associated with either the fibrous growth mode or the spherulitic growth mode of gel network formation. It has been proven that crystallographic mismatch nucleation of SAFiNs can be controlled by supersaturation and the rate of cooling of the system. Under low supersaturation of gelators, fibers grow one-dimensionally to form less branched fibrils, whose entanglement results in a fibrous mode of network (Figure 2.3). On the other

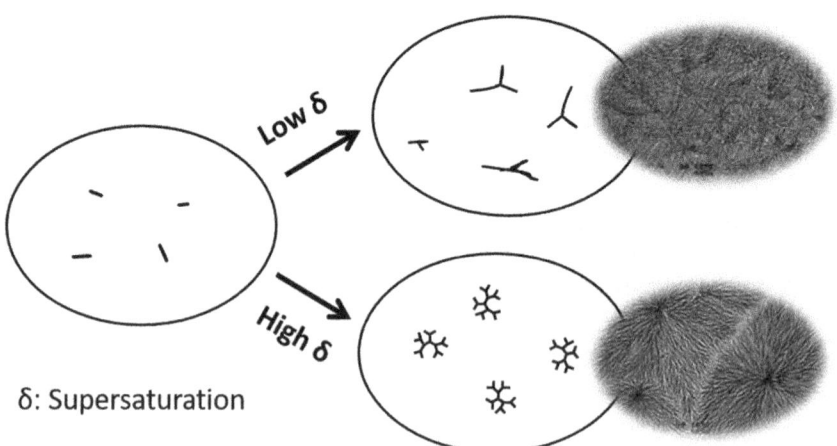

Figure 2.3   Effect of gelator saturation on the crystallization phenomenon of gelators (SEM images are reprinted with permission from ref. 78. Copyright 2006 American Chemical Society).

hand, high supersaturation favours the formation of dense network of fibrils arranged in highly branched spherulite domains or rosette-like structure. Thus, the early stage of gelation, *i.e.* the stage following nucleation, is crucial in determining the SAFiNs network in molecular gels. This shows that the branching behaviour of SAFiNs in molecular gels can be controlled by changing the processing parameters during the synthesis of molecular gels. Thus, molecular gels can be fine-tuned or engineered at all the levels of structures (gelator and solvent type can influence the primary and secondary structures) depending on the applications.

Hydroxylated fatty acids including, 12-hydroxy stearic acid (12HSA) and ricinelaidic acid (*trans* isomer of 12HSA) are capable of gelling edible oils (*e.g.* canola oil) at concentrations as low as 0.3% and 0.5%, respectively.[82] They are natural fatty acids of castor oil. The organogelation ability of 12HSA has been well known since 1946 as it is used in greases like lithium grease. 12HSA has also been used in waste disposal of engine oils and vegetable oils. The high structuring ability of 12HSA can be attributed to its structural diversity. The hydroxyl group at the C_{12} position and the carboxylic acid group play crucial roles in structuring the oils at such a low concentration. During gelation, 12HSA molecules stack in a zig-zag fashion as lamellar plates owing to intermolecular hydrogen bonding between the hydroxyl groups. The adjacent lamellae adhere together *via* head-to-head hydrogen bonds between carboxyl groups and extend unidirectionally as fiber-shaped crystals along a single axis, called the fiber axis. The fibrillar crystals interact with each other to form SAFiNs after reaching a critical parameter called the percolation threshold. At the molecular level, imperfect dimerization of the carboxylic heads leads to CMB, which results in branching of fibers. Thus, due to CMB, a single fiber branches to multiple daughter fibers with a fractal Cayley tree substructure (Scheme 2.5).[83] This kind of gelator network (SAFiNs) is also associated with fatty acids and the fractal Cayley tree substructure of stearic acid shown in Scheme 2.5. The SAFiNs formed in this fashion possesses a high surface area for contact with the solvent, which facilitates the

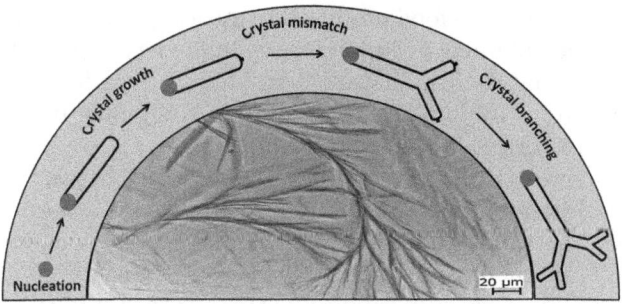

Scheme 2.5   The schematic growth of crystals *via* CMB and the insert shows the optical microstructure of stearic acid fibers formed as a fractal Cayley tree substructure in a soybean oil oleogel.

entrapment of a large volume of solvent by capillary forces. Hence, 12HSA has shown a lot of promise as a molecular gelator. However, its application is limited owing to the lack of regulatory approval for use in food products. The possible laxative effect of castor oil derivatives has been the major concern for 12HSA approval in food applications. Though 12HSA is not a GRAS (generally recognized as safe) material, it is one of the thoroughly explored molecular gelators. It has helped food researchers unearth the crystallization phenomenon of fatty acids and their derivatives.

With regard to fatty acids, an increase in chain length reduces the minimum gelator concentration (MGC) required to structure the vegetable oils.[27] The order of fatty acids in terms of their MGC is lauric acid (15%) > myristic acid (6%) > palmitic acid (4%) > stearic acid (2%). The fatty acids beyond C_{18} chain length possess approximately the same MGC (2%). Besides chain length, functional groups attached to the carbon chain also play a major role in structuring the vegetable oils. Dicarboxylic acids like adipic, suberic and sebacic acids are more efficient in structuring vegetable oils than the monocarboxylic acids of equal chain length.[27] The presence of an additional hydroxyl group at C_{12} of 12HSA, compared to stearic acid, has resulted in over 100% increase in gelation efficiency. On the other hand, with a methyl group instead of a hydroxyl group at C_{12} of stearic acid, the gelation efficiency of the resulting gelator is reduced five-fold. Similar to fatty acids, fatty alcohols can also structure vegetable oils but their efficiency is lower than that of fatty acids. However, a synergistic effect is seen when fatty acids and fatty alcohols of the same length are mixed to structure edible oils.[28] When stearic acid and stearyl alcohol are mixed at a ratio of 7:3, harder and more elastic oleogels are obtained in comparison to the gels obtained using them alone. The small crystal size (high surface area to volume ratio) of needle-like crystals formed by the fatty acid–alcohol mixture is responsible for the increased strength of the oleogels.

Waxes are fatty substances with long hydrocarbon chains with or without a functional group. The functional group can be an alcohol, aldehyde, ester, or ketone. Waxes have been extracted either from plants, animals, insects, or petroleum for various industrial applications. Waxes are widely used in lubricants, cosmetics and food products. In most of the food products, waxes are used as either adjuncts or supplements. In the recent past, the cosmetic, pharmaceutical and food industries have shifted their focus towards bio-derived waxes from mineral waxes. The most commonly used bio-derived waxes for structuring edible oils are: candelilla wax (CLW), carnauba wax (CRW), rice bran wax (RBW), sunflower wax (SFW) and beeswax (BZW) (insect origin). In addition to them, berry wax (BW), fruit wax (FW) and jojoba wax (JW) can also be used. The structuring ability of waxes in vegetable oils depends on their chemical composition, purity and solubility in vegetable oils. The solubility of waxes in canola oil is in the order of RBW < SFW < CLW < CRW. Correspondingly CLW and CRW showed an MGC of 1% (w/w) and the others (RBW, SFW) showed an MGC of 0.5% (w/w).[54] Henceforth, to achieve better gelation efficiency, gelators should be neither completely soluble nor insoluble in vegetable oils. Morphological differences in the microstructure of waxes also contribute to their gelation efficiency. RBW and SFW form long needle-like fibrous crystals, while on the other hand CLW and CRW form grain-like crystals and dendrite crystals, respectively.[54] The high aspect ratio of fibers provides a high surface area of contact for hydrophobic interactions between gelators and vegetable oil. In general, the highest surface area of contact is provided by fibers, followed by platelets/sheets and spheres. The morphological differences in the crystals are owing to the composition of waxes. Chemically, RBW and SFW are homogeneous (>90% esters) as compared to CLW and CRW. This homogeneous nature allows the crystals to grow anisotropically to form fibers. Furthermore, minor components present in CLW and CRW act as crystal habit modifiers and prevent the crystal growth, which results in the formation of platelets and grain-like morphologies. This kind of behaviour was also seen in BW and FW oleogels. The heterogeneous nature of BW and FW results in the formation of spherulitic crystals, which in turn yields high MGC values for both the waxes. The formation of large spherulitic crystals is also assisted by the lack of long chain hydrocarbons and the presence of small and medium chain fatty acids in their composition. Since these fatty acids are not free and linked to glycerol, lateral packing of non-linear molecules resulted in the formation of large crystals. Hence, the composition plays a significant role in the gelation behaviour of wax-based oleogels.

2.3.2 Polymeric Gelators (Cellulose Derivatives)

An enormous amount of research has been conducted on polymer-based hydrogels but the research on polymeric oleogels and gelators is limited, providing huge scope for exploring new molecules as polymeric oleogelators. The major challenge being faced by researchers is that the food polymers are hydrophilic and immiscible in edible oils. Cellulose derivatives, *viz.* ethyl cellulose (EC) and hydroxypropyl methylcellulose (HPMC), have shown gelation in vegetable oils.

EC is a hydrophilic biopolymer with ethoxyl substituents at the hydroxyl groups of carbons 2, 3 or 6 of the cellulose backbone (1,4-β-D linked glucose units). The ethoxyl substitution should be 2.5/3 for the dissolution of EC in vegetable oils and organic solvents.[84] According to the guidelines framed by the WHO (World Health Organization) and the FAO (Food and Agriculture Organization) of the United Nations, EC can be used as a food additive in 60 different food applications. Hence, EC is considered as a food-grade or near food-grade material for culinary applications. Depending on the type of application, EC with different molecular weight and viscosity can be manufactured. EC grades with molecular weights 24, 57 and 74 kDa and corresponding viscosities of 10, 45 and 100 cP, respectively, have formed oleogels in canola oil, soybean oil and flax seed oil. The microstructure of the oleogels was composed of interconnected polymeric chains with 3.0–4.5 μm size oil filled pores.[69] The pore size and polymer interconnectivity can be controlled by varying the polymer grade, concentration and oil unsaturation.

HPMC is also a hydrophilic biopolymer, prepared by substituting the hydroxyl groups of cellulose with hydroxypropyl and methyl groups. The substituted groups are hydrophobic in nature, which imparts amphiphilicity to the molecule. The US FDA (Food and Drug Administration) has approved HPMC as a GRAS material for human consumption up to 0.047 mg kg^{-1} body weight day^{-1}. The maximum tolerable intake capacity of HPMC for humans is 5 mg kg^{-1} body weight day^{-1}, which is 100 times higher than the prescribed amount. It has been used as a food additive in the form of an emulsifier and a thickening agent. Unlike EC, HPMC is used in the preparation of oleogels by following an indirect approach. In the direct approach, EC is dissolved in oil by heating and then upon cooling, disperses as interconnected polymeric chains while structuring. HPMC oleogels are formed by following indirect approaches, namely foam-template or emulsion-template approaches.[85,86] In these approaches, the first step involves the hydration of HPMC to open the polymer chains and form a structural network in water. In the second step, the absorbed water is removed from the 3D polymeric network by heating or freeze drying and then vegetable oil is immobilized by soaking or shearing the dried polymeric strands in the oil.

In addition to the structuring agents mentioned above, proteins (β-lacto-globulin, gelatin),[64,65] gums (xanthan gum)[65] and resins (shellac from *Laccifer lacca*)[66] can also be used to structure edible oils.

2.4 Multifunctional Molecular Gelators as Next-generation Oil Structuring Agents: Design, Synthesis and Self-assembly

Though there are several commercially available oil thickening agents and gelators, there is still an ongoing quest and demand for new and improved thickening agents that can mimic the food products given in Table 2.1, which are synthesized based on traditional edible oil structuring methods. All the food products mentioned in Table 2.1 suggest that synthesis of a desired food product undergoes multiple processing protocols. Since the advent of the Molecular gastronomy approach to food making and food processing, the existing multiple-step edible oil structuring methodology can be simplified by designing a simple and efficient edible oil structuring technology using bioderived molecular gelators (next-generation oil structuring agents). The developed technology may become crucial for the synthesis of food products in the near future. The design and development of next-generation edible oil structuring technology is represented schematically in Scheme 2.6. The success of this technology is largely dependent on the identification of precursors for the design of structuring agents from natural sources. Briefly, the technology can be explained as follows. The established precursors from natural resources are mainly sugars (*e.g.* trehalose), sugar alcohols (*e.g.* sorbitol, mannitol) and glycosides (*e.g.* raspberry ketone glucoside). The precursors are enzymatically esterified with fatty acids using lipase B by following a

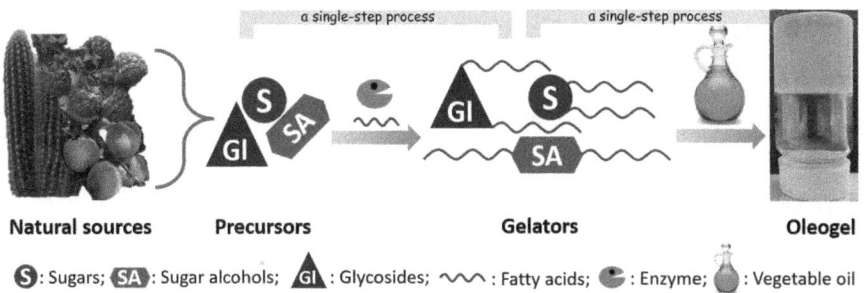

Scheme 2.6 Design of next-generation edible oil structuring technology.

simple, one-step regiospecific heterogeneous catalysis. Thermal processing of fatty acid esters of precursors in vegetable oils lead to the formation of oleogels, structured vegetable oils.

The use of carbohydrate derivatives to structure edible oils proves very promising. Sugars, sugar alcohols and glycosides esterified with fatty acids have shown excellent gelation in edible oils at very low concentrations. The self-assembling properties of the carbohydrate derivatives facilitate their use as edible oil structuring agents. For example, trehalose, an α-D-glucopyranosyl-$(1 \rightarrow 1)$-β-D-glucopyranoside, an α-linked disaccharide, is extensively used as a non-reducing sugar in food, pharmaceutical and cosmetic industries. Specifically, fatty acids are esterified onto trehalose to synthesize trehalose diesters or trehalose-based oleogelators using a lipase-catalyzed regiospecific process.[87,88] The enzymatic catalysis is driven under anhydrous conditions using lipase B from *Candida antarctica* at 45 °C. In this process, long chain fatty acids are appended at the 6 and 6′ positions of trehalose to yield diesters, namely, trehalose distearate, trehalose dimyristate and trehalose didecanoate, which form oleogels when dissolved in edible oils.[76] Trehalose didecanoate, for example, has been used to structure olive oil at extremely low concentrations, as low as 0.09 wt%/v.[76] On the other hand, trehalose diesters with short chain fatty acids, such as trehalose dibutyrate, are soluble in water and insoluble in olive oil, and do not form oleogels. The chemical structures of trehalose diesters are shown in Figure 2.4. The gelation ability of carbohydrate derivatives and trehalose derivatives can be attributed to their amphiphilic nature. Hence, hydrophilic lipophilic balance (HLB) plays a crucial role in structuring liquid solvents. Since trehalose is a disaccharide with eight hydroxyl groups, long chain fatty acids (C_{10}, C_{14}, C_{18}) are able to balance the large hydrophilic head in order to create an appropriate HLB for the formation of oleogels. Trehalose diesters with short chain fatty acids (C_4) cannot attain the HLB required for their solubility in vegetable oils. Self-assembly of the gelators is mainly governed by the high hydroxyl group density of trehalose, which participates in intermolecular hydrogen bonding. Furthermore, both gelator–gelator and gelator–solvent interactions are crucial in forming a stable oleogel. During self-assembly, the long fatty acid chains of trehalose

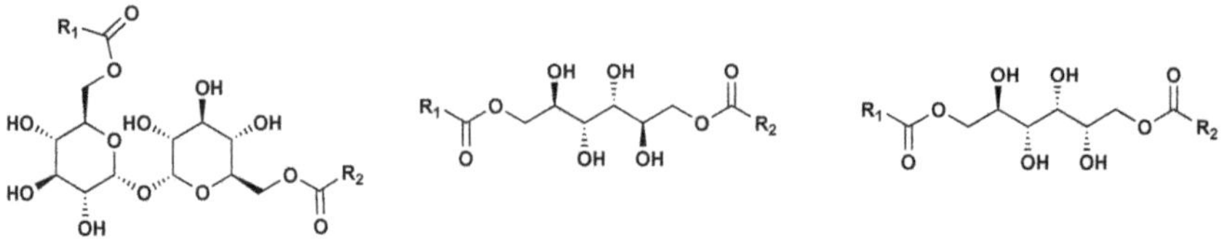

Trehalose dialkanoate   **Mannitol dialkanoate**   **Sorbitol dialkanoate**

$$R_1, R_2 = C_nH_{2n+1}; n = 1 \text{ to } 17$$

Figure 2.4   The chemical structures of molecular gelators.

diesters are involved in hydrophobic interactions amongst themselves and with the solvent. These self-assembled gelators are further stabilized by extensive hydrogen bonding between the hydroxyl groups of trehalose.[89] In the case of diesters with short chain fatty acids, gelator–gelator interactions *via* hydrogen bonding dominate the fatty acid hydrophobic interactions. This imbalance is responsible for the precipitation of trehalose dibutyrate in edible oils. Thus, both gelator–gelator and gelator–solvent interactions are crucial to form a stable oleogel. Furthermore, trehalose diester-based gels consist of three-dimensional entangled fiber-like aggregates with lengths in the micrometer scale and diameters around 10 to 500 nm. Such a high length-to-diameter aspect ratio suggests that inter-gelator interactions are highly anisotropic.

Similar to sugar derivatives, sugar alcohol derivatives have also been synthesized by regiospecific enzymatic catalysis and checked for their ability to gelate edible oils. Sugar alcohols are non-reducing and have lower calorific/glycemic indices than sugars.[90] In this regard, the use of sugar alcohols improves the nutritional value of oleogels and their corresponding food products. Mannitol and sorbitol are six-carbon open-chain sugar alcohols that have been used as sugar substitutes in the food industry.[91] Mannitol reduces blood glucose and insulin levels, and is helpful in cough clearance related to hypersecretary diseases like asthma.[92] Mannitol is also an osmotic diuretic with both antioxidant and renovascular properties that could improve renal function.[93] Mannitol has been listed in the "List of essential medicines" by the WHO. Sorbitol is a stereoisomer of mannitol, with a difference in the orientation of one hydroxyl group at the C_2 position. To create gelators based on these sugar alcohols, medium chain fatty acids, namely, butyric acid (C_4) and caprylic acid (C_8), were esterified onto the hydroxyl groups at the first and sixth carbons of sugar alcohols *via* lipase-mediated regiospecific catalysis. Caprylic acid does not easily re-esterify to form triglycerides; its use in gelation has an added advantage of its known nonhypercholesterolemic property.[8] All the as-synthesized molecules, *viz.* sorbitol dibutyrate, sorbitol dicaprylate, mannitol dibutyrate and mannitol dicaprylate, have the versatility to gelate a wide range of oils. The list includes canola oil, grape seed oil, olive oil, sesame oil, soybean oil and sunflower oil.[46] However, their gelation potentials differ from each other in terms of the type of sugar alcohol and fatty acid chain. Mannitol-based gelators have been found to have better gelation potential (lower minimum gelator concentration) and form strong gels (higher gel-to-sol transition temperature). This is owing to the subtle difference in the chemical structure of hydrophilic sugar alcohols (stereoisomers). Additionally, caprylic acid derivatives are able to gel edible oils at very low concentrations compared to butyric acid derivatives. Their better gelling ability can be attributed to strong gelator–solvent interactions, facilitated by the long fatty acid chains of caprylic acid. Thus, the gelling behaviour as well as the physical and mechanical properties of the oleogels can be modulated by altering either the hydrophilic head or hydrophobic tail of the gelators. The stereochemical difference between the molecules also affects the orientation of the gelator network and the opacity of the oleogels. Mannitol dioctanoate-based oleogels are opaque owing to a densely packed long fibrillar gelator network in their microstructure (Figure 2.5). On the other hand, sorbitol dioctanoate-based oleogels are translucent and their morphology consists of randomly dispersed clusters with needle-like microcrystallites in the oil matrix. Moreover, the mechanical strength (storage modulus, yield strain) and thermal strength (gel-to-sol transition temperature) of mannitol dioctanoate-based oleogels are several times higher compared to sorbitol dioctanoate-based oleogels. The difference in the strength is attributed to the loosely connected and spatially dispersed microcrystallites in the microstructure of sorbitol dioctanoate-based oleogels.

Glycosides, which are natural surfactants of plant origin, have been used in foods, perfumes and shampoos. Raspberry ketone glucoside (RKG) is available in fruits and was granted GRAS status by the US FDA in 1968. RKG has been coupled with saturated and unsaturated fatty acids *via* lipase-mediated regiospecific esterification.[47] RKG esters have shown excellent gelation behaviour in edible oils. Comparatively, saturated glucoside esters (RKG caprylate, RKG stearate) have better gelation potential (<0.5 wt%) than unsaturated glucoside esters (>0.5 wt%, RKG oleate). Similarly, cardanol-based glycolipids have shown their oleogelation capability in linseed oil, sesame oil, hazelnut oil and olive oil.[94]

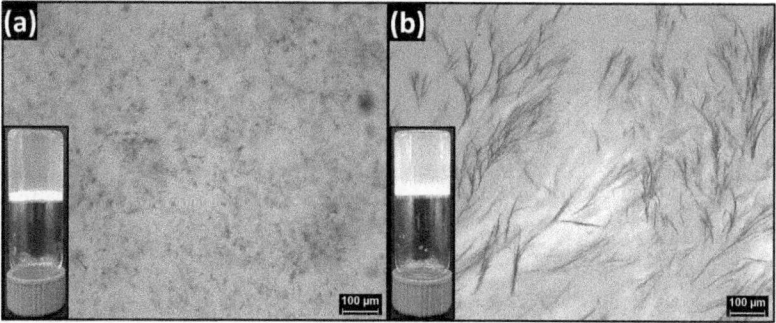

Figure 2.5   Microstructure of (a) translucent sorbitol dioctanoate oleogels and (b) opaque mannitol dioctanoate oleogels.

2.5   Conclusions

In conclusion, this chapter enumerates and exemplifies the role and neces-
sity of bioderived molecular structuring agents designed for edible oils.
Though solidification of edible oils is possible *via* hydrogenation, interesterification and blending of fats, the need to generate solid oils without any
adverse side effects has been growing. The structuring of edible oils using
biobased gelators is possible by combining the principles of green chemistry
and supramolecular aggregation under the given conditions. The gelation
mechanism of bioderived gelators is mostly driven by non-covalent interactions, *viz.* hydrogen bonding between the hydroxyl groups, π–π interactions
of aromatic rings and van der Waals forces amongst the fatty acid chains and/
or lipids in edible oils. The bioderived oleogelators have excellent gelling abilities in different edible oils, ranging from highly unsaturated linseed oil to
saturated palm oil. In a nutshell, bioderived gelators have proven their potential as molecular structuring agents. The results are intriguing for generating
and expanding the scope of bioderived molecular gelators towards sustainable soft materials in the future. Currently, extensive research is in progress
to develop food-grade oleogels with bio-active agents (curcuminoids, pharmaceutical drugs) as nutraceuticals.[95] Though exemplary research has been
carried out in designing biobased structuring agents, their industrial usage
is limited. The industrial application of molecular gelators can be enhanced
by designing molecular gelators as multifunctional gelators and nutraceuticals. The multifunctional structuring agents can be efficiently designed by
incorporating antioxidant, anti-aging and anti-inflammatory capabilities in
the targeted molecular gelator *via* enzyme catalysis.

Acknowledgements

This research was made possible in part by Grants to G. J. from NIFA, the
United States Department of Agriculture (GRANT11890945) and the Institute
of Innovation and Nutrition, GrupoBimbo sponsored through Forent/Barcel
S. A. de C. V. (RFCUNY project # 76600-00 01). M. S. thanks the RISE Program
at the City College of New York, under the National Institute of Health (NIH-
5R25GM056833-15), for financial support *via* a research fellowship. We thank
Uduak Thomas for the help in reviewing the manuscript for publication.

References

1. A. R. Patel, P. S. Rajarethinem, A. Gredowska, O. Turhan, A. Lesaffer,
   W. H. De Vos, D. Van de Walle and K. Dewettinck, *Food Funct.*, 2014, **5**,
   645–652.
2. S. S. Sagiri, B. Behera, R. R. Rafanan, C. Bhattacharya, K. Pal, I. Banerjee
   and D. Rousseau, *Soft Mater.*, 2013, **12**, 47–72.
3. C. M. O'Sullivan, S. Barbut and A. G. Marangoni, *Trends Food Sci. Technol.*, 2016, **57**(Pt A), 59–73.

4. R. R. Allen, *J. Am. Oil Chem. Soc.*, 1981, **58**, 166–169.
5. M. T. Tarrago-Trani, K. M. Phillips, L. E. Lemar and J. M. Holden, *J. Am. Diet. Assoc.*, 2006, **106**, 867–880.
6. A. Philippaerts, P. A. Jacobs and B. F. Sels, *Angew. Chem., Int. Ed.*, 2013, **52**, 5220–5226.
7. N. C. Acevedo and A. G. Marangoni, *Food Biophys.*, 2014, **9**, 368–379.
8. C. C. Akoh and D. B. Min, *Food Lipids: Chemistry, Nutrition, and Biotechnology*, CRC press, 2008.
9. A. H. Hite, R. D. Feinman, G. E. Guzman, M. Satin, P. A. Schoenfeld and R. J. Wood, *Nutrition*, 2010, **26**, 915–924.
10. R. Chowdhury, S. Warnakula and S. Kunutsor, *et al.*, *Ann. Intern. Med.*, 2014, **160**, 398–406.
11. P. W. Siri-Tarino, Q. Sun, F. B. Hu and R. M. Krauss, *Am. J. Clin. Nutr.*, 2010, **91**, 535–546.
12. L. Hooper, N. Martin, A. Abdelhamid and G. Davey Smith, *Cochrane Database Syst. Rev.*, 2015, **6**, CD011737.
13. U. Schwab, L. Lauritzen, T. Tholstrup, T. I. Haldorsson, U. Riserus, M. Uusitupa and W. Becker, *Food Nutr. Res.*, 2014, **58**, 25145.
14. P. W. Siri-Tarino, S. Chiu, N. Bergeron and R. M. Krauss, *Annu. Rev. Nutr.*, 2015, **35**, 517–543.
15. G. Parfene, V. Horincar, A. K. Tyagi, A. Malik and G. Bahrim, *Food Chem.*, 2013, **136**, 1345–1349.
16. B. Martínez-Vallespín, W. Vahjen and J. Zentek, *Cytotechnology*, 2016, **68**, 1925–1936.
17. S. Rial, A. Karelis, K.-F. Bergeron and C. Mounier, *Nutrients*, 2016, **8**, 281.
18. F. Gunstone, *Vegetable Oils in Food Technology: Composition, Properties and Uses*, John Wiley & Sons, New Jersy, 2011.
19. M. Kellens, V. Gibon, M. Hendrix and W. De Greyt, *Eur. J. Lipid Sci. Technol.*, 2007, **109**, 336–349.
20. E. J. Pérez-Monterroza, C. J. Márquez-Cardozo and H. J. Ciro-Velásquez, *LWT–Food Sci. Technol.*, 2014, **59**, 673–679.
21. A. Lopez-Martínez, M. A. Charó-Alonso, A. G. Marangoni and J. F. Toro-Vazquez, *Food Res. Int.*, 2015, **72**, 37–46.
22. S. Da Pieve, S. Calligaris, E. Co, M. Nicoli and A. Marangoni, *Food Biophys.*, 2010, **5**, 211–217.
23. E. Yılmaz and M. Öğütcü, *J. Am. Oil Chem. Soc.*, 2014, **91**, 1007–1017.
24. C. H. Chen, I. Van Damme and E. M. Terentjev, *Soft Matter*, 2009, **5**, 432–439.
25. N. Realdon, E. Ragazzi and E. Ragazzi, *Drug Dev. Ind. Pharm.*, 2001, **27**, 165–170.
26. A. López-Martínez, J. A. Morales-Rueda, E. Dibildox-Alvarado, M. A. Charó-Alonso, A. G. Marangoni and J. F. Toro-Vazquez, *Food Res. Int.*, 2014, **64**, 946–957.
27. J. Daniel and R. Rajasekharan, *J. Am. Oil Chem. Soc.*, 2003, **80**, 417–421.
28. F. Gandolfo, A. Bot and E. Flöter, *J. Am. Oil Chem. Soc.*, 2004, **81**, 1–6.

29. J. F. Toro-Vazquez, J. Morales-Rueda, A. Torres-Martínez, M. A. Charó-Alonso, V. A. Mallia and R. G. Weiss, *Langmuir*, 2013, **29**, 7642–7654.
30. S. S. Sagiri, V. K. Singh, K. Pal, I. Banerjee and P. Basak, *Mater. Sci. Eng., C*, 2015, **48**, 688–699.
31. F. R. Lupi, D. Gabriele, V. Greco, N. Baldino, L. Seta and B. de Cindio, *Food Res. Int.*, 2013, **51**, 510–517.
32. H. M. Schaink, K. F. van Malssen, S. Morgado-Alves, D. Kalnin and E. van der Linden, *Food Res. Int.*, 2007, **40**, 1185–1193.
33. A. Bot and W. M. Agterof, *J. Am. Oil Chem. Soc.*, 2006, **83**, 513–521.
34. A. Bot, R. den Adel, C. Regkos, H. Sawalha, P. Venema and E. Flöter, *Food Hydrocolloids*, 2011, **25**, 639–646.
35. M. A. Rogers, A. J. Wright and A. G. Marangoni, *Soft Matter*, 2009, **5**, 1594–1596.
36. T.-M. Wang and M. A. Rogers, *Lipid Technol.*, 2015, **27**, 175–178.
37. F. R. Lupi, D. Gabriele, D. Facciolo, N. Baldino, L. Seta and B. de Cindio, *Food Res. Int.*, 2012, **46**, 177–184.
38. F. Lupi, D. Gabriele and B. de Cindio, *Food Bioprocess Technol.*, 2012, **5**, 2880–2888.
39. I. F. Almeida and M. F. Bahia, *Int. J. Pharm.*, 2006, **327**, 73–77.
40. R. Sánchez, J. M. Franco, M. A. Delgado, C. Valencia and C. Gallegos, *Chem. Eng. Res. Des.*, 2008, **86**, 1073–1082.
41. S. Murdan, G. Gregoriadis and A. T. Florence, *J. Pharm. Sci.*, 1999, **88**, 608–614.
42. D. K. Shah, S. S. Sagiri, B. Behera, K. Pal and K. Pramanik, *J. Appl. Polym. Sci.*, 2013, **129**, 793–805.
43. M. Pernetti, K. van Malssen, D. Kalnin and E. Flöter, *Food Hydrocolloids*, 2007, **21**, 855–861.
44. B. Behera, V. Patil, S. S. Sagiri, K. Pal and S. S. Ray, *J. Appl. Polym. Sci.*, 2012, **125**, 852–863.
45. S. R. Jadhav, P. K. Vemula, R. Kumar, S. R. Raghavan and G. John, *Angew. Chem., Int. Ed.*, 2010, **49**, 7695–7698.
46. S. R. Jadhav, H. Hwang, Q. Huang and G. John, *J. Agric. Food Chem.*, 2013, **61**, 12005–12011.
47. J. R. Silverman and G. John, *J. Agric. Food Chem.*, 2015, **63**, 10536–10542.
48. S. Jana and S. Martini, *J. Agric. Food Chem.*, 2014, **62**, 10192–10202.
49. M. Öğütcü, N. Arifoğlu and E. Yılmaz, *Int. J. Food Sci. Technol.*, 2015, **50**, 404–412.
50. E. Ylmaz and M. Ogutcu, *Food Funct.*, 2015, **6**, 1194–1204.
51. E. Ylmaz and M. Ogutcu, *RSC Adv.*, 2015, **5**, 50259–50267.
52. M. Öğütcü, N. Arifoğlu and E. Yılmaz, *J. Am. Oil Chem. Soc.*, 2015, **92**, 459–471.
53. A. Jang, W. Bae, H.-S. Hwang, H. G. Lee and S. Lee, *Food Chem.*, 2015, **187**, 525–529.
54. A. Blake, E. Co and A. Marangoni, *J. Am. Oil Chem. Soc.*, 2014, **91**, 885–903.

55. J. F. Toro-Vazquez, J. A. Morales-Rueda, E. Dibildox-Alvarado, M. Charó-Alonso, M. Alonzo-Macias and M. M. González-Chávez, *J. Am. Oil Chem. Soc.*, 2007, **84**, 989–1000.

56. J. F. Toro-Vazquez, R. Mauricio-Pérez, M. M. González-Chávez, M. Sánchez-Becerril, J. d. J. Ornelas-Paz and J. D. Pérez-Martínez, *Food Res. Int.*, 2013, **54**, 1360–1368.

57. F. M. Alvarez-Mitre, J. A. Morales-Rueda, E. Dibildox-Alvarado, M. A. Charó-Alonso and J. F. Toro-Vazquez, *Food Res. Int.*, 2012, **49**, 580–587.

58. F. M. Alvarez-Mitre, J. F. Toro-Vázquez and M. Moscosa-Santillán, *J. Food Eng.*, 2013, **119**, 611–618.

59. J. C. B. Rocha, J. D. Lopes, M. C. N. Mascarenhas, D. B. Arellano, L. M. R. Guerreiro and R. L. da Cunha, *Food Res. Int.*, 2013, **50**, 318–323.

60. H.-S. Hwang, M. Singh, E. Bakota, J. Winkler-Moser, S. Kim and S. Liu, *J. Am. Oil Chem. Soc.*, 2013, **90**, 1705–1712.

61. D. C. Zulim Botega, A. G. Marangoni, A. K. Smith and H. D. Goff, *J. Food Sci.*, 2013, **78**, C1845–C1851.

62. M. Öğütcü and E. Yılmaz, *Grasas Aceites*, 2014, **65**, e040.

63. D. C. Zulim Botega, A. G. Marangoni, A. K. Smith and H. D. Goff, *J. Food Sci.*, 2013, **78**, C1334–C1339.

64. A. I. Romoscanu and R. Mezzenga, *Langmuir*, 2006, **22**, 7812–7818.

65. A. R. Patel, P. S. Rajarethinem, N. Cludts, B. Lewille, W. H. De Vos, A. Lesaffer and K. Dewettinck, *Langmuir*, 2015, **31**, 2065–2073.

66. A. R. Patel, D. Schatteman, W. H. D. Vos and K. Dewettinck, *RSC Adv.*, 2013, **3**, 5324–5327.

67. M. Davidovich-Pinhas, A. J. Gravelle, S. Barbut and A. G. Marangoni, *Food Hydrocolloids*, 2015, **46**, 76–83.

68. A. K. Zetzl, A. G. Marangoni and S. Barbut, *Food Funct.*, 2012, **3**, 327–337.

69. A. K. Zetzl, A. J. Gravelle, M. Kurylowicz, J. Dutcher, S. Barbut and A. G. Marangoni, *Food Struct.*, 2014, **2**, 27–40.

70. A. R. Patel, N. Cludts, M. D. B. Sintang, A. Lesaffer and K. Dewettinck, *Food Funct.*, 2014, **5**, 2833–2841.

71. A. R. Patel, N. Cludts, M. D. Bin Sintang, B. Lewille, A. Lesaffer and K. Dewettinck, *ChemPhysChem*, 2014, **15**, 3435–3439.

72. G. John, M. Masuda, Y. Okada, K. Yase and T. Shimizu, *Adv. Mater.*, 2001, **13**, 715–718.

73. V. S. Balachandran, S. R. Jadhav, P. K. Vemula and G. John, *Chem. Soc. Rev.*, 2013, **42**, 427–438.

74. N. K. O. Ojijo, I. Neeman, S. Eger and E. Shimoni, *J. Sci. Food Agric.*, 2004, **84**, 1585–1593.

75. M. Law, *BMJ [Br. Med. J.]*, 2000, **320**, 861–864.

76. G. John, G. Zhu, J. Li and J. S. Dordick, *Angew. Chem., Int. Ed.*, 2006, **45**, 4772–4775.

77. P. K. Vemula, J. Li and G. John, *J. Am. Chem. Soc.*, 2006, **128**, 8932–8938.

78. R.-Y. Wang, X.-Y. Liu, J. Narayanan, J.-Y. Xiong and J.-L. Li, *J. Phys. Chem. B*, 2006, **110**, 25797–25802.

79. A. Marangoni, *J. Am. Oil Chem. Soc.*, 2012, **89**, 749–780.
80. M. George and R. G. Weiss, *Acc. Chem. Res.*, 2006, **39**, 489–497.
81. M. A. Rogers and A. G. Marangoni, *Cryst. Growth Des.*, 2008, **8**, 4596–4601.
82. M. A. Rogers, A. J. Wright and A. G. Marangoni, *Food Res. Int.*, 2008, **41**, 1026–1034.
83. R. Lam, L. Quaroni, T. Pedersen and M. A. Rogers, *Soft Matter*, 2010, **6**, 404–408.
84. T. A. Stortz, A. K. Zetzl, S. Barbut, A. Cattaruzza and A. G. Marangoni, *Lipid Technol.*, 2012, **24**, 151–154.
85. A. R. Patel, D. Schatteman, A. Lesaffer and K. Dewettinck, *RSC Adv.*, 2013, **3**, 22900–22903.
86. A. R. Patel and K. Dewettinck, *Eur. J. Lipid Sci. Technol.*, 2015, **117**, 1772–1781.
87. T. Higashiyama, *Pure Appl. Chem.*, 2002, **74**, 1263.
88. A. B. Richards, S. Krakowka, L. B. Dexter, H. Schmid, A. P. M. Wolterbeek, D. H. Waalkens-Berendsen, A. Shigoyuki and M. Kurimoto, *Food Chem. Toxicol.*, 2002, **40**, 871–898.
89. G. John, B. Vijai Shankar, S. R. Jadhav and P. K. Vemula, *Langmuir*, 2010, **26**, 17843–17851.
90. P. K. Vemula and G. John, *Acc. Chem. Res.*, 2008, **41**, 769–782.
91. M. Grembecka, *Eur. Food Res. Technol.*, 2015, **241**, 1–14.
92. E. Daviskas, S. D. Anderson, S. Eberl and I. H. Young, *Respir. Med.*, 2010, **104**, 1645–1653.
93. K. Kalimeris, N. Nikolakopoulos, M. Riga, K. Christodoulaki, K. G. Moulakakis, C. Dima, C. Papasideris, T. Sidiropoulou, G. Kostopanagiotou and A. Pandazi, *J. Cardiothorac. Vasc. Anesth.*, 2014, **28**, 954–959.
94. K. Muthusamy, V. Sridharan, C. U. Maheswari and S. Nagarajan, *Green Chem.*, 2016, **18**, 3722–3731.
95. H. Yu, K. Shi, D. Liu and Q. Huang, *Food Chem.*, 2012, **131**, 48–54.

CHAPTER 3

Biomimicry: An Approach for Oil Structuring

MICHAEL A. ROGERS

University of Guelph, Department of Food Science, 50 Stone Rd, Guelph, N3C3X9, Canada
*E-mail: mroger09@uoguelph.ca

3.1 Introduction

Ubiquitously found in nature, self-assembled materials have a remarkable ability to spontaneously engineer complex supramolecular assemblies, often using very simple small molecules.[1] Molecular self-assembly, inspired by nature, has driven an entirely new platform, 'bottom-up' nano-fabrication, for constructing materials with precision, integrating both order and dynamics, to achieve functionality ranging from stimuli responsiveness and recognition to catalysis.[1] The supramolecular structures are assembled molecule-by-molecule, whereby the coding for assembly is embedded in the structural motifs of the molecule and is based on chemical complementarity and structural compatibility, facilitating weak non-covalent interactions to stabilize the hierarchical resultant structures.[2] Most famously, spider silk relies on self-assembly of small proteins, spidroins, that form intermolecular interactions between repeat regions of the protein, giving rise to its strength and elasticity.[3] Biological materials that are examples of 'bottom-up' nano-fabrication range from hagfish slime,[4] collagen,[5] lipid bilayers,[6] and DNA[7] to the stratum corneum.[8] Of interest in designing a novel hardstock fat replacer,

Food Chemistry, Function and Analysis No. 3
Edible Oil Structuring: Concepts, Methods and Applications
Edited by Ashok R. Patel
© The Royal Society of Chemistry 2018
Published by the Royal Society of Chemistry, www.rsc.org

the stratum corneum (SC) seems an ideal target since it is composed of molecules that are universally found in our food supply.

The primary benefit of mimicking biological systems that self-assemble is that rational design of small molecular gelators has remained elusive, despite the vast body of literature devoted to such gels over the past decade, making it exceedingly difficult to design new oleogelators.[9] Unfortunately, finding small molecules capable of self-assembling into fibrillar networks is an empirical science and relies more on serendipity to discover new oleogelators than their rational design. In part, this is because self-assembled fibrillar networks (SAFiNs) rely on an intricate balance of weak non-covalent interactions (*i.e.*, hydrogen-bonding, π–π stacking and van der Waals interactions) stabilizing their hierarchical assembled structures and processes that govern their dissolution/solubility in each solvent. It is well-established that SAFiNs form 1-dimensional (1D) fibers spontaneously *via* an aggregation-nucleation-growth pathway.[10-14] However, the molecular features that govern this pathway are more complex and researchers do not yet have a 'toolkit' to build new gelators. As a case in point, one of the most widely studied molecular gelators, 12-hydroxyoctadecanoic acid (*i.e.*, 12 hydroxystearic acid), loses its ability to self-assemble when chirality at position 12 is lost, when the hydroxyl group is moved too close to the carboxylic acid, and if the electronegativity is reduced of the functional group at either position 1 (*i.e.*, the carboxylic acid) or at position 12 (*i.e.*, the hydroxyl group).[15-20] In other words, minor changes in the chemical structure have drastic consequences on the assembly and physical properties of an oleogelator.

3.2   The Stratum Corneum

Although skin is comprised of numerous layers it is the outermost layer, the SC, that is particularly of interest, as its primary function is to regulate moisture migration and desiccation. The SC is comprised of approximately 60% protein (*i.e.*, corneocytes), 20% water and 20% lipids.[21] The barrier properties of the SC arise from a subcellular organization of epidermal proteins and lipids, which are arranged in a continuous extracellular lipid matrix surrounding lipid-depleted, protein-enriched cells (corneocytes).[21] The impermeability of the SC arises due to the specific location and organization of the lipid species within the cellular interstices.[21] Unlike most lipid organizations in the human body, the SC is essentially devoid of phospholipids and instead is enriched in sphingolipids, free sterols and fatty acids. Remarkably, most common skin ailments, such as psoriasis, eczema, and ichthyosis, result in impaired barrier function and excess desiccation as a result of a disproportionate imbalance of lipid species within the lipid domains.[22,23] This suggests that the unique composition and relative ratios of the lipid species in the SC alter the supramolecular assemblies, thereby introducing a diseased state in skin. Thus, by working with ceramides, sterols and fatty acids, in different proportions, an effective network capable of structuring liquids oils in foods may be engineered.

The stratum corneum is thought of as a two-compartment system: the protein-enriched and lipid-depleted cornified cells are embedded in an expanded extracellular matrix of highly nonpolar lipids. The extracellular lipid matrix acts as the permeability barrier and plays major roles in cohesion and desquamation. It is well established that removal of the non-polar lipids from the stratum corneum has very little influence on the barrier properties; however, removal of polar lipids, including cholesterol and ceramides, causes profound barrier damage.[21] Upon disruption of the lamellar bodies the lipid bilayers are either absent or defective allowing excess desiccation. Work characterizing the lipid domains show a 4.1 Å spacing, suggesting a tight hydrocarbon packing that functions as the water barrier.[24]

3.3 Ceramides

Sphingolipids, a class of polar lipid, are composed of sphingosines, ceramides and sphingomyelins, which all share a common long-chain base.[25] Herein, the common long-chain base is sphingosine (Figure 3.1). Ceramides contain the sphingosine base attached to a fatty acid of differing chain lengths and degrees of saturation *via* an amide bond (Figure 3.1).

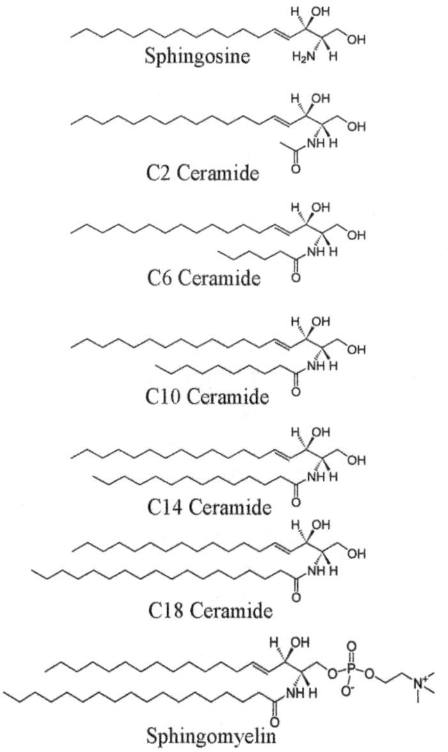

Figure 3.1 Chemical structure of sphingolipids.

Sphingomyelin contains a phosphocholine and a ceramide, which is covalently bound to the primary alcohol group. The ability of sphingolipids (*i.e.*, ceramides) to form an oleogel in canola oil is critically linked to their chemical structure.

3.3.1   Health Aspects of Ceramides

Dietary ceramides pass through the gut with minimal hydrolysis and are enzymatically broken down into sphingosine in the intestinal tract.[26,27] These hydrolytic enzymes are mainly expressed in the small and large intestine, allowing ceramides to pass through the gut giving rise to their potential as delivery vehicles for bioactive delivery. Sphingolipids have several physiological effects, which include influencing cell growth, differentiation, and cell death[26] as well as modifying cholesterol profiles.[28]

Sphingolipids are lipid secondary messengers for growth factors and cytokines [*e.g.*, tumor necrosis factor α (TNF-α)].[29] During platelet-derived growth factor activity, ceramides are converted enzymatically to sphingosine-1-phosphate and a potent mutagen and an inhibitor of apoptosis. Conversely, when TNF-α is active, elevated ceramide levels lead to the inhibition of growth and/or apoptosis, thus they are referred to as tumor-suppressing lipids.[30] Ceramides stimulate production of interleukin 1 beta, leading to apoptosis in tumor cells. Ceramides play an important role in inhibition of colon carcinogenesis.[30] During a first 4 week study, where colon cancer was chemically induced in CF1 mice, followed by treatment with and without the ingestion of sphingomyelin, a 70% reduction of aberrant crypt foci (ACF) was observed when mice were fed sphingomyelin.[27] ACF is the earliest morphological change in the colon and it is thought that adenomas and adenocarcinomas develop from these early lesions. After 40 weeks of consuming the sphingomyelin-enriched diet, malignant adenocarcinomas were reduced.[27] Comparisons of the abilities of C2 and C18 ceramides, as well as sphingosine and sphingomyelin, to cause cell death in common cancer cell lines have illustrated the potential of some of these compounds to inhibit cancer cell growth. A panel of colon cell lines (Figure 3.2) was evaluated to determine their sensitivity to increasing concentrations of ceramides (C2 and C18), sphingomyelin and sphingosine.[31] Across all cancer cell lines, sphingosine and C2 ceramide displayed the greatest effect at reducing colon cancer cell viability.[31] In contrast, C18 ceramide and sphingomyelin had little impact on colon cancer cell viability, as measured by the cell assays.

Sphingolipids have also been shown to reduce cholesterol by 30% and to quench free radicals, inhibiting lipid peroxidation in VLDL.[26] Thus, sphingomyelin may directly and/or indirectly reduce the risk factors for atherosclerosis.[26] Several reports indicate that sphingomyelin aids in the reduction of atherosclerosis. Chatterjee found that LDL had different binding and utilization capabilities in the presence of sphingomyelin.[32] Further, when sphingomyelinase is enriched in atherosclerotic lesions, there is an alteration in the aggregation state that promotes the activity of macrophages.[33]

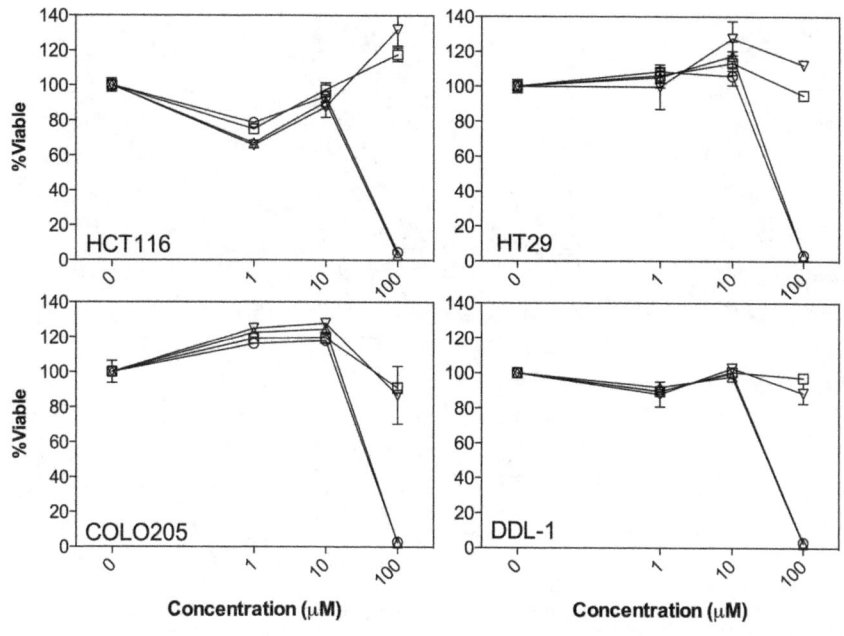

-○- 18:1 Sphingosine -△- C2-Ceramide -▽- C-18 Ceramide -□- 18:0 Sphingomyelin

Figure 3.2 Impact of increasing sphinoglipid concentrations on colon cancer cell line viability. Reproduced from Rogers, Spagnuolo, Wang, Angka, *Food Science & Nutrition*, 2016, 1–8 DOI: 10.1002/fsn3.433 with permission from Wiley Open Access. © 2016 The Authors, *Food Science & Nutrition*, published by Wiley Periodicals, Inc.

3.3.2   Ceramide Oleogels

Neither sphingosine nor sphingomyelin can form oleogels in canola oil; however, short-chain ceramides (*i.e.*, C2, C6, and C10) immobilize the oil, preventing its flow when the vial is inverted.[31] As the chain length of the ceramide increases, the turbidity of the gel transitions from a translucent to an opaque gel, suggesting that the supramolecular network of the gel is coarsening or that different microstructural elements are produced (*i.e.*, a transition from fibrillar to platelet or spherulitic crystals).[34,35] Microstructural elements, along with the supramolecular network established by ceramides, imaged under polarized light clearly show that ceramides that are capable of limiting flow are assembled into fibrillar crystals and the viscous solutions (*i.e.*, ceramides with a fatty acid chain length greater than 10 carbons) produced spherulitic crystals that were responsible for impeding light causing them to be opaque (Figure 3.3).[31]

Most commercially available sphingolipids are a mixture of sphingomyelins with varying chain lengths. For example, egg-derived sphingomyelin, converted to the ceramide using phospholipase c, resulted in a mixed chain ceramide that did not produce a stable gel at 5 wt% (Figure 3.4).[36]

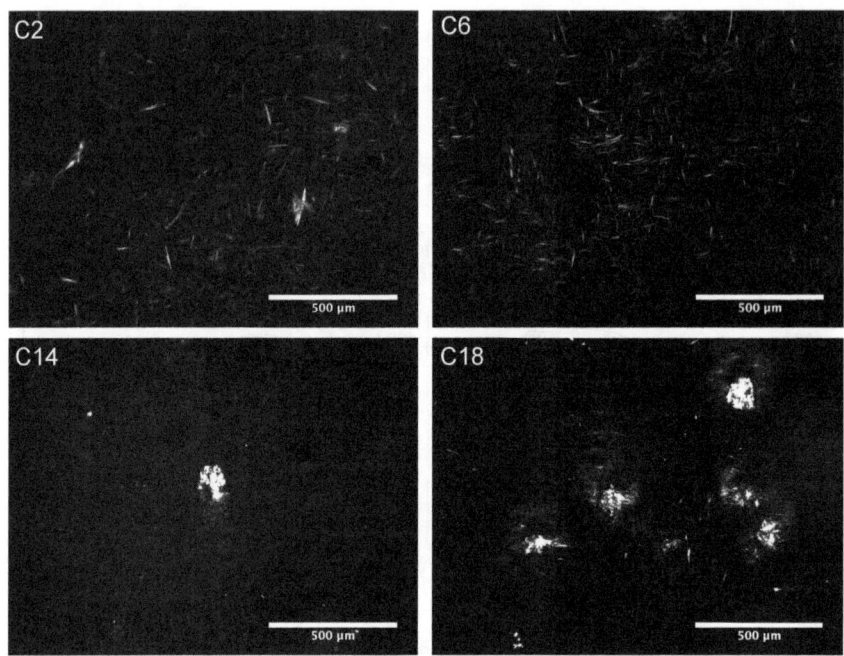

Figure 3.3   Polarized light micrographs of 5 wt% C2, C6, C14, and C18 in canola oil after being stored for 24 hr at 20 °C. Reproduced from Rogers, Spagnuolo, Wang, Angka, *Food Science & Nutrition*, 2016, 1–8 DOI: 10.1002/fsn3.433 with permission from Wiley Open Access. © 2016 The Authors, *Food Science & Nutrition*, published by Wiley Periodicals, Inc.

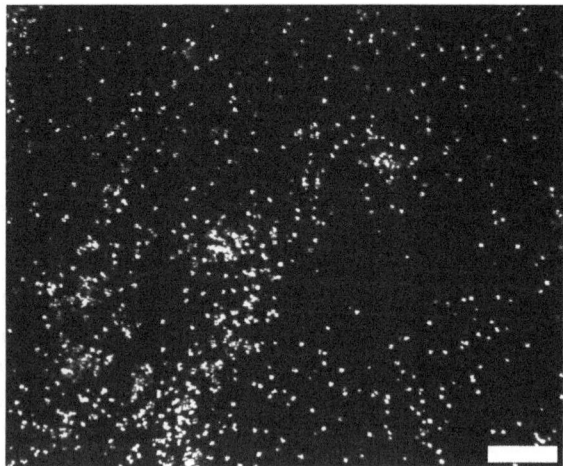

Figure 3.4   10 wt% milk ceramide, converted enzymatically from sphingomyelin, in canola oil oleogels imaged using polarized light [magnification bar = 100 μm]. Adapted from ref. 35 with permission from The Royal Society of Chemistry.

Ceramides derived from egg contains 86% palmitic acid, 6% stearic acid, 3% behenic acid, 2% nervonic acid and 2% make up all other fatty acids; whereas ceramides derived from milk contain only 16% palmitic acid, 20% behenic acid, 34% ticosylic acid, 21% lignoceric acid, 3% nervonic acid and 6% make up all other fatty acids.[36] The high degree of fatty acid chain length variation in the sphingomyelin-derived ceramides promoted small dendritic crystals that interact forming flocs, requiring higher concentrations of ceramides to form oleogels in canola oil. The heterogeneous nature of ceramide sidechains may disrupt the unidirectional growth of the fibrillar aggregates and promote a three-dimensional crystallization, similar to colloidal fat crystal networks.

Not only are the gel-forming abilities and visual appearance altered by chain length but so are the elastic modulus (G') and yield point.[31] Increasing the chain length of the fatty acid decreased the ability of the network to form a continuous network, resulting in lower elastic modulus (Figure 3.5) and yield points.[31] Previous rheological assessments, based on frequency sweeps, at 2 wt% ceramide in canola oil were frequency-independent. Using Clark and Ross-Murphy's classical definition of gels[37] they are classified as strong gels. At 5 wt% ceramide in canola oil, irrespective of the fatty acid chain length, all combinations formed oleogels and G' and G'' were both frequency-independent. Although ceramides with fatty acid chain lengths of

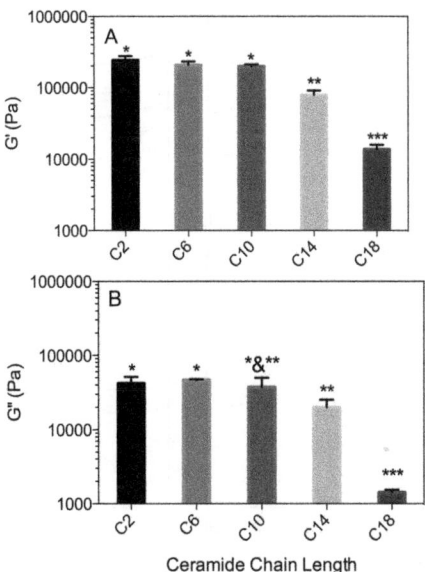

Figure 3.5 (A) Mean storage (G') and (B) loss modulus (G'') of 5 wt% ceramide oleogels in canola oil obtained from the frequency sweep. Asterisk indicate statistical significance at $p < 0.05$. Reproduced from Rogers, Spagnuolo, Wang, Angka, *Food Science & Nutrition*, 2016, 1–8 DOI: 10.1002/fsn3.433 with permission from Wiley Open Access. © 2016 The Authors, *Food Science & Nutrition*, published by Wiley Periodicals, Inc.

six and ten carbons did not exhibit flow when inverted, their elastic and loss modulus (G'') were both frequency-dependent, suggesting that the supramolecular structure is an entangled network.[31] This decrease in hardness is attributed to a combination of effects, including changes in solubility and differing arrangement of fibers *versus* spherulites.

The sol and gel transition, including both the crystallization and melting transition temperatures, increased with increasing chain length (Figure 3.6).[31] A single melting and crystallization peak were observed, except for C2 in canola oil. The C2-canola oil oleogel had two exothermic events during crystallization and two endothermic transitions during the melting process, suggesting a polymorphic solid–state transition or that a liquid crystal transition occurred followed by a transition to a 3D crystalline arrangement.[31] At 2 wt% the C2-canola oil oleogel had a higher crystallization compared to the C6-canola oil oleogel, which was also observed for the melting profiles.[31] At 2 wt%, after an initial melt of the C2 oleogel, there is a recrystallization event leading to a more stable polymorph followed by a second melting peak. At 5 wt% a weak crystallization peak is observed at ~40 °C, followed by another crystallization event. This transitions to a much larger melting peak

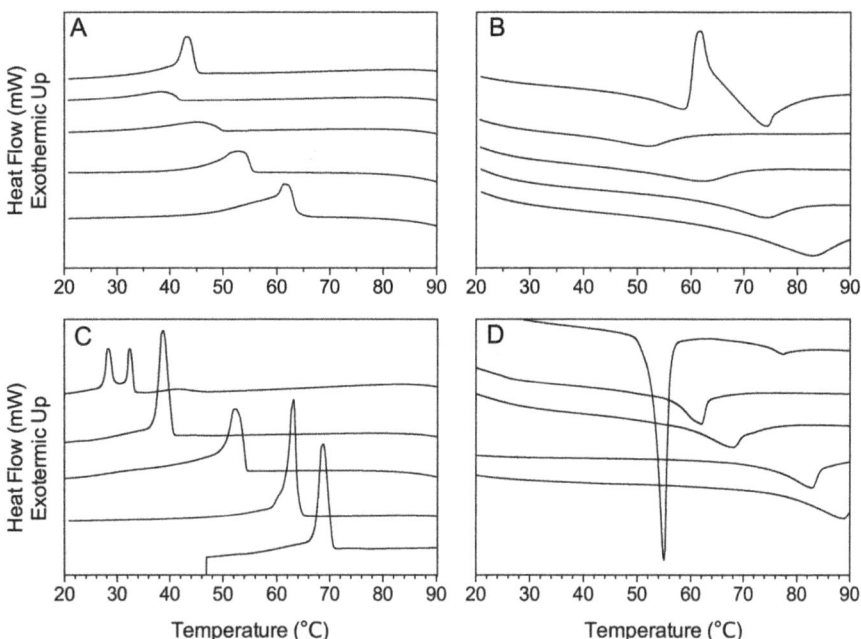

Figure 3.6 Crystallization (a, c) and melting (b, d) profiles for 2 wt% (a, b) and 5 wt% (c, d) ceramide gels in canola oil. The order from top to bottom for the ceramides is C2, C6, C10, C14, and C18. Reproduced from Rogers, Spagnuolo, Wang, Angka, *Food Science & Nutrition*, 2016, 1–8 DOI: 10.1002/fsn3.433 with permission from Wiley Open Access. © 2016 The Authors, *Food Science& Nutrition* published by Wiley Periodicals, Inc.

compared to the other chain length ceramides. The shorter chain length ceramides have an enhanced ability to self-assemble into low dimensionality crystals that are more adept at preventing the flow of the continuous oil phase compared to long-chain ceramides.[31]

3.4   Mimicking the Stratum Corneum Lipid Domains

Neither β-sitosterol, long-chain ceramides (ceramide III or phytosphingosine acylated with stearic acid) nor stearic acid are effective at forming a solid at 5 wt% in canola oil. At 5 wt%, stearic acid and ceramide III form platelets, while β-sitosterol forms a mixture of fibers and spherulites (Figure 3.7). Although fatty acids have been widely studied for their ability to form oleogels in edible oils, this is usually done with added fatty alcohols.[38] The crystal structure of fatty acids varies between one-dimensional (*i.e.*, lamellar structures in liposomes, micelles and bilayers) and three-dimensional crystals (*i.e.*, platelet and spherulitic crystals). Meanwhile, cholesterol has also been reported to form oleogels when incorporated in canola oil with δ-oryzanol.[39–45] Mixtures of β-sitosterol with fatty acids reveal a plethora of crystal structures depending on the fatty acid to cholesterol ratio. Changing the ratio of cholesterol and stearic acid showed a crystal transition from rosettes to ribbons in the case of β-sitosterol (Figure 3.8).[46] A 2:3 ratio of stearic acid to β-sitosterol forms platelets arranged in rosettes, while a 1:4 ratio still presents platelet crystals. When the β-sitosterol is replaced with cholesterol, 1:4 stearic acid to β-sitosterol changes the crystal morphology from platelets, observed at a 2:3 ratio, to fibers where the aspect ratio of the fibers increases to ~1000:1. These fibers may not be desirable in foods as they often have poor sensory attributes. However, other applications in structuring cosmetics and delivering pharmaceutics may be feasible.

Increasing the complexity of the system to contain β-sitosterol, ceramide III and stearic acid requires the formation of a ternary phase diagram to illustrate the different crystal morphologies when the sum of the structurants totaled 8 wt%. For the phase diagram, structurants were varied at 1 wt% intervals in canola oil for a total of 21 different combinations (Figure 3.9).[46] All 21 combinations had different crystal morphologies and physical properties; however, only eight of the 21 systems were capable of structuring the liquid oil into a solid-like fat matrix.[46] In other words, eight combinations of β-sitosterol, ceramide III and stearic acid did not flow upon vial tube inversion. Samples with less than 3 wt% stearic acid flowed upon inverting the vial and at 8 wt% structurants the G'' (*i.e.*, loss modulus) was greater than the G' (*i.e.*, storage modulus), apart from [2 wt% stearic acid, 1 wt% β-sitosterol and 5 wt% ceramide III (SBC215)] and SBC314. In five ratios (SBC314, SBC413, SBC512, SBC521 and SBC611), fibers were observed. Platelet-like crystals were observed in SBC215, short fibril crystals were seen in SBC422, and rosette-like crystals for SBC431 were observed in the remaining solid-like samples.[46]

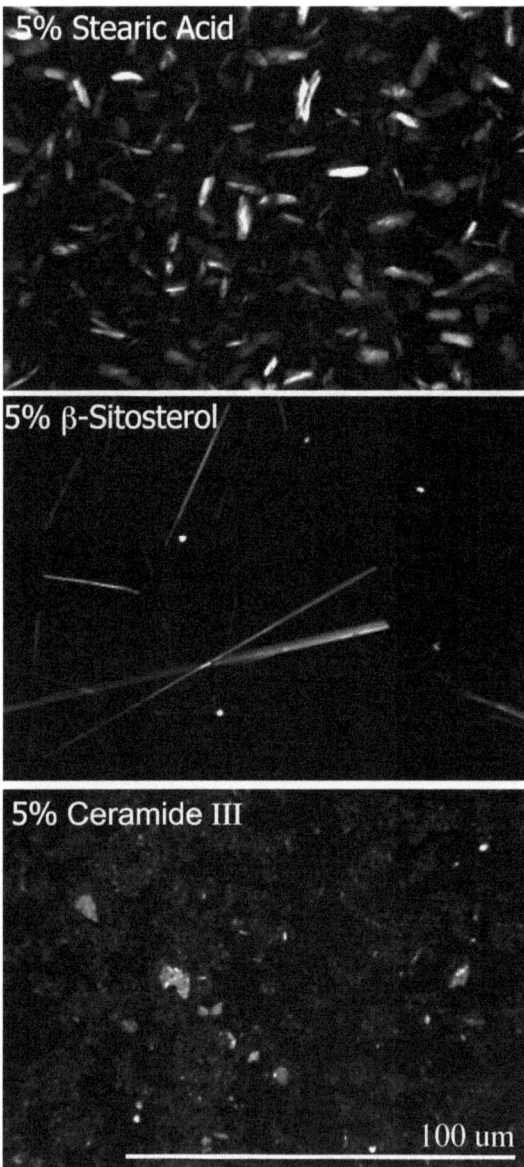

Figure 3.7 Polarized light micrographs of viscous solutions of 5 wt% stearic acid, β-sitosterol and ceramide III in canola oil. Reproduced from Wang & Rogers, *Lipid Technology*, 2015, 27, 175–178 with permission from Wiley-VCH Cerlag GmbH & Co. Copyright © 2015 WILEY-VCH Verlag GmbH & Co. KGaA, Weinheim.

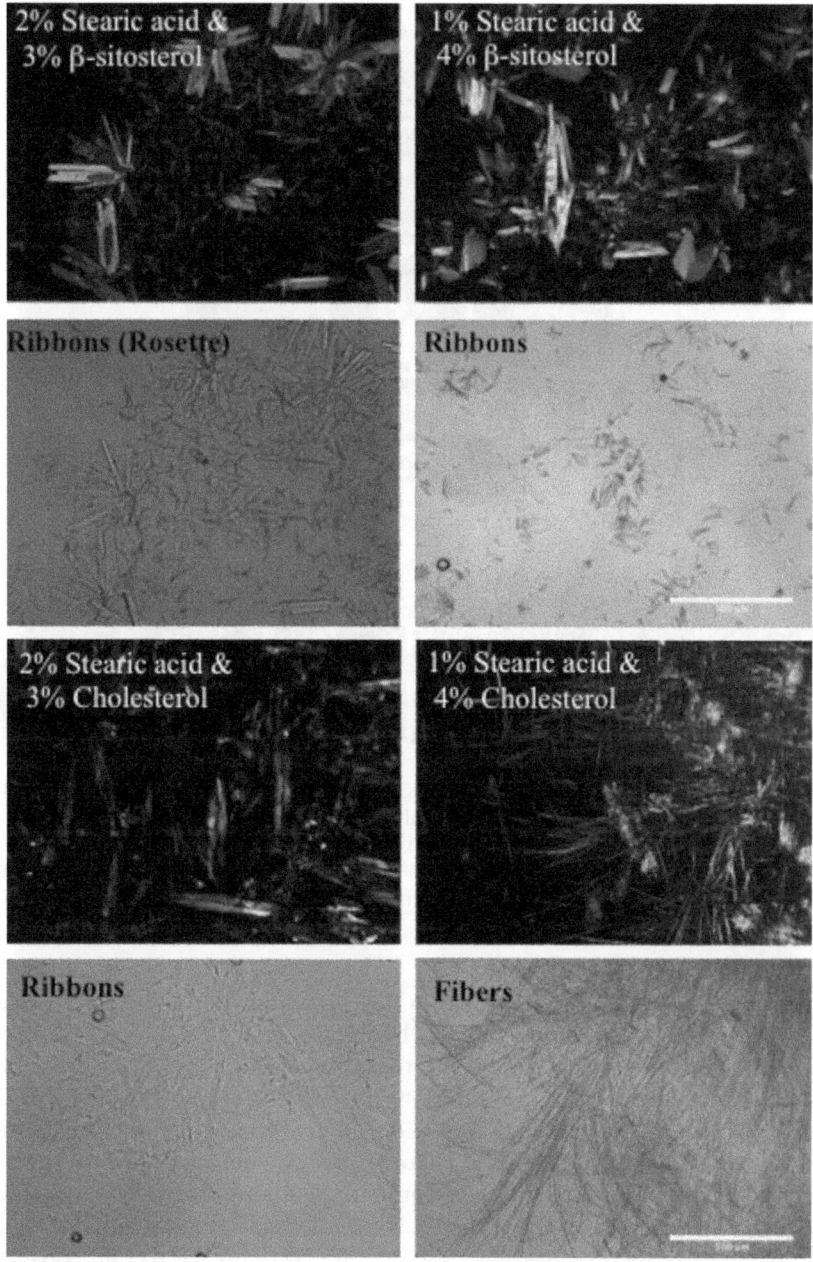

Figure 3.8 Polarized light micrographs (top) and brightfield micrographs (bottom) of 5 wt% combinations of stearic acid and β-sitosterol or cholesterol in canola oil. Adapted from Wang & Rogers, *Lipid Technology*, 2015, 27, 175–178 with permission from Wiley-VCH Cerlag GmbH & Co. Copyright © 2015 WILEY-VCH Verlag GmbH & Co. KGaA, Weinheim.

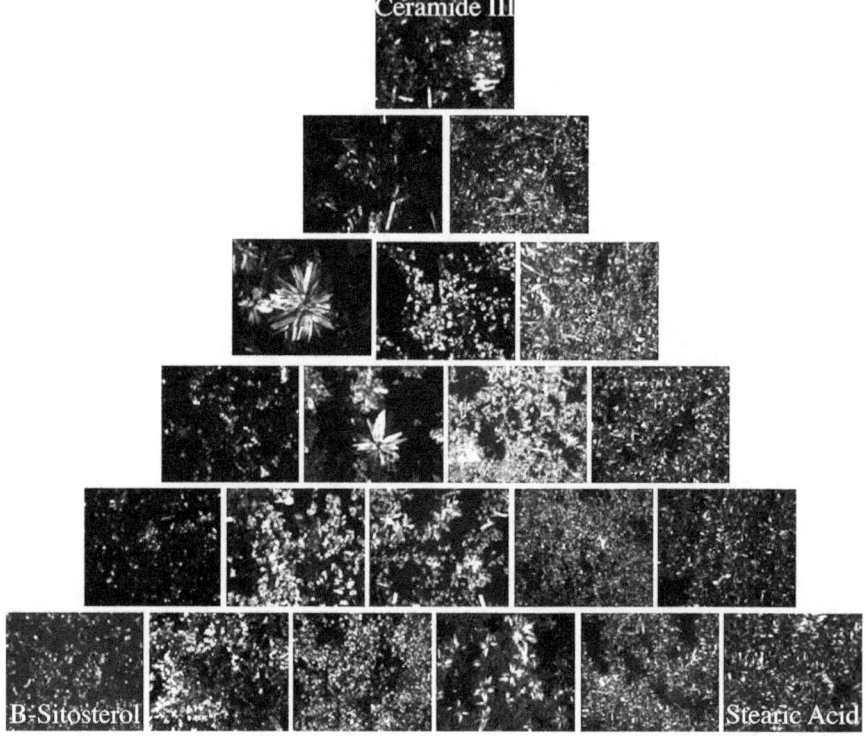

Figure 3.9   Ternary phase diagram consisting microscopy images for 21 samples.
The composition of each point is expressed as the stearic acid:β-sitos-
terol:ceramide III ratio in canola oil. Reproduced from Wang & Rogers,
Lipid Technology, 2015, 27, 175–178 with permission from Wiley-VCH
Cerlag GmbH & Co. Copyright © 2015 WILEY-VCH Verlag GmbH & Co.
KGaA, Weinheim.

Although SBC215, SBC422, and SBC431 appeared solid and did not flow
under vial inversion, more advanced rheological techniques highlighted the
frequency dependence of the samples that contained platelets, short fibers or
rosette-like crystals.[46] These combinations were not true gels as they exhibited frequency dependence of G' and G''. The five specific ratios that formed
crystal fibers each exhibited solid rheological behaviors and each of the gels
contained similar compositions where the β-sitosterol concentration was
either 1 or 2 wt% [*i.e.*, on the right side of the ternary phase diagram (Figure
3.9)]. The ratios SBC314, SBC413, SBC512, SBC521 and SBC611 all consisted
of fibers, all had an elastic modulus greater than the loss modulus, and were
frequency independent between 0.1 and 10 Hz.[46] The elastic modulus ranged
from 10^2 Pa to 10^5 Pa at 8 wt% depending on the ratio of the three structurants.[46] Within these combinations, there is a wide range of tailorability of
the final physical properties depending on the desired application. The yield

stresses of SBC611, SBC521 and SBC512 were 4.65 ± 2.36, 1.22 ± 0.73 and 3.43 ± 2.07 Pa, respectively, indicating that they are easily spread. Not only is the hardness of the gels tailorable, using different structurant compositions, but so are their melting properties. The onset of melting was between 42 and 48 °C, and when there was a greater concentration of stearic acid compared to β-sitosterol the melting temperature was higher.[46]

3.5   Conclusions

Although studies on this system are still in their infancy, they have tremendous potential to facilitate the incorporation of more unsaturated fats into processed foods while reducing the *trans* and saturated fat content. Studies on phytosterol, ceramide and fatty acid structures may reveal more synergistic combinations that produce novel materials that could be incorporated into our food and feed supplies. Preliminary work has shown that when mixed in ideal ratios the elastic modulus of 8 wt% structurants can reach 10^5 Pa, making it an ideal hardstock fat replacer. Fiber-forming combinations were observed at $3:1:4$, $4:1:3$, $5:1:2$, and $6:1:1$ stearic acid:β-sitosterol:ceramide III.

Acknowledgement

This research was undertaken, in part, thanks to funding from the Canada Research Chairs program.

References

1. A. C. Mendes, E. T. Baran, R. L. Reis and H. S. Azevedo, *Wiley Interdiscip. Rev.: Nanomed. Nanobiotechnol.*, 2013, **5**, 582–612.
2. M. Boncheva and G. M. Whitesides, *MRS Bull.*, 2005, **30**, 736–742.
3. G. Askarieh, M. Hedhammer, K. Nordling, A. Saenz, C. Casals, A. Rising, J. Johansson and S. D. Kinight, *Nat. Lett.*, 2010, **465**, 236–239.
4. L. Böni, P. Fischer, L. Böcker, S. Kuster and P. A. Rühs, *Sci. Rep.*, 2016, **6**, 30371–30379.
5. F. H. Silver, J. W. Freeman and G. P. Seehra, *J. Biomech.*, 2003, **36**, 1529–1553.
6. J. N. Israelachvili, D. J. Mitchell and B. W. Ninham, *J. Chem. Soc., Faraday Trans. 2*, 1976, **72**, 1525–1568.
7. A. Reinhardt and D. Frenkel, *Phys. Rev. Lett.*, 2014, **112**, 238103.
8. P. M. Elias, *J. Gen. Intern. Med.*, 2005, **20**, 183–200.
9. Y. Lan, M. G. Corradini, R. G. Weiss, S. R. Raghavan and M. A. Rogers, *Chem. Soc. Rev.*, 2015, **44**, 6035–6058.
10. J. Gao, S. Wu, T. Emge and M. A. Rogers, *CrystEngComm*, 2013, **15**, 4507–4515.

11. R. S. H. Lam and M. A. Rogers, *CrystEngComm*, 2010, **13**, 866–875.
12. R. Y. Wang, P. Wang, J. L. Li, B. Yuan, Y. Liu, L. Li and X. Y. Liu, *Phys. Chem. Chem. Phys.*, 2013, **15**, 3313–3319.
13. J. L. Li and X. Y. Liu, *Adv. Funct. Mater.*, 2010, **20**, 3196–3216.
14. X. Y. Liu and P. D. Sawant, *Adv. Mater.*, 2002, **14**, 421–426.
15. M. A. Rogers and R. G. Weiss, *New J. Chem.*, 2015, **39**, 785–799.
16. A. Pal, S. Abraham, M. A. Rogers, J. Dey and R. G. Weiss, *Langmuir*, 2013, **29**, 6467–6475.
17. M. A. Rogers, S. Abraham, F. Bodondics and R. G. Weiss, *Cryst. Growth Des.*, 2012, **12**, 5497–5504.
18. M. A. Rogers, S. Abraham, F. Bodondics and R. G. Weiss, *Cryst. Growth Des.*, 2012, **12**, 5497–5504.
19. S. Abraham, Y. Lan, R. S. H. Lam, D. A. S. Grahame, J. J. H. Kim, R. G. Weiss and M. A. Rogers, *Langmuir*, 2012, **28**, 4955–4964.
20. D. A. S. Grahame, C. Olauson, R. S. H. Lam, T. Pedersen, F. Borondics, S. Abraham, R. G. Weiss and M. A. Rogers, *Soft Matter*, 2011, 7, 7359–7365.
21. S. M. Jackson, M. L. Williams, K. R. Feingold and P. M. Elias, *West. J. Med.*, 1993, **158**, 279–285.
22. M. J. Choi and H. I. Maibach, *Am. J. Clin. Dermatol.*, 2005, **6**, 215–223.
23. L. Coderch, O. Lopez, A. de la Maza and J. L. Parra, *Am. J. Clin. Dermatol.*, 2003, **4**, 107–129.
24. T. J. McIntosh, *Biophys. J.*, 2003, **85**, 1675–1681.
25. K. Miazek, S. Lebecque, M. MHamaidia, A. Paul, S. Danthine, L. Willems, M. Frederich, E. De Pauw, M. Deleu, A. Richel and D. Goffin, *Biotechnol., Agron., Soc. Environ.*, 2016, **20**, 321–336.
26. E. M. Schmelz, *Nutr. Bull.*, 2000, **25**, 135–139.
27. E. M. Schmelz and A. H. Merrill, *Nutr. Bull.*, 1998, **14**, 717–719.
28. T. Kobayashi, T. Shimizugawa, T. Osakabe, S. Watanabe and H. Okuyama, *Nutr. Res.*, 1997, **17**, 111–114.
29. A. E. Cremesti and A. S. Fischl, *Lipids*, 2000, **35**, 937–945.
30. P. W. Parodi, *Aust. J. Dairy Technol.*, 2001, **56**, 65–71.
31. M. A. Rogers, P. A. Spagnuolo and T.-M. Wang, *Wiley Food Sci. Nutr.*, 2017, **5**, 579–587.
32. S. Chatterjee, *J. Biol. Chem.*, 1999, **268**, 3401–3406.
33. S. Marathe, S. L. Schissel, M. J. Yellin, N. Beatini, R. Mintzer, K. J. Williams and I. Tabas, *J. Biol. Chem.*, 1998, **273**, 4081–4088.
34. M. A. Rogers, A. J. Wright and A. G. Marangoni, *Soft Matter*, 2008, **4**, 1483–1490.
35. M. A. Rogers, A. J. Wright and A. G. Marangoni, *Curr. Opin. Colloid Interface Sci.*, 2009, **14**, 223.
36. M. A. Rogers, A. J. Wright and A. G. Marangoni, *Soft Matter*, 2009, **5**, 1594–1596.
37. A. H. Clark and S. B. Ross-Murphy, *Adv. Polym. Sci.*, 1987, **83**, 55–192.
38. J. Daniel and R. Rajasekharan, *J. Am. Oil Chem. Soc.*, 2003, **80**, 417–421.
39. A. Bot and W. G. M. Agterof, *J. Am. Oil Chem. Soc.*, 2006, **83**, 513–521.

40. A. Bot, R. den Adel, E. Roijers and C. Regkos, *Food Biophys.*, 2009, **4**, 266–272.
41. A. Bot, R. den Adel and E. C. Roijers, *J. Am. Oil Chem. Soc.*, 2008, **85**, 1127–1134.
42. A. Bot and E. Flöter, in *Edible Oleogels: Structure and Health Implications*, ed. A. G. Marangoni and N. Garti, AOCS Press, Urbana, IL, 2011, pp. 49–79.
43. M. A. Rogers, *Food Res. Int.*, 2009, **42**, 747–753.
44. M. A. Rogers, *CrystEngComm*, 2011, **13**, 7049–7057.
45. M. A. Rogers, A. Bot, R. S. H. Lam, T. Pedersen and T. May, *J. Phys. Chem.*, 2010, **114**, 8278–8295.
46. T.-M. Wang and M. A. Rogers, *Lipid Technol.*, 2015, **27**, 175–178.

Section III

Structuring Units: Crystalline Particles and Self-assembled Structures

CHAPTER 4

New Insights into Wax Crystal Networks in Oleogels

K. D. MATTICE AND A. G. MARANGONI*

Department of Food Science, University of Guelph, 50 Stone Road E,
Guelph, Ontario, Canada N1G 2W1
*E-mail: amarango@uoguelph.ca

4.1 Introduction

Wax-based oleogels are a type of structured oil system where waxes act as the gelator of a liquid oil phase. These waxes are capable of forming gels at low concentrations (as low as <1% w/w) by trapping large volumes of liquid oil within a three-dimensional network. Waxes are composed of a complex mixture of molecules, including esters of long-chain carboxylic acids and long chain alcohols derived from fatty acids and fatty alcohols hydrocarbons, free fatty acids, free fatty alcohols and other components in lesser quantities.[1-3] A wax dispersed in an oil will crystallize with decreasing temperature, forming three-dimensional crystals of characteristic shape. These crystals will assemble to form a three-dimensional structure, known as a wax crystal network, which is stabilized through intermolecular forces at transient or permanent junction zones.[4-6] The exact forces that cause the self-assembly of these three-dimensional structures are unknown, but it is this phenomenon that allows for gelation by waxes.

Properties of wax oleogels depend on many critical variables. Each type of wax takes on a characteristic morphology and crystal size. The compositional

Food Chemistry, Function and Analysis No. 3
Edible Oil Structuring: Concepts, Methods and Applications
Edited by Ashok R. Patel
© The Royal Society of Chemistry 2018
Published by the Royal Society of Chemistry, www.rsc.org

diversity between waxes causes these differences, which subsequently affect the overall properties of the three-dimensional crystal network that is formed. Because these structural properties are so unique to the type of wax, analysis of a formed wax oleogel can actually identify the type of wax contained in the system.[5] This structural variability is only amplified when waxes contain impurities. As highly heterogeneous materials, the endless arrangements of major and minor components increase the potential for diversity in crystallization behaviour and crystal morphologies. In addition to the wax composition, the presence of impurities, such as surfactants, can impact the physical properties of a gel.[7] The type of oil used in gelation will also affect gel functionality, as the oil acts as the solvent for the wax gelator, making the solubility of the wax in the oil a determinant factor. There have also been studies published that demonstrate the use of different cooling rates or the use of shear during cooling as a means of modifying gel functionality.[8-11] Modification of these parameters produces changes in the nucleation and crystal growth rates, which are the foundations of the microstructure and physical properties.

It is reasonable to consider the gelation of oils using waxes on a large-scale basis given the considerable annual global production of waxes, the small amount of wax required to gel a liquid oil, often between 1 and 4% (w/w), and the cost of these waxes as raw materials.[5] When compared with the many discovered oleogelators, the uniqueness here stems from the waxes' natural origin. The incorporation of wax oleogels containing common sources of waxes (rice bran oil, sunflower oil, *etc.*) into food and cosmetics satisfies the growing demand for natural ingredients. As an ingredient in cosmetics, wax oleogels can impart texture and stability to topical products. As a food ingredient, wax oleogels are used as an alternative solid fat source with an improved fatty acid profile. Significant research is now being done to examine the practicality of using wax oleogels as a solid unsaturated fat source in foods.[12-15] The fact that the functionality and physical properties can be altered by modification of certain variables, combined with the current trends favouring the use of natural ingredients in food, make waxes of great interest for this application. The use of oleogels in foods is quite novel, therefore all oleogelators must be approved for use in foods. At this point, the US FDA has determined that rice bran wax, candelilla wax, and carnauba wax are generally recognized as safe (GRAS) in foods under regulations 21 CFR 172.890, 184.1976, and 184.1978, respectively.[16-18]

4.2   Natural Waxes

Natural waxes used for the gelation of oils are obtained from primarily plant sources, extracted either directly from the plant itself, or from the oil that was previously extracted from the plant. There are also waxes derived from animal sources, the most notable being beeswax. While the umbrella of natural waxes includes a large number of species, all types of waxes vary greatly in melting and crystallization behaviour, chemical composition and

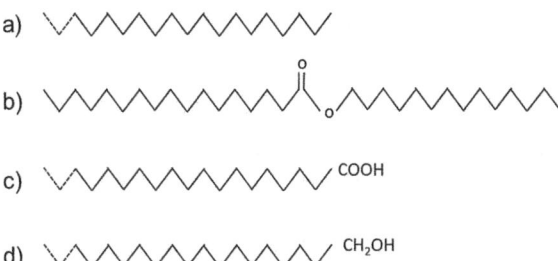

Figure 4.1   General structure of major components found in waxes: (a) hydrocarbons (*n*-alkanes) of formula $H_3C[CH_2]_nCH_3$ where n = 22–36; (b) wax
esters (mainly mono esters) of formula R^1COOR^2 where R^1 and R^2 are
10–20 carbon atoms or more in length; (c) fatty acids (mainly alkanoic
acids) of formula RCOOH where R is 12–34 carbon atoms in length; and
(d) fatty alcohols (mainly primary alcohols) of formula RCH_2OH where
R is 12–34 carbon atoms in length.

morphology, making no type of wax identical to another. Waxes are composed of a mixture of wax esters, free fatty acids, free fatty alcohols and
hydrocarbons, each of varying carbon chain length. These are considered the
major components, but waxes will also contain a variety of minor components. The general structure of each of the major components can be found
in Figure 4.1. It is the wax ester content that is generally considered responsible for the gelation behaviour of natural waxes.[5,6,19] The combination of polar
and non-polar components makes natural waxes slightly soluble in liquid
oils, evident by a lower experimental solid fat content (SFC) as compared to the
theoretical values based on the amount of wax added to each oleogel.[5] This fits
the requirements for oleogelation, where an insoluble gelator molecule will
simply precipitate out of solution, and a highly soluble species would prevent the formation of a crystal network, thus no gelation would occur.[20]

The gelation efficiency of a wax has been shown to vary depending on the
composition of the wax itself. Variations in the ratio of major to minor components, as well as variation in the chain length of these components, will
affect the gelation properties of a wax. It is therefore necessary to identify
the precise composition of each wax in order to truly understand how that
wax behaves when used as a gelator and the properties of the subsequent gel
that would be formed. The general compositions of common wax types have
been well known and available for some time, but the extremely complex
composition of waxes make determining the exact composition profile very
difficult.[3,5] With wax crystal networks becoming more prevalent in research
and industry, determining the detailed composition was vital to explain
the behaviour of different types of waxes. Separation of complex mixtures,
such as waxes, is commonly done by thin layer chromatography, which can
then identify the presence of a class of compounds to determine their general composition.[21,22] However, a study published by Doan *et al.* described
updated methods employing high-performance liquid chromatography

Table 4.1 General chemical composition of the waxes analyzed through HPLC-ELSD on an Alliance System. Reprinted from *Food Chemistry*, **214**, C. D. Doan, C. M. To, M. De Vrieze, F. Lynen, S. Danthine, A. Brown, K. Dewettinck and A. R. Patel, Chemical profiling of the major components in natural waxes to elucidate their role in liquid oil structuring, 717–725, Copyright 2017, with permissions from Elsevier.[a]

(%)	RBX	SFX	CLX	CRX	BEX	BSX
WE	93.49 ± 2.63	96.23 ± 0.19	15.76 ± 0.35	62.05 ± 3.03	0.02 ± 0.02	58.00 ± 0.68
FFA	6.00 ± 2.12	3.29 ± 0.16	9.45 ± 1.14	6.80 ± 0.76	95.70 ± 1.11	8.75 ± 0.75
FAL	0.22 ± 0.22	0.32 ± 0.38	2.20 ± 1.02	30.74 ± 2.48	4.24 ± 1.10	6.42 ± 0.90
HC	0.29 ± 0.29	0.17 ± 0.15	72.92 ± 2.23	0.41 ± 0.30	0.03 ± 0.01	26.84 ± 1.04

[a]WE: wax ester; FFA: free fatty acid; FAL: free fatty alcohol; HC: hydrocarbon; RBX: rice bran wax; SFX: sunflower wax; CLX: candelilla wax; CRX: carnauba wax; BEX: berry wax; BSX: beeswax. Source: Doan *et al.*[3]

(HPLC) with an evaporative light scattering detector (ELSD) as more sensitive to provide results with much greater resolution.[3] Then, the separated fractions from HPLC were subjected to gas chromatography and mass spectrometry (GC-MS) as a method to identify specific compounds in the wax species. The authors saw distinct retentions times for free fatty acids, free fatty alcohols, hydrocarbons and wax esters when six different waxes were analyzed (Table 4.1). GC-MS was then used to identify specific species within each class of compound, the results of which are presented in Table 4.2, where the absence of a value indicates that the amount of that compound was below the detection limit. The wax ester fraction is presented separately as fatty acid and fatty alcohol moieties because the saponification required for the analysis of wax esters caused fractionation of these entities. While general interpretations for this analysis can be made, true conclusions regarding the wax ester content are limited by the inability to determine which fatty acid and fatty alcohol moieties were once bound together.

4.3 The Gelation of Oil by Waxes

The entrapment of liquid oils within a crystal network is not a new concept, and is in fact the method by which hard-stock fats exist, where the higher melting, solid TAGs form a crystal network, which structures the lower melting fractions. Wax crystal networks function by the same concept, where an oleogelator acts to form a three-dimensional crystal network to entrap and structure liquid oils. Many types of natural waxes have demonstrated their ability as gelators, and can be differentiated by their critical concentrations (C^*), defined as the minimum amount of gelator required to cause gelation of liquid oil. The C^* is an indicator of gelation efficiency, where a more efficient gelator can be identified by the appearance of structured oil at a lower concentration of gelator, corresponding to a lower C^* value. However, the C^* can vary for the same type of wax when gelling different of oil, resultant from changes in the oil composition. Blake and Marangoni demonstrated the effect of oil type on the critical concentration of a wax, stating that differences result from changes in wax solubility when incorporated into different types of oil.[5]

To form a wax-based oleogel, a determined amount of wax is dispersed into a selected type of liquid oil. While the C^* of a specific wax can be used to determine the amount of wax to be added, in many cases, this produces gels that are too weak for realistic use in many applications so prior determination is required. The mixture is then heated to 90 °C for approximately 30 minutes in order to completely melt and dissolve the wax crystals into the oil. Upon cooling, the wax begins to form crystals of characteristic size and shape for each particular wax. These crystals then form a network *via* transient or permanent junction zones, which entraps the oil liquid phase within the solid structure and an oleogel is formed.[5,23] In most cases, gelation does not occur simultaneously with the formation of crystals, as it depends on the reorganization and assembly of these crystals into the three-dimensional microstructure at low temperatures.[19] Confirmation of gelation can simply be done by visual inspection; any sign of separation or flow upon inversion indicates that the oil has not been gelled. It is also possible to assign certain rheological measurements as minimum requirements to be able to quantitatively classify a system as a gel.

As the amount of gelator in the system is increased beyond the C^*, the melting and gelation points are also increased.[4,6,8] Figure 4.2 displays the relationship between wax concentration and gel melting temperature for rice bran wax (RBX), sunflower wax (SFX), carnauba wax (CRX) and candelilla wax (CLX), where a higher temperature of melting (T_m) and temperature of crystallization (T_c) are observed with increasing gelator concentration in all cases, not reaching a maximum value until the concentration reaches 100% (w/w).[5] Understanding this relationship can be useful for tailoring wax oleogels to specific applications when a specific melting temperature is desired. Accompanying these changes is an exponential increase in gel firmness with increasing gelator concentration. This is, therefore, also a means of modifying the physical properties of a wax oleogel. However, it has been suggested that adding wax in concentrations far above the corresponding C^* is detrimental to the physical properties of a gel. Blake and Marangoni proposed that a gel prepared with a wax concentration of even 5% (w/w) when the C^* is 1–2% (w/w) would have too much "solid" character such that the physical properties of the system no longer resemble a gel state.[11] Another observation is that a higher gelator concentration results in the development of larger crystals.[5] This result is likely caused by the presence of more wax material at higher gelator concentrations, leading to greater crystalline growth.

A study by Hwang *et al.* described the contribution of the chemical composition to the gelation efficiency of waxes.[6] Specifically, the authors related the observed efficiency of a number of different waxes to their composition in terms of wax ester content. Overall, a lower gelation efficiency was found when waxes were composed predominantly of other components, explained

Table 4.2 Content of specific hydrocarbon, free fatty acid, free fatty alcohol and the fatty acid and fatty alcohol moieties of wax esters species in common waxes by GC-MS. Reprinted from *Food Chemistry*, **214**, C. D. Doan, C. M. To, M. De Vrieze, F. Lynen, S. Danthine, A. Brown, K. Dewettinck and A. R. Patel, Chemical profiling of the major components in natural waxes to elucidate their role in liquid oil structuring, 717–725, Copyright 2017, with permission from Elsevier.[a]

	RBX	SFX	CLX	CRX	BEX	BSX
Hydrocarbons						
C27	-	-	0.15 ± 0.13	-	-	40.28 ± 3.43
C29	-	-	6.26 ± 2.57	-	-	25.63 ± 1.90
C31	-	-	82.48 ± 1.95	-	-	18.05 ± 0.85
C33	-	-	7.68 ± 0.85	-	-	3.12 ± 1.19
Free fatty acids						
C16	20.76 ± 2.55	33.18 ± 2.90	17.84 ± 1.78	16.17 ± 1.85	82.13 ± 2.49	30.11 ± 5.57
C18	5.54 ± 0.61	20.42 ± 1.83	1.90 ± 0.29	17.78 ± 1.79	11.26 ± 1.67	6.84 ± 1.54
C20	8.70 ± 1.58	22.12 ± 2.05	1.31 ± 0.29	10.57 ± 2.14	0.39 ± 0.19	1.74 ± 0.64
C22	17.52 ± 3.12	9.83 ± 2.26	1.60 ± 0.17	6.28 ± 2.40	-	4.33 ± 1.39
C24	28.33 ± 3.91	4.00 ± 0.95	0.64 ± 0.19	15.47 ± 3.49	-	27.80 ± 2.84
C26	3.36 ± 0.76	2.26 ± 0.40	0.90 ± 0.29	6.17 ± 1.15	-	6.55 ± 0.76
C28	3.34 ± 0.71	2.85 ± 0.09	3.13 ± 0.24	10.58 ± 0.18	-	7.20 ± 0.30
C30	3.03 ± 0.80	0.99 ± 0.73	31.38 ± 0.82	4.70 ± 2.39	-	6.60 ± 0.31
Free fatty alcohols						
C24	25.30 ± 2.08	25.25 ± 1.13	1.77 ± 0.12	-	-	12.01 ± 2.11
C26	26.15 ± 0.36	24.11 ± 2.91	4.53 ± 1.25	-	-	9.46 ± 2.60
C28	16.16 ± 1.41	17.74 ± 1.68	11.48 ± 1.45	6.38 ± 1.64	-	12.95 ± 1.35
C30	15.39 ± 1.36	10.46 ± 3.00	42.28 ± 2.30	13.94 ± 1.22	-	29.23 ± 0.20
C32	9.47 ± 2.50	5.92 ± 0.63	14.57 ± 1.97	65.39 ± 2.71	-	20.40 ± 2.74
C34	4.70 ± 0.15	-	2.83 ± 0.20	11.17 ± 1.05	-	3.94 ± 0.28

Wax ester fractions

Fatty acid moieties						
C16	8.63 ± 1.47	4.89 ± 1.85	34.16 ± 3.91	15.16 ± 5.19	79.89 ± 4.55	80.27 ± 4.50
C18	4.97 ± 0.69	4.84 ± 1.49	17.44 ± 1.01	16.69 ± 3.91	13.25 ± 2.38	9.54 ± 1.57
C20	22.95 ± 1.25	48.65 ± 6.87	6.31 ± 3.23	16.02 ± 0.11	-	0.51 ± 0.15
C22	26.75 ± 2.19	24.14 ± 1.55	10.87 ± 2.56	12.19 ± 0.77	-	0.43 ± 0.17
C24	27.83 ± 1.17	5.87 ± 1.27	1.20 ± 0.70	22.38 ± 5.35	-	0.47 ± 0.19
Fatty alcohol moieties						
C18	2.80 ± 0.93	1.82 ± 0.60	20.70 ± 7.09	13.69 ± 3.10	66.02 ± 2.42	2.93 ± 0.85
C22	6.92 ± 1.04	9.48 ± 1.43	-	3.08 ± 0.09	-	-
C24	29.46 ± 4.24	34.41 ± 4.06	2.05 ± 0.08	3.61 ± 1.66	8.77 ± 2.09	25.19 ± 3.50
C26	23.94 ± 1.10	30.07 ± 1.10	8.18 ± 1.16	1.73 ± 0.87	3.82 ± 0.29	14.20 ± 0.94
C28	13.70 ± 1.31	12.51 ± 1.98	18.72 ± 0.79	3.15 ± 1.98	3.32 ± 0.07	13.45 ± 0.51
C30	11.16 ± 2.92	4.01 ± 1.59	26.64 ± 3.05	11.95 ± 0.74	3.05 ± 0.08	26.12 ± 1.70
C32	5.31 ± 2.68	2.20 ± 0.28	7.99 ± 2.48	51.74 ± 6.57	3.53 ± 0.06	13.57 ± 1.73

[a]HC: hydrocarbon; FFA: free fatty acid; WE: wax ester; FAL: free fatty alcohol; FA moieties: fatty acid moieties; FAL moieties: free fatty alcohol moieties; RBX: rice bran wax; SFX: sunflower wax; CLX: candelilla wax; CRX: carnauba wax; BEX: berry wax; BSX beeswax. Source: Doan et al.[3]

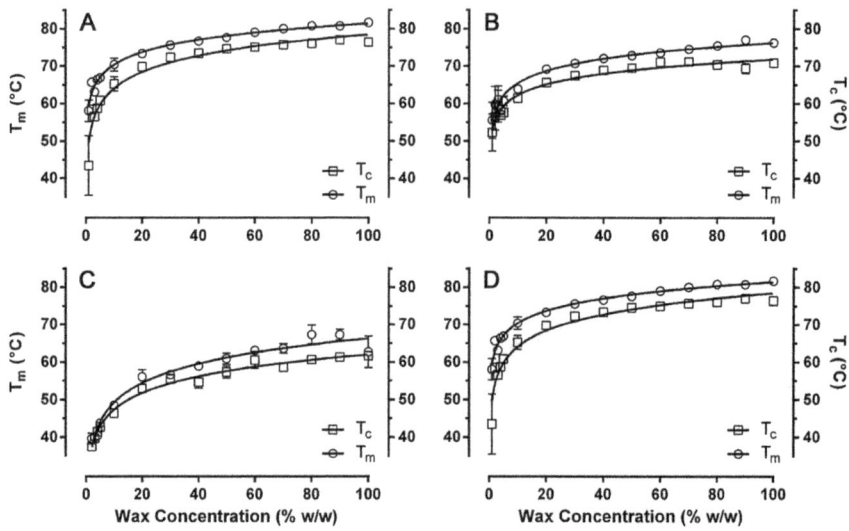

Figure 4.2   Peak melting temperatures (T_m) and peak crystallization temperatures (T_c) as functions of wax concentration for (a) rice bran wax, (b) sunflower wax, (c) candelilla wax, and (d) carnauba wax each in canola oil.[5] *Journal of the American Oil Chemists' Society*, Structure and Physical Properties of Plant Wax Crystal Networks and Their Relationship to Oil Binding Capacity, **91**, 2014, A. I. Blake, E. D. Co and A. G. Marangoni. © AOCS 2014. With permission from Springer.

simply due to the lack of wax esters. The nature and amount of wax esters or *n*-alkanes have been proposed as the main factors responsible for the gelation of oil. One particular characteristic necessary for efficient gelation is the prevalence of saturated esters. Waxes that contain a significant amount of unsaturated esters, which are highly soluble in unsaturated oils, require larger amounts of wax for gelation, making them inefficient. This is especially true if the wax has a similar composition to that of the unsaturated alkyl chains in the oil in which the wax is dispersed. Additionally, waxes that contain longer alkyl chains of wax esters demonstrate better gelling abilities than those that contain shorter alkyl chains.

Another compositional consideration is the presence of minor components and impurities. Essentially, a greater heterogeneity equates to a less ordered structure that is less capable of entrapping oils. Hwang *et al.* obtained RBX from three different suppliers and compared the respective oleogels.[6] The results confirmed varying gelation results between the three waxes, including C^* values and melting behaviour determined by differential scanning calorimetry (DSC). This is a significant indication that the even minor changes in the composition of a wax can cause its gelation efficiency to be altered. Further, the presence of impurities in CLX was shown to impede the growth of crystals causing an overall reduction in crystal size.[20] This puts great emphasis on the need for precise analysis of wax composition.

4.4 Wax Crystal Network Microstructure

Wax crystal networks are three-dimensional structures with particular characteristics at all length scales. The microstructure of a wax crystal network is unique to the type of wax used and has been deemed an important characteristic in terms of the physicochemical properties of the formed gels. To characterize the nanoscale structure, powder X-Ray diffraction is used, which reveals the crystalline sub-cell arrangement. Wax crystal networks tend to have wide angle XRD patterns similar to those of β′ polymorphic form triacylglycerols, while the small angle peaks will vary among different waxes exhibiting different lamellar packing.[4,5] A notable difference between the intensity of the peaks at wide and small angles has also been observed, indicating anisotropic (unidirectional) growth along directions perpendicular and parallel to the lamellar planes. This results in crystals that are thin and needle-like fibres or platelets.[4,5] A more complex system that experiences co-assembly of multiple components will undergo more complex crystal growth depending on the intermolecular interactions that exist between these components.[19] This leads to the structural features of a larger scale, including crystal size, morphology, and other properties of the assembled three-dimensional network. The microstructure is often studied using optical light microscopy as it can distinguish between solid and liquid phases, where the wax crystals appear bright, and contrasts the liquid oil phase, which appears dark.[24] Micrographs of eight different wax crystal networks are shown in Figure 4.3.

Hwang *et al.* proposed that greater gelation efficiency would be observed for waxes possessing a highly homogeneous chemical composition, as well as having high compatibility between major and minor components, as this would allow for tighter packing and the formation of fibrillar crystals.[6] This was supported by the findings of Morales-Rueda *et al.*, who demonstrated that the heterogeneity of CLX and CRX allows for the co-crystallization of foreign components within their structure, subsequently hindering the tight molecular packing and preventing the formation of an ordered crystal structure.[25] It has been hypothesized that the gelation efficiency of a wax depends on the morphology of the crystals, where a needle-like or fibrillar morphology allows for easy entrapment of liquid oil, as compared to a wax that displays spherulitic or granular morphology. It was theorized that the fibrous nature produced a highly entangled and dense network that easily entrapped the surrounding oil phase within.[4] The high gelation efficiency of RBX and SFX, which display fibrillar morphology, compared with the lower efficiency of CRX and CLX, which display spherical morphology, support these conclusions.

This theory of gelation differences based on fibrillar *vs.* non-fibrillar morphology was examined by Blake and Marangoni.[26] The authors scrutinized the common technique of optical light microscopy owing to being limited by resolution and length scale, meaning wax crystals that are smaller than the lower limit of this method, including CLX crystals, would not be

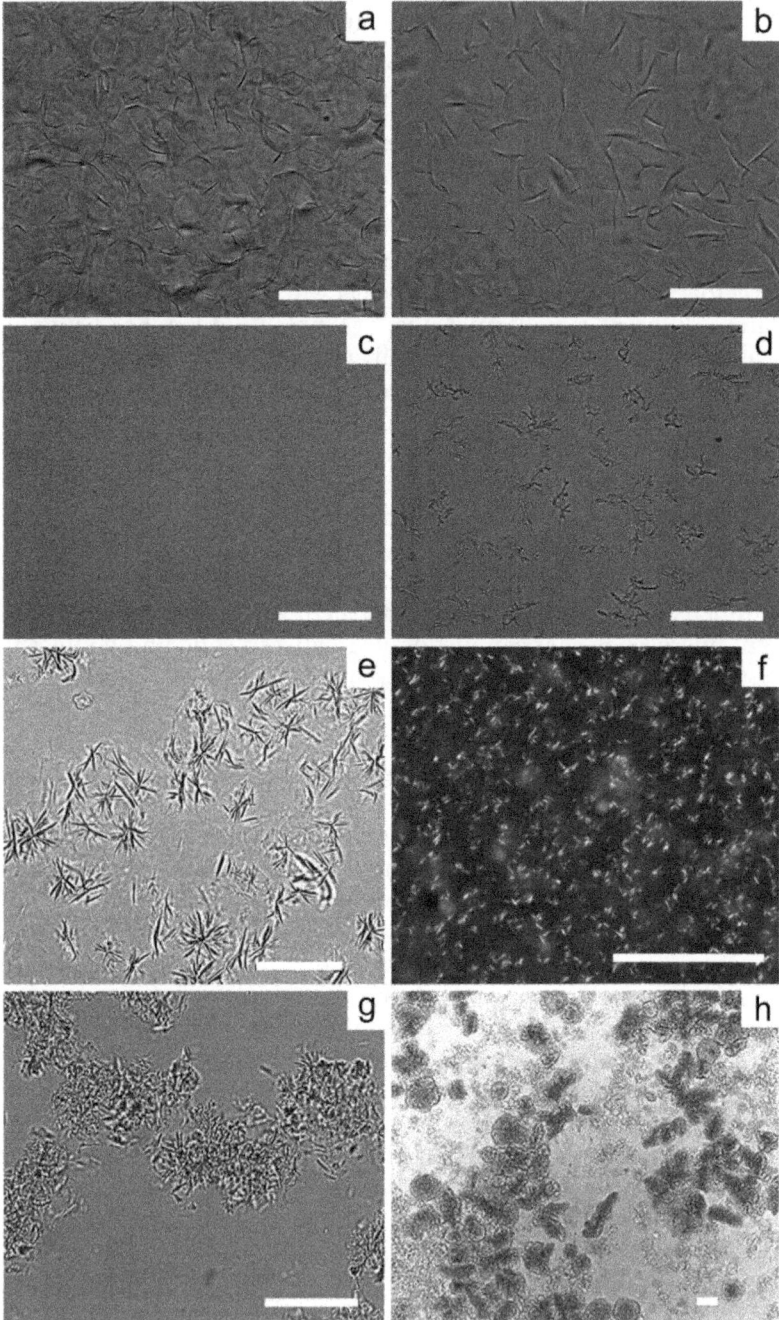

Figure 4.3   Optical light micrographs (under normal and polarized light) of oleogels prepared with: (a) 1% rice bran wax in canola oil; (b) 1% sunflower
wax in canola oil; (c) 2% candelilla wax in canola oil; (d) 4% carnauba
wax in canola oil; (e) 1% beeswax in high oleic sunflower oil; (f) 4%

pictured clearly. The liquid oil or solvent phase also creates challenges, as distinguishing the crystals from this phase can be difficult. The authors suggested the use of cryogenic scanning electron microscopy (cryo-SEM) as an alternative method. In this method, samples must be de-oiled prior to analysis, and the wax crystals are immobilized by flash freezing in liquid nitrogen (−210 °C). This makes the analysis much more rigorous, but can provide a clearer image of the microstructure owing to the removal of the oil phase and greater resolution. Using this method, the authors concluded that RBX, SFX and CLX wax crystals actually exist in platelet form. Knowing that all research on mineral oil waxes reported platelet-like morphology, they suggested that the needle-like shapes observed were actually the protruding edges of platelets, not visible using the usual optical light microscopy methods.[27] Differences in gelation efficiency between waxes were then explained using visual differences in the crystal size, dispersion and pore size of the networks. It was observed that CLX crystal networks contained smaller and more homogenously dispersed crystals when compared to both RBX and SFX. The larger and less homogenously dispersed crystals of the RBX and SFX crystal networks resulted in large pores. These larger pores can hold more oil, meaning less material is needed to gel oil and therefore a greater gelation efficiency can be expected. A summary of reported morphologies and crystal size for each type of wax is presented in Table 4.3. These findings led to the conclusion that that traditional optical light microscopy is a truly imprecise method, and cryo-SEM is a more suitable method for the study of wax crystal morphology. This conclusion, however, has not yet been universally accepted, with studies still reporting non-platelet morphologies for these waxes from cryo-SEM images.[19] A comparison of optical light micrographs and cryo-SEM images for RBX, SFX and CLX is presented in Figure 4.4.

Microscopy is also used to determine the fractal dimension and pore area fraction. Determining the fractal dimension is a method of characterizing the uniformity of the mass distribution in a crystal network by assigning a numerical classification. In the method of box-counting fractal

---

sugar cane wax in soybean oil; (g) 6% berry wax in high oleic sunflower oil; and (h) 7% fruit wax in high oleic sunflower oil. Scale bars = 50 μm (a–g), or 20 μm (h). (a) *Journal of the American Oil Chemists' Society*, Structure and Physical Properties of Plant Wax Crystal Networks and Their Relationship to Oil Binding Capacity, **91**, 2014, A. I. Blake, E. D. Co and A. G. Marangoni. © AOCS 2014. With permission from Springer. (b) Reproduced with permission from A. R. Patel, M. Babaahmadi, A. Lesaffer and K. Dewettinck, *J. Agric. Food Chem.*, 2015, **63**, 4862–4869. Copyright 2015 American Chemical Society. (c) Reprinted from *Food Research International*, **50**, J. C. B. Rocha, J. D. Lopes, M. C. C. N. Mascarenhas, D. B. Arellano, L. M. R. Guerreiro and R. L. da Cunha, Thermal and rheological properties of organogels formed by sugarcane or candelilla wax in soybean oil, Copyright 2013, with permission from Elsevier.

Table 4.3   Summary of observed morphology and crystal size of different waxes.[a]

Wax type	Description of observed morphology	Approximate crystal size	Aggregate size
Rice bran wax	Needle-like/platelet-like	10–20 μm	–
Sunflower wax	Needle-like/platelet-like/ rod-like	15–25 μm	–
Candelilla wax	Spherical/grain-like	<10 μm	–
Carnauba wax	Spherical/dendritic	10–20 μm	50–100 μm
Beeswax	Fibrous/"sea urchin"-like	20–30 μm	–
Sugar cane wax	Needle-like	<10 μm	–
Berry wax	Platelet-like in spherical crystalline aggregates	<10 μm	40–70 μm
Fruit wax	Flat crystals radiating from spherical units/spherulites	30–50 μm	–

[a]Sources: Dassanayake *et al.*,[4] Blake and Marangoni,[5] Blake and Marangoni,[26] Patel *et al*,[19] Rocha *et al.*[32] and Doan *et al.*[33]

dimension, a value that is closer to 2 indicates a homogenous mass distribution. Conversely, a value that is farther from 2 indicates a network with uneven distribution. Discussion of fractal dimension is entertained to compare the mass distribution of two different wax crystal networks. Pore area fraction is used to discuss the percent area of pores within a crystal network, where a larger value indicates that pores are taking up a larger area in the network overall. From a micrograph, the number of black (pores) to white (crystals) pixels can be used to calculate the pore area fraction. The fractal dimension of a gel has been shown to have a negative correlation with the pore area fraction, which is easily explained by the fact that a uniformly distributed network of smaller crystals will contain smaller pores, while a network of large, uneven crystals will contain larger pores that take up a greater fraction of the area.

4.5   Types of Natural Wax Gelators

4.5.1   Rice Bran Wax (RBX)

RBX is obtained by extraction from rice bran oil. Rice bran oil is estimated to contain somewhere between 0.4 and 4% wax, and based on the annual production of rice bran, there is the potential for an annual production of approximately 35 100 metric tons of RBX.[5,28] Its chemical composition is highly homogenous, with commercially available RBX containing primarily long chain esters of saturated C46–C62 derived from C20–C36 fatty alcohols and C20–C26 fatty acids. The homogeneity and high melting point of these components gives rise to this wax's sharp melting point at high temperatures. RBX forms long fibrous crystals, which likely contribute to its high gelation efficiency, where RBX has a $C*$ of just 1% (w/w) in canola oil.[4,5] A $C*$ as low as 0.50 (w/w) has been observed in soybean oil when specific cooling rates were used.[6]

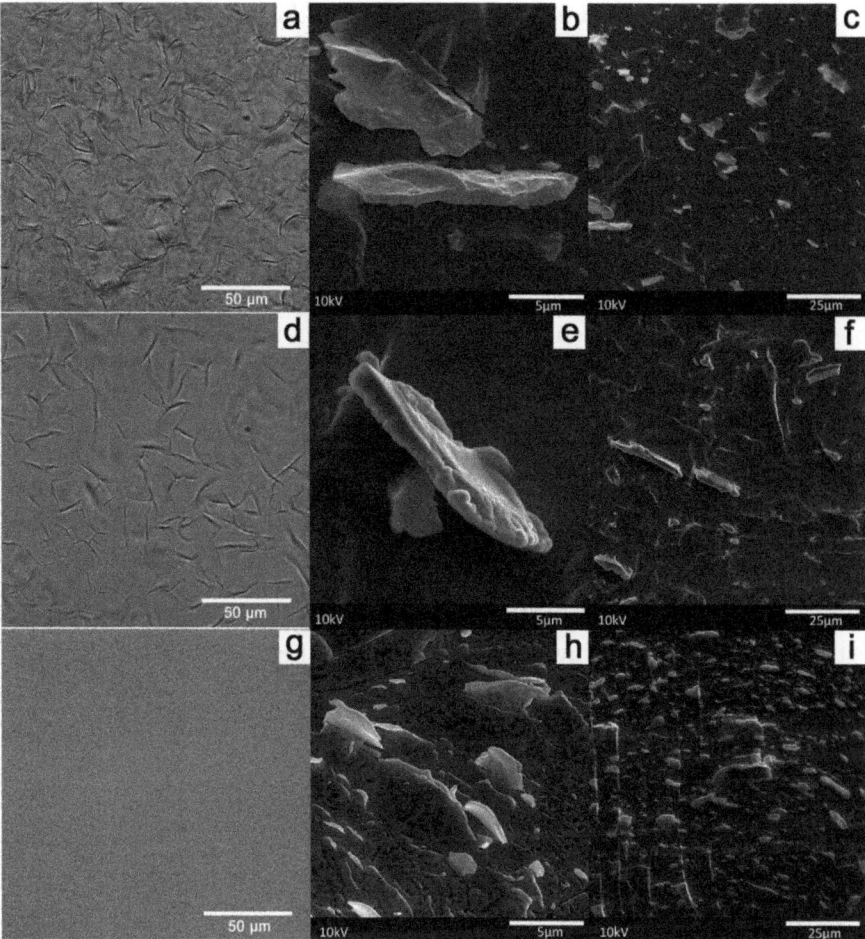

Figure 4.4   Optical light micrographs (left) and cryo-SEM images at high (middle) and low (right) magnification of rice bran wax (a–c), sunflower wax (d–f) and candelilla wax (g–i). (Adapted from *Journal of the American Oil Chemists' Society*, Structure and Physical Properties of Plant Wax Crystal Networks and Their Relationship to Oil Binding Capacity, **91**, 2014, A. I. Blake, E. D. Co and A. G. Marangoni. © AOCS 2014 with permission from Springer[5] and *Food Structure*, 3, A. I. Blake and A. G. Marangoni, Plant wax crystals display platelet-like morphology, 30–34, copyright 2015, with permission from Elsevier[26]).

RBX has been approved for use in foods by the FDA (21 CFR, 172.890) under the conditions that the refined wax has specific properties, including a melting point from 75 to 80 °C and a free fatty acid content less than or equal to 10%.[16] The intended food uses for RBX are as a component in candy coatings, coatings for fresh fruits and vegetables and as a plasticizer in chewing gum. Other non-food uses for RBX include applications in the cosmetic industry, where it is used to create texture in products such as lipsticks and mascaras, in pharmaceuticals and in the leather industry.[1,29]

4.5.2 Sunflower Wax (SFX)

SFX is a hard, crystalline, high melting wax obtained from sunflower oil. Crude sunflower oil contains 0.35–1.0% wax, and based on the annual production of sunflower oil, there is the potential for production of SFX up to 42 300 metric tons annually.[5,30] Production of sunflower wax can be considered sustainable as it is a by-product of sunflower oil production. Its composition is highly homogenous, with commercially available SFX containing >95% wax esters ranging from C42 to C60 primarily from C20–C32 fatty alcohols and C20–C28 fatty acids.[3,19] This homogeneity causes SFX to have single sharp melting and crystallization points. The high content of wax esters has been noted as the cause of the rod-like or fibrillar morphology of SFX crystals.[19] The combination of these two factors explains the low C^* at 1% (w/w) in canola oil, given both of their association with excellent gelation properties. Another study also reported successful gelation at concentrations as low as 0.50% (w/w) for SFX in soybean oil when specific cooling rates were used.[6] Currently SFX can only be used in cosmetic applications, as it has not yet been deemed GRAS by the FDA. These applications include lipsticks, lip balms, and other cosmetics owing to added benefits of thickening and lubricity.[31]

4.5.3 Candelilla Wax (CLX)

CLX is obtained from the candelilla plant, native to Mexico and the southern United States, and is a yellowish-brown and opaque to translucent, hard wax. CLX has a moderately efficient C^* of 2% (w/w) in canola oil. It has been reported to form weak gels in a 1% concentration formulation. However, these gels do not demonstrate adequate stability during rheological testing at temperatures greater than 40 °C, meaning this cannot be considered the C^* value.[5] Identified major and minor components have determined that CLX is a very chemically diverse wax. Generally, CLX contains approximately 73% hydrocarbons, 16% wax esters and 9% fatty acids.[19] The major component has been determined to be hentriacontane, a long chain hydrocarbon, and the two most prevalent minor components of interest have been identified as the *n*-alkanes nonacosane and tritriacontane.[8,25] The minor components in CLX have been found to co-crystallize and form mixed molecular structures with major alkane components, causing interference in molecular packing and a broad melting point at mid-range temperatures.[4,5,8] The reduced crystalline order caused by the heterogeneity of this wax causes and also hinders crystal growth, causing formation of smaller spherulitic, grain-like or worm-like crystals.[5,6] The spherical crystals exhibit greater oil binding capacities, presumed to be caused by the greater surface area of the crystals, allowing for greater oil surface absorption. However, they exhibit a reduced gelation efficiency (as compared with other waxes) because the smaller, less ordered crystals are less soluble in oil.[26]

CLX can be used in foods under conditions outlined in the FDA regulation 21 CFR 184.1976.[18] Currently, food uses of CLX are as an ingredient in chewing gum and hard candies. Toro Vazquez *et al.* reported stable CLX gels that experienced no phase separation over three months of storage at room temperature and suggested that this stability, combined with the achieved textures, could make CLX gels practical for use in the food industry.[8] However, large scale use of CLX is not currently feasible, as the many inexpensive, synthetic waxes available as alternatives in non-food applications cause production of this wax to be limited.[5]

4.5.4 Carnauba Wax (CRX)

CRX is a high melting wax obtained from the leaves and buds of the Brazillian wax palm, *Copernicia cerifera* Maritius, and is a very hard, brittle wax. Compositionally, CRX is not chemically homogenous, comprising approximately 60% wax esters and 30% fatty acid alcohols.[3,19] The main components of commercially available CRX are C24–C32 fatty acids and primary alcohols.[3] In terms of gelation efficiency, many studies have classified CRX as one of the least efficient gelators due to having a high C^* of 4% (w/w) in canola oil.[4,5] It has been suggested that this lower efficiency results from the characteristic formation of spherical, grain-like crystals that are less ordered owing to the heterogeneity of this wax, and therefore capable of trapping less oil.[5] In addition, the primary presence of aliphatic esters and aliphatic diesters in place of wax esters is also thought to contribute as these species do not possess the same gelation abilities.[6] Use of CRX in foods must comply with the FDA regulation 21 CFR 184.1978.[17] Current food uses of CRX are as an ingredient in baked goods, chewing gum, confections and frostings, fresh fruit and fruit juices, gravies and sauces, processing fruits and fruit juices, and soft candies.

4.5.5 Other Natural Wax Gelators

There has been extensive research into the gelation mechanisms of the waxes listed above, but studies are now beginning to include other wax gelator species, with more research still to come. These include sugar cane wax (SCX), beeswax (BSX), berry wax (BEX) and fruit wax (FRX). While none of these waxes have been given GRAS status as of yet, their natural sources make it reasonable to assume that it is in the realm of possibility for the future. SCX has many similar physical properties to the commonly used waxes, including a high melting point and needle-like morphology. Rocha *et al.* conducted an investigation into the use of SCX as an oleogelator in soybean oil.[32] The authors determined that SCX was able to produce a gel at 2% (w/w) concentration, similar to the C^* of CLX. However, all gels were significantly softer than their CLX counterparts despite having very similar physical properties. This result again validates the fact that even minor variations in wax composition and microstructure will influence the traits of the oleogels created. The use of BSX as an oleogelator is also promising. It has a melting point in the mid-range of that common in waxes, and is composed of approximately 58% wax esters, 27% hydrocarbons and 9% fatty acids. A C^* ranging from <1% to 1.5% (w/w) has been reported in the gelation of high oleic sunflower, olive oil and rice bran oil.[19,33,34] Despite this equivalent gelation efficiency, BSX has been shown to produce gels that are more brittle when compared to other wax crystal networks unless added at a much greater level of 7% (w/w), potentially limiting their application.[19,34] BSX also has a "sea-urchin" like morphology, completely unique from all other wax oleogelators studied. BEX and FRX differ from all other sources in that they have low melting points resulting from the high content of lower melting, short chain fatty acids and free fatty alcohols. These components also have a higher solubility in oil and cause the C^* of BEX and FRX to be much greater than other waxes, determined to be as high as 6% (w/w) and 7% (w/w), respectively, in high oleic sunflower oil.[19] In agreement with these findings, both of these waxes form spherical crystals, which have been described to be less effective for the entrapment of oil. However, one study reported gel formation in rice bran oil with a drastically different BEX concentration as low as 1% (w/w).[33] Interestingly, BEX and FRX oleogels required minimal crystalline assembly for gel formation, evident by the simultaneous occurrence of crystallization and gelation different from other waxes. While the compositional differences between the oil phases in these systems can be used to explain these contrasting C^* results, this is a clear indication that more research is required to better understand the gelation mechanisms of BEX and FRX.

4.6 Oil Binding Capacity of Wax Crystal Networks

The oil binding capacity describes the ability of a system to retain its liquid oil phase and is an important parameter to study in order to characterize the stability and therefore, the potential applications, of an oleogel. The binding of liquid oil within fat crystal networks depends on microstructural factors and intermolecular interactions, and has been attributed to two main mechanisms: the entrapment of oil in pores within the surrounding crystalline network, and the absorption of oil onto the surface of crystalline triacylglycerol nanoplatelets.[35,36] Essentially, it is considered that oil trapped within a pore is weakly bound and more likely to be exuded and lost in a shorter amount of time.[10] The more oil that is tightly adsorbed to the surface of wax crystals, the higher the oil binding capacity of a gel will be. Therefore, a network of smaller, evenly distributed crystals with greater surface area will have a greater oil binding capacity than one with large crystal masses. The oil binding capacity can be quantified by observing and measuring the oil that is lost by a system during the destabilization period as grams lost per hour. Testing allows for the determination of both the amount of oil that is lost (a long-term consequence), as well as the rate at which it is lost (a short-term consequence). Gravimetric tests are often used, which employ kinetic forces to induce destabilization, including gravitational or centrifugal forces.[10]

Blake and Marangoni performed an extensive study of the oil binding capacity of wax oleogels.[5,10] First, their work considered the oil binding capacity of wax based oleogels prepared with waxes at their respective $C*$ in canola oil in static conditions. According to this work, CLX gels were found to have the greatest oil binding capacity, attributed to the small crystal size and therefore largest surface area. RBX gels experienced the greatest oil loss, which can be explained by the larger crystal size. However, further considerations led to the conclusion that further investigation of the oleogels in non-static conditions was required. This analysis would provide more information on the how the gels would behave in industrial processing conditions and shed additional light on the potential use of these gels in different applications. In the case of food applications, a gel with a low oil binding capacity would not be suitable as the release of oil would compromise the integrity of the food matrix, as well as the sensory and texture properties. The relationship between oil binding capacity and the application of shear on wax oleogels will be described in a later section.

Given the dependence of oil binding capacity on pore size and crystal surface area, the fractal dimension and pore area fraction are common parameters that can be used to predict the extent of oil loss a gel is expected to exhibit. This was observed by Blake and Marangoni, who reported a negative relationship between box-counting fractal dimension and oil loss.[10] The authors suggested that modification of the microstructure could be a method of engineering a gel with desirable oil binding properties. This includes increasing the homogeneity in crystal mass distribution, otherwise known as increasing the fractal dimension and surface area for oil adsorption, or by decreasing the pore area fraction, thereby decreasing the opportunity for oil to be weakly bound within the pores. The authors plotted the fractal dimension *vs.* pore area fraction for individual gels made with three different gelators (RBX, SFX, and CLX) at two different cooling rates (1.5 °C min^{-1} and 5 °C min^{-1}) (Figure 4.5). The significance of including these six different gels was the fact that each one had a different fractal dimension and pore area fraction, with gelator type and cooling rate both having an effect on the microstructure of the oleogels. The effect of cooling rate will be discussed in a later section. The resultant curve revealed that the relationship between these parameters is easily modeled using an exponential growth function such that one could achieve the desired oil binding capacity by manipulating the pore area fraction of the wax crystal network. The exponential model also reveals that there is a limit to the extent that pore area can be reduced before the oil binding capacity is negatively affected.

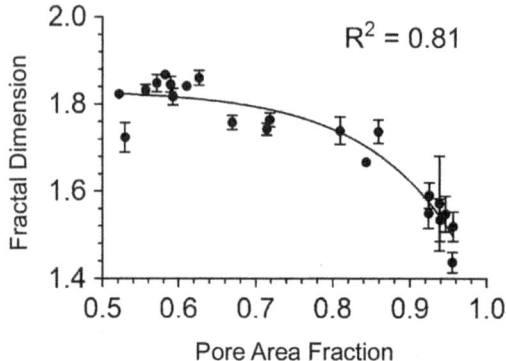

Figure 4.5   Fractal Dimension as a function of pore area fraction for rice bran wax, sunflower wax, and candelilla wax gels formed at cooling rates of 1.5 °C min^{-1} and 5 °C min^{-1}. Curve modeled using Fractal Dimension (FD) = $A + Be^{-A\,Fp/C}$ where A, B, and C are fitting parameters. *Food Biophysics*, The use of cooling rate to engineer the microstructure and oil binding capacity of wax crystal networks, **10**, 2015, 456–465, A. I. Blake and A. G. Marangoni, © Springer Science+Business Media New York 2015. With permission from Springer.

4.7   Rheological Profiling of Wax Crystal Networks

Based on the large-scale interest in the use of wax crystal networks in food applications, an extensive fundamental understanding of the macroscopic properties of these systems is required. The macroscopic functionality of a wax crystal network describes parameters including elasticity, mechanical strength and oil binding capacity. Rheological measurements compare the storage modulus (G'), which describes the elastic or solid properties, to the loss modulus (G''), which describes the viscous or liquid properties. Given the solid state of these gelled systems, it is expected that $G'' \ll G'$ for all wax based oleogels. These measurements allow for differentiation of the wax crystal networks in that a system with greater mechanical strength would have a greater G'. These properties are dependent on the microstructure of the network, including crystal size, shape and distribution, as well as degree of branching, and the proportion of transient and permanent junction zones.[10] A network containing a greater proportion of permanent junction zones is associated with high mechanical strength, while elasticity is associated with a network containing transient junction zones.

The many variables have been shown to cause changes in the rheological properties of a wax crystal network among gels prepared with different wax types at both their respective C^* as well as at equal concentrations. The fatty acid composition of the oils has also been suggested to have an impact on the final properties of oleogels, including their viscosity and hardness, where oleogels prepared with oils containing saturated fatty acids and high melting fatty acids cause the production of viscous and hard oleogels.[37] Overall, wax oleogels are characterized as viscoelastic gels, which vary in type of deformation (either ductile or brittle) when force is applied and in stiffness based on the bond strength between the particles of the gel. The use of different types of oils must also be considered, as a varying triacylglycerol composition could cause differing degrees of interactions such that different rheological properties result. The strong relationship between the rheological properties of a gel and the aforementioned variables means the functionality of a wax oleogel can be modified by means of altering these parameters.

It has been shown that the stiffness of any gel will increase as the solid volume fraction increases, as the fractal dimension decreases and as the particle size decreases.[5,38,39] In the case of wax crystal networks, a wax with a greater C^* will form a stiffer gel based on the greater concentration of solids present in the gel. This correlation makes estimation of gel stiffness easy. While it is assumed that the determined fractal dimension and particle size of gels could also be used to predict stiffness, it is necessary to consider the cumulative effect from both of the factors as the expected stiffness of certain gels based on one factor only may not follow the usual trend. For example, Blake *et al.* found that gels prepared with CLX were stiffer than gels prepared with RBX and SFX, despite having the greatest fractal dimension. This result is theorized to be due to the combined effect of its small particle size and higher solid phase content.[5,8]

A comprehensive study by Patel *et al.* compared the rheological properties of six different wax-based oleogels.[19] Each type of wax oleogel was found to display shear thinning behaviour, which is a reduction in viscosity with applied shear. They also displayed low thixotropic recovery, meaning that once the shear stress is removed, the consistency is not completely recovered. Both CRX and SFX gels were found to have the greatest strength. Interestingly, SFX, which has the highest proportion of wax esters, did not form a stronger gel than CRX. Despite the established excellent gelling properties of wax esters, the combination of wax esters and fatty alcohols in CRX formed the strongest gel. Gels produced by both BEX and FRX were found to be weak in comparison with other wax oleogels. Overall, the authors concluded that morphology strongly influences the rheological behaviour of a wax oleogel. Given the dependence of morphology on chemical composition, this in turn can also be related to the rheological behaviour.

Gels prepared with the critical concentration in rice bran oil were used by Doan *et al.* to determine correlations between the rheology of the wax oleogels and the chemical composition of the wax.[3] The authors found correlations that suggested gel strength increases with decreasing amounts of free fatty alcohols, and increasing amounts of hydrocarbons, free fatty acids and wax esters. A higher content of wax esters was correlated to producing a strong but brittle gel, while those with a greater amount of free fatty acids and hydrocarbons exhibited smoothness and plasticity.

4.8   The Effect of Cooling Rate on the Properties of Wax Crystal Networks

The use of different cooling rates in wax oleogels impacts the size of the crystals formed, and the development of transient or permanent junction zones. In general, faster cooling rates have been shown to increase the rate of nucleation and induce formation of smaller crystals. This causes an increase in the number of crystals formed, in addition to an increase in crystallographic mismatches and branching. The rapid crystallization occurring at faster cooling rates also causes a reduced level of molecular organization of the rotator phases. On the contrary, a slower cooling rate can be used to obtain larger and less branched crystals. The manipulation of cooling rate can therefore be used to modify the microstructure of the gels, which in turn can affect rheological behaviour and oil binding capacity.[10]

One study by Toro-Vazquez *et al.* investigated the use of different cooling rates on the many properties of wax oleogels.[8] They determined that a faster cooling rate causes a lower gelation temperature, and a lower heat of gelation (ΔH). Gels prepared with CLX in safflower oil were cooled at rates of 1 °C min^{-1} and 10 °C min^{-1}. The gels that were cooled at a slower rate were found to achieve the structural alignment required for gelation at a higher temperature, while the gels that were cooled faster had less time to organize into the necessary structures, therefore gelling at a temperature approximately 5 °C lower than the slowly cooled gel. The decreased molecular organization from rapid cooling means that gels formed at a faster cooling rate would have a corresponding lower ΔH of melting. The authors also concluded that the changes in microstructure (size and organization) resulting from a faster cooling rate caused CLX gels to have lower G' values. Similar results were reported by Hwang *et al.* when investigating SFX gels.[6] Again, a faster cooling rate caused a smaller crystal size. The authors also observed that a slower cooling rate allowed for the development of uniformly shaped crystals, while a faster cooling rate caused irregularity in the crystal shapes. Interestingly, the smaller, incongruent shaped crystals formed a denser network, causing an increase in gel firmness.

Blake and Marangoni discussed the effect of cooling rate in the preparation of RBX, SFX and CLX gels.[10] The authors used their results to relate the affected microstructure to a gel's oil binding capacity. It was determined that a gel cooled at a faster rate, and therefore a gel with decreased crystal size, experienced an increase in fractal dimension and an decrease in pore area fraction. This corresponded to gels with a greater oil binding capacity. Essentially, the larger surface area and smaller pore size of a network with smaller crystals has a greater ability to tightly bind oil.

4.9 The Effect of Shear on the Properties of Wax Crystal Networks

The use of shear in oleogels can alter the crystallization and microstructure of a gel. Fundamentally, applying shear causes an increase in heat and mass transfer. As a result, the frequency of molecular collisions increases causing crystals to break or to form many smaller crystal entities.[11,40,41] It has also been suggested that shear can change the rate of nucleation and the alignment of nuclei formed, inducing the formation of specific crystal orientations.[41,42] There is an upper limit for the application of shear, as extensive shear forces can cause network disruption, compromising the structure and oil binding capacity. Something else to consider is that increased heat transfer resultant from applying shear will also cause an inherently faster cooling rate; therefore, temperature must also be monitored when examining the effect of shear.[11] Similar to other variables, there is a critical shear rate for each type of wax oleogel, above which the effects of aggregation, orientation and the like are no longer observed.

The effect of shear is dependent on the extent and duration of the shear applied and on the type of wax employed as the gelator in each case. Additionally, different results arise when shear is applied to a formed crystal network or at different points during gel formation (nucleation or pre-crystallization). In studies by Alvarez-Mitre *et al.* and Chopin-Doroteo *et al.*, shearing until the nucleation point potentially impeded network formation, while shearing only until metastable conditions, defined by the authors as a specific temperature at least 5 °C above the nucleation temperature, prevented crystal breakage and allowed for the development of permanent junction zones and thus a stronger gel.[7,40]

In a study by Blake and Marangoni, the effect of shear on RBX, SFX and CLX gels was examined.[11] In general, shear stress was found to cause the formation of smaller wax crystals, except in the case of gels that contained very low concentrations (0.5% w/w). They concluded that this result was caused by the increase in nucleation rate in favour of crystal growth during shear or to some extent by the breaking of crystals caused by shear. The latter is less likely to contribute significantly as the authors describe the use of shear only for a short period during gel setting. Shear under slow cooling conditions induced crystal aggregation in RBX and SFX gels. In the case of RBX gels, the aggregation was correlated to a decrease in oil loss, suggesting that the application of shear could produce RBX oleogels with a greater range of applications. This was not the case for SFX gels, where crystal aggregation actually resulted in increased oil loss. This was attributed to the dependence of SFX oil binding capacity on surface adsorption, which decreases when crystal surface area decreases. In contrast, CLX gels experienced an increase in platelet size rather than aggregation, an occurrence also observed in other studies.[7,10,40,43] In this case, shear caused molecular alignment and the formation of flow-induced mesophase structures, enhancing the nucleation and crystallization process. The larger crystals also caused a decrease in oil binding capacity, again likely because of surface area reduction. Because CLX is a chemically heterogeneous wax, shear increased the potential crystallization combinations resulting in the formation of mixed molecular packing structures. As RBX and SFX are highly homogenous, this would explain the different effects observed from shear. Blake and Marangoni also determined a combined effect of shear and cooling rate as the same shear rate applied to gels that had been cooled at different rates did not demonstrate properties that could be explained by previous theories.[11] The authors described the combination of shear and rapid cooling as a complex and dynamic environment that is difficult to interpret.

4.10 Conclusions

Wax crystal networks are very efficient gelled oil systems due to their strong oil-binding capacity at lower wax concentrations. The difficulty in achieving widespread understanding of wax based oleogels stems from the complex chemical composition of waxes that varies between types of wax, with a degree of variation even observable between the same types of wax produced by different companies. With extensive research and the discovery of new analytical techniques, much has been uncovered about wax crystal networks. However, even more is yet to be revealed. In order to truly understand the fundamentals of these gelled systems and how to manipulate their many properties, the involvement of the macroscopic structure at all length scales within the overall system must be considered. Further investigation into their ability to produce stable crystal networks in unsaturated, liquid oils is required to allow for the eventual large-scale production for the food industry. Doing so will enable the use of oleogels as fat substitutes, producing food products with reduced *trans* and saturated fat contents. However, the shear sensitivity and low thixotropic recovery of these gels currently poses a barrier to their direct incorporation into food systems. Benefits associated with the particular use of natural waxes as gelators include their natural origins, the low critical concentration for gelation and the ability to manipulate the physical properties. RBX and SFX have the greatest potential for economical large-scale oleogel production owing to their low critical concentrations (1% w/w) as well as having the largest possible production annually, relative to the other available waxes. Specifically, RBX has been suggested to be the most suitable wax for the creation of highly functional oleogels capable of acting as a solid fat replacement owing to the desirable properties that can be achieved under specific conditions, including a high oil binding capacity, at low concentrations of wax.

References

1. S. R. Vali, Y. H. Ju, T. N. B. Kaimal and Y. T. Chern, *J. Am. Oil Chem. Soc.*, 2005, **82**, 57–64.
2. D. C. Zulim Botega, A. G. Marangoni, A. K. Smith and H. D. Goff, *J. Food Sci.*, 2013, **78**, 1334–1339.
3. C. D. Doan, C. M. To, M. De Vrieze, F. Lynen, S. Danthine, A. Brown, K. Dewettinck and A. R. Patel, *Food Chem.*, 2017, **214**, 717–725.

4. L. S. K. Dassanayake, D. R. Kodali, S. Ueno and K. Sato, *J. Am. Oil Chem. Soc.*, 2009, **86**, 1163–1173.
5. A. I. Blake, E. D. Co and A. G. Marangoni, *J. Am. Oil Chem. Soc.*, 2014, **91**, 885–903.
6. H. S. Hwang, S. Kim, M. Singh, J. K. Winkler-Moser and S. X. Liu, *J. Am. Oil Chem. Soc.*, 2012, **89**, 639–647.
7. M. Chopin-Doroteo, J. a. Morales-Rueda, E. Dibildox-Alvarado, M. a. Charó-Alonso, A. de la Peña-Gil and J. F. Toro-Vazquez, *Food Biophys.*, 2011, **6**, 359–376.
8. J. F. Toro-Vazquez, J. A. Morales-Rueda, E. Dibildox-Alvarado, M. Charó-Alonso, M. Alonzo-Macias and M. M. González-Chávez, *J. Am. Oil Chem. Soc.*, 2007, **84**, 989–1000.
9. J. A. Morales-Rueda, E. Dibildox-Alvarado, M. a. Charó-Alonso and J. F. Toro-Vazquez, *J. Am. Oil Chem. Soc.*, 2009, 765–772.
10. A. I. Blake and A. G. Marangoni, *Food Biophys.*, 2015, **10**, 456–465.
11. A. I. Blake and A. G. Marangoni, *Food Biophys.*, 2015, **10**, 403–415.
12. N. Hughes, A. G. Marangoni, A. J. Wright, M. A. Rogers and J. W. E. Rush, *Trends Food Sci. Technol.*, 2009, **20**, 470–480.
13. D. Zulim Botega, *Application of Rice Bran Wax Organogel to Substitute Solid Fat and Enhance Unsaturated Fat Content in Ice Cream*, MSc Thesis, Univeristy of Guelph, 2012.
14. H. S. Hwang, M. Singh and S. Lee, *J. Food Sci.*, 2016, **81**, 1045–1054.
15. F. C. Wang, A. J. Gravelle, A. I. Blake and A. G. Marangoni, *Curr. Opin. Food Sci.*, 2016, **7**, 27–34.
16. FDA, 2016, https://www.accessdata.fda.gov/scripts/cdrh/cfdocs/cfcfr/CFR-Search.cfm?fr=172.890, accessed February 2017.
17. FDA, 2016, https://www.accessdata.fda.gov/scripts/cdrh/cfdocs/cfcfr/CFR-Search.cfm?fr=184.1978, accessed February 2017.
18. FDA, 2016, https://www.accessdata.fda.gov/scripts/cdrh/cfdocs/cfcfr/CFR-Search.cfm?fr=184.1976, accessed February 2017.
19. A. R. Patel, M. Babaahmadi, A. Lesaffer and K. Dewettinck, *J. Agric. Food Chem.*, 2015, **63**, 4862–4869.
20. E. D. Co and A. G. Marangoni, *J. Am. Oil Chem. Soc.*, 2012, **89**, 749–780.
21. P. J. Holloway and S. B. Challen, *J. Chromatogr.*, 1966, **25**, 336–346.
22. T. C. S. Kanya, L. J. Rao and M. C. S. Sastry, *Food Chem.*, 2007, **101**, 1552–1557.
23. R. Wang, X.-Y. Liu, J. Xiong and J. Li, *J. Phys. Chem. B*, 2006, **110**, 7275–7280.
24. N. C. Acevedo and A. G. Marangoni, *Cryst. Growth Des.*, 2010, **10**, 3327–3333.
25. J. A. Morales-Rueda, E. Dibildox-Alvarado, M. A. Charó-Alonso, R. G. Weiss and J. F. Toro-Vazquez, *Eur. J. Lipid Sci. Technol.*, 2009, **111**, 207–215.
26. A. I. Blake and A. G. Marangoni, *Food Struct.*, 2015, **3**, 30–34.
27. Y. Miyazaki and A. G. Marangoni, *Mater. Res. Express*, 2014, **1**, 1–12.
28. R. M. Saunders, *Food Rev. Int.*, 1985, **1**, 465–495.
29. Koester Keunen, http://www.koster-wax.com/us/wax-products/organic-and-natural/rice-bran-wax, accessed February 2017.
30. S. Martini and M. C. Añón, *J. Am. Oil Chem. Soc.*, 2000, **77**, 1087–1092.

31. Koester Keunen, http://www.koster-wax.com/us/waxes-by-industry/candle-waxes/sunflower-wax–258g, accessed February 2017.
32. J. C. B. Rocha, J. D. Lopes, M. C. C. N. Mascarenhas, D. B. Arellano, L. M. R. Guerreiro and R. L. da Cunha, *Food Res. Int.*, 2013, **50**, 318–323.
33. C. D. Doan, D. Van de Walle, K. Dewettinck and A. R. Patel, *J. Am. Oil Chem. Soc.*, 2015, **92**, 801–811.
34. E. Yilmaz and M. Öğütcü, *J. Food Sci.*, 2014, **79**, E1732–E1738.
35. C. MacDougall, M. Razul, E. Papp-Szabo, F. Peyronel, C. Hanna, A. Marangoni and D. Pink, *Faraday Discuss.*, 2012, **158**, 425–433.
36. M. S. G. Razul, C. J. MacDougall, C. B. Hanna, A. G. Marangoni, F. Peyronel, E. Papp-Szabo and D. a. Pink, *Food Funct.*, 2014, **5**, 2501–2508.
37. L. S. K. Dassanayake, D. R. Kodali, S. Ueno and K. Sato, *J. Oleo Sci.*, 2012, **61**, 1–9.
38. S. S. Narine and A. G. Marangoni, *Food Res.*, 1999, **32**, 227–248.
39. D. Tang and A. G. Marangoni, *J. Am. Oil Chem. Soc.*, 2006, **83**, 377–388.
40. F. M. Alvarez-Mitre, J. A. Morales-Rueda, E. Dibildox-Alvarado, M. A. Charó-Alonso and J. F. Toro-Vazquez, *Food Res. Int.*, 2012, **49**, 580–587.
41. N. C. Acevedo, J. M. Block and A. G. Marangoni, *Faraday Discuss.*, 2012, **158**, 171.
42. G. Mazzanti, S. E. Guthrie, E. B. Sirota, A. G. Marangoni and S. H. J. Idziak, *Cryst. Growth Des.*, 2003, **3**, 721–725.
43. F. M. Alvarez-Mitre, J. F. Toro-Vázquez and M. Moscosa-Santillán, *J. Food Eng.*, 2013, **119**, 611–618.

CHAPTER 5

Structuring Edible Oil Phases with Fatty Acids and Alcohols

ARJEN BOT*[a] AND ECKHARD FLÖTER[b]

[a]Unilever R&D Vlaardingen, Olivier van Noortlaan 120, NL-3133 AT Vlaardingen, The Netherlands; [b]Food Process Engineering, Department of Food Technology and Food Chemistry, Technical University Berlin, Seestraße 13, D-13353 Berlin-Wedding, Germany
*E-mail: arjen.bot@unilever.com, eckhard.floeter@tu-berlin.de

5.1 Introduction

Structuring of edible oil phases to achieve certain textural characteristics is not new. Long ago, nature invented triglycerides (TAGs) that are especially rich in saturated and relatively long-chain fatty acids for this purpose, which structure as a result of their partial solubility in other shorter and less saturated triglycerides (TAGs). Such triglycerides form crystals which have a number of interesting properties.

First of all, these triglycerides form crystallites that can aggregate to form a network in the oil phase. The size of these particles varies with crystallisation conditions. Typically, slow cooling to temperatures relatively close to the dissolution temperature will result in big crystals. On the other hand, rapid cooling to very low temperatures compared to the dissolution temperature tends to result in smaller crystallites. Generally, smaller crystals tend to form firmer networks, and thus processing conditions affect the properties of the

Food Chemistry, Function and Analysis No. 3
Edible Oil Structuring: Concepts, Methods and Applications
Edited by Ashok R. Patel
© The Royal Society of Chemistry 2018
Published by the Royal Society of Chemistry, www.rsc.org

resulting network. The crystallisation conditions and triglyceride composition may also affect the shape, or morphology, of the crystals.

A complicating factor is that the crystals that are initially formed are not necessarily the crystals that ultimately form the network. For triglycerides, it is common that recrystallization occurs of the initially formed crystals to a differently ordered but energetically more favourable molecular arrangement. This is because it is often easier for molecules to assemble in a conformation that is closer to the conformation in the liquid melt than to immediately arrange in the energetically lowest conformation. The occurrence of multiple arrangements is called polymorphism. In its simplest incarnation, the crystal transforms through three common conformations: α, β′ and β crystals, the latter being the most ordered and stable. Depending on the triglyceride composition and purity, some stages may be skipped, not reached or the number of polymorphs may be much bigger than the three mentioned above.

The most important property of crystallising triglycerides in the context of network formation is the fact that the molecules in the crystals are in dynamic equilibrium with the triglycerides in the melt. Because some of the molecules migrate from solid to liquid phase and *vice versa*, new connections between crystals in the network can be formed, which has a major impact on the texture of the gel.[1] Partly this happens during the initial formation of the network, but the same process also results in curing of the network after intense shearing. Such behaviour would be desirable for oleogel networks too.[2]

Therefore, it seems logical to first investigate very similar compounds to TAGs in order to identify potential oleogelling structurants: fatty acids, and their somewhat more distant cousins alcohols (for naming conventions, see Tables 5.1 and 5.2). Fatty acids can be obtained from triglycerides by hydrolysis. Fatty alcohols can be derived from triglycerides by transesterification

Table 5.1   Naming of saturated fatty acids and alcohols with even chain length.

Fatty acid chain length	Systematic chain name	Trivial fatty acid name	Trivial fatty alcohol name
10	Decane	Capric	Capric alcohol
12	Dodecane	Lauric	Lauryl alcohol
14	Tetradecane	Myristic	Myristyl alcohol
16	Hexadecane	Palmitic	Cetyl alcohol
18	Octadecane	Stearic	Stearyl alcohol
20	Cosane	Arachidic	Arachidyl alcohol
22	Docosane	Behenic	Behenyl alcohol
24	Tetracosane	Lignoceric	Lignoceryl alcohol
26	Hexacosane	Cerotic	Ceryl alcohol
28	Octacosane	Montanic	Montanyl alcohol
30	Triacontane	Melissic	Melissyl alcohol
32	Dotriacontane	Lacceroic	Lacceryl alcohol
34	Tetratriacontane	Geddic	Geddyl alcohol

Table 5.2 Naming of a number of saturated fatty di-acids with even chain length.

Fatty acid chain length	Systematic chain name	Trivial fatty di-acid name
6	Hexanedioic	Adipic
8	Octanedioic	Suberic
10	Decanedioic	Sebacic

with methanol followed by high-pressure hydrogenation, although earlier processes involved splitting of natural waxes.[3] Note that typically restrictions will apply to the use of these ingredients in food applications.

5.2 Fatty Acids (FA)

For the use of fatty acids in oleogelling, two classes of materials should be distinguished: saturated non-branched long-chain fatty acids on the one hand, and other long-chain fatty acids on the other hand (see Tables 5.1 and 5.2). The latter group includes unsaturated fatty acids, fatty acids with hydroxyl groups or multiple acid groups.

The first extensive report on the use of saturated fatty acids in oil structuring was made by Eini and Tamarkin, who identified a number of alternative applications for *e.g.* arachidic, behenic and montanic acid in oil spreads and flavour delivery systems.[4] More systematic studies were done by Daniel and Rajasekharan[5] and Gandolfo *et al.*[6] Gandolfo *et al.* investigated the structuring potential of palmitic, stearic, arachidic and behenic acid in sunflower oil. Daniel and Rajasekharan demonstrated that the minimum concentration of fatty acid necessary to establish oil gelling at room temperature seems to level off at a fatty acid chain length of ~30, probably because the system is well below the gelling temperature and most of the high melting material can participate in structuring. The exact polymorph of the fatty acid crystals is still a point of discussion as different groups have found slightly different X-ray diffraction data for the stearic acid oleogels.[7–9] This points to a dependence of the final molecular arrangement on the process that was followed to create the oleogel, while less impact on the results is expected from the oil types used in these studies: sunflower,[6,7] canola,[9] or sesame or soybean oil.[10] Schaink *et al.* suggest that the arrangement of stearic acid molecules in sunflower oil oleogels follows an orthorhombic structure similar to β′-type triglyceride crystals.[7] Sagiri *et al.* and Blach *et al.* found support in their data for a double layer structure with a d-spacing of 4.02 nm,[9,10] consistent with the C-form found in pure stearic acid.[11]

Next to the saturated fatty acids, the structuring potential of a number of other fatty acids has also been investigated. The addition of a second acid group, as in adipic, suberic and sebacic acid, improves the efficiency of the structurant compared to the saturated fatty acids with a single acid group mentioned above, although it also leads to an increase of the melting point of the oleogel to 80–145 °C.[5] Another feature that enhances the structuring efficiency is the presence of a hydroxyl group in the fatty acid. Actually, one of the earliest examples of non-triglyceride structuring of oil phases that can be found in the literature employs 12-hydroxystearic acid to structure peanut butter.[12] Extensive studies have been performed on oleogelling by means of 12-hydroxystearic acid and 12-hydroxy-9-*trans*-octadecenoic acid (ricinelaidic acid).[13,14] Both systems create fibrillar networks that form quite firm gels when sufficient structurant is used.

5.3 Fatty Alcohols

For fatty alcohols, literature is available for the oleogelling properties of saturated fatty alcohols. Again, Eini and Tamarkin identified potential applications of cetyl, stearyl and behenyl alcohol in products like oil spreads, shortening, cocoa butter substitutes, fatty coatings, flavour delivery systems, or milk and yoghurt alternatives based on structured vegetable oil.[4] In some cases the oleogelling system was supplemented with high melting triglycerides, such as palm oil flakes. More systematic studies were performed by Gandolfo *et al.* and Valoppi *et al.*, who showed for palmityl, stearyl, arachidyl and behenyl alcohol that the firmness of a fatty alcohol oleogel increases approximately proportionally with structurant concentration above ~2%.[6,15] Daniel and Rajasekharan investigated fatty alcohols series from behenyl alcohol up to melissyl alcohol, a range in which the disintegration temperature of the respective oleogels increases from about 40–45 °C to 55–60 °C.[5] For these systems structurant concentrations of 3–4% are typically required, although structuring can be more efficient in more saturated triglyceride oils. Valoppi *et al.* showed that the crystal platelets decrease in size with increasing chain length.[15] Furthermore, for the fatty alcohols there is discussion on the exact polymorph that is obtained, again as a result of small differences in X-ray diffraction data.[7,9,16] Schaink *et al.* favour two polymorphs with β'-like orthorhombic packing, one double layer and the other triple layer,[7] Valoppi *et al.* identified two double-layer β'-like orthorhombic packing variations in which a slightly thicker metastable bilayer structure converts to a more compact bilayer structure during storage,[16] and Blach *et al.* identified a sub-α and a β form in their data[9] (*cf.* Kolb *et al.*[17]).

In contrast to the fatty acids, no information seems to be available on how hydroxyl moieties affect the oleogelling capability of fatty alcohols.

5.4 Fatty Acids + Fatty Alcohols

Mixtures of structurants are usually more interesting in food applications than pure components because it is easier to tailor the properties of the structurant to the application. For example, pure triglycerides have a sharp melting point (and a relatively sharp dissolution temperature when dissolved in oil), but mixtures of triglycerides melt over a temperature range and the triglyceride composition can be varied deliberately to obtain a blend that satisfies a pre-defined melting profile.[18] The same principle applies to mixtures of fatty acids and fatty alcohols, though we will limit ourselves here to mixtures of a pure fatty acid with a pure fatty alcohol.

Mixtures of long-chain fatty acids and fatty alcohols often structure edible oils more effectively than each of the individual components.[6,7] The gels form at structurant concentrations in excess of ~2%[6] and break at strains of ~10%.[7] The gels are particularly strong for a structurant mixture of 30% stearic acid + 70% stearyl alcohol if compared to other mixtures, but generally oleogels are formed preferentially in mixtures with a fatty acid and the fatty alcohol of the same chain length (16 to 22) and in mixtures of stearic acid with fatty alcohols of different chain length (16 to 22).[6,7]

Casual observation of differential scanning calorimetric (DSC) curves for mixtures of stearic acid and stearyl alcohol, either with or without oil, might suggest that the mixture forms an ideal liquid solution combined with a complete immiscibility in the solid phase.[9,19] A more detailed assessment, in particular of the melting/crystallisation enthalpy, reveals deviations that point towards a solid–solid transition at compositions not too close to the neat components, which can be attributed to the formation of mixed crystals (Figure 5.1).[19] X-ray scattering also indicates that mixtures of fatty acid and fatty alcohol do indeed form such crystals.[7,9] Schaink *et al.* attributed the enhanced firmness of the oleogels based on 30% stearic acid + 70% stearyl alcohol mixtures to these differently packed crystals.[7] Building on the data available today, a diagram involving these more complex intermediate mixed crystalline phases as shown in Figure 5.2 can be proposed. The basis for this new interpretation is the consistent combination of the DSC data[19] and

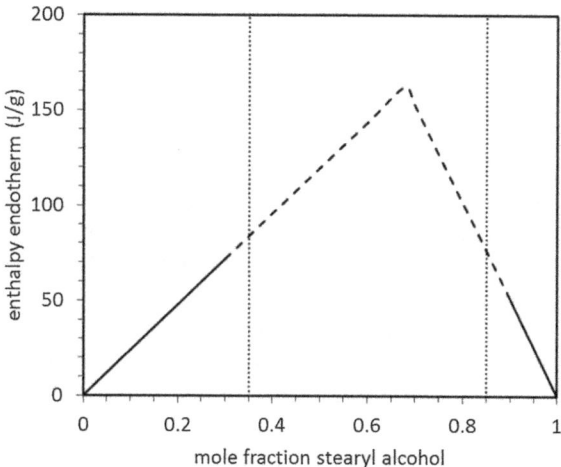

Figure 5.1 Schematic enthalpy endotherm for the binary stearic acid + stearyl alcohol system, as obtained from DSC heating curves. The experimental enthalpy data follow the solid lines, but deviate from linear behaviour in the concentration range indicated by the dashed line.[7] This coincides with the observation of X-ray diffraction patterns consistent with mixed crystals in the composition range between the dotted vertical lines.[9]

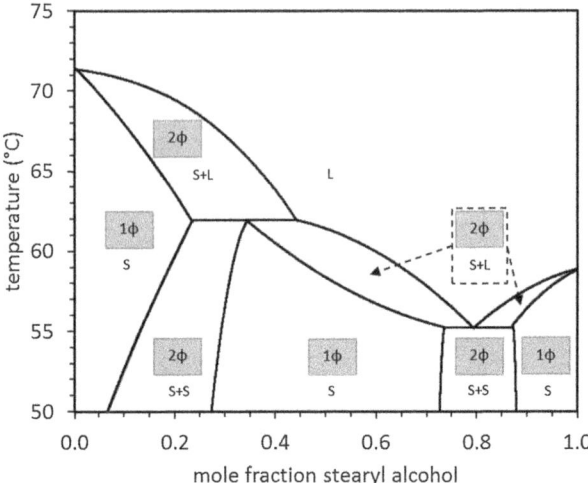

Figure 5.2   Proposed schematic phase diagram for the binary stearic acid + stearyl alcohol system, as derived from DSC heating curves and X-ray scattering. Single phase regions are indicated by 1φ, two phase regions by 2φ, S = solid, L = liquid.

enthalpy data[19] with the X-ray data of the dilute systems.[9] The re-evaluation is built on the understanding that DSC peaks expand over two complete phase regions, that shoulders have to be considered, and that the X-ray data allow the identification of two-phase regions because of superimposed structural signals. Figure 5.2 suggests that the system has an intermediate mixed solid solution that expands over a composition from somewhere between 20 and 30% to somewhere between 70 and 80% of stearyl alcohol. By introduction of this mixed solid phase, two solid–solid two-phase regions automatically emerge. These ranges of demixed solid phases are well represented in the SAXS pattern disclosed by Blach *et al.*[9] for diluted systems. It should be noted that the phase behaviour shown Figure 5.2 does not take into account possible polymorphic transitions, which, for example, seem to occur at concentrations close to pure stearyl alcohol, as indicated by the DSC data of Gandolfo *et al.*[19] and X-ray data of Blach *et al.*[9]

Blach *et al.* observed that these mixed crystal do not always lead to enhanced firmness, and therefore no direct relation should be assumed between mixed crystal formation and enhanced firmness. Low-resolution nuclear magnetic resonance indicates that the presence of mixed crystals coincides with higher amounts of solid phase in the oleogel.[9] Light microscopy suggests that the mixed crystals form small needle-like crystals whereas the neat components crystallise as platelets,[6,7,9] although the latter does depend on the crystallisation conditions applied to the samples used for microscopy.[9] The previously conjectured relation between small crystallite size and oleogel firmness[19] does not hold as softer gels seem to contain more fine crystals.[9] However, recent scanning electron microscopy results show that the needle-like crystals are likely artefacts caused by sample preparation, and that the crystals always have a platelet morphology.[9] The platelets become gradually thinner with increasing stearic acid content in the mixture.

Blach *et al.* suggest that the enhanced firmness of the fatty alcohol-rich oleogels is best explained by a combination of two effects.[9] Obviously, the amount of solid phase is important, but not sufficient. Earlier explanations that identified an increased nucleation rate due to a reduced interfacial energy leading to smaller crystallites[6] as the second important element fall short in view of the newer data. Instead, Blach *et al.* implicate the network structure, which reveals itself by the concentration dependence of the network firmness on the concentration of the structurant: The firmness of fatty alcohol-rich oleogels varies slightly steeper than linear with structurant concentration, whereas the firmness for fatty acid-rich oleogels varies slightly less steep than with the square of the structurant concentration. This difference is explained as a transition from a closed to an open cellular structure, which is corroborated with the amount of oil exudation observed for these oleogels, which is higher for many of the fatty acid-rich oleogels.[9]

5.5 Potential Applications

Next to oil structuring being of scientific interest, the current attention is driven by its promise of having applications in food and other products. This is particularly true of structuring using mixtures of fatty acids and alcohols, which has been considered in a multitude of applications in the patent literature. As patents go, some of these applications seem more feasible than others, but the sheer number of ideas makes it hard to think that the system would not serve as a solution in some instances.

To select such opportunities, Bot and Flöter listed a number of important criteria that a potential oil structurant(s) should satisfy to be considered for product structuring:[20]

- has a food-grade status
- acts as an oil structurant
- allows manipulation of melting temperature/level of structuring
- has no adverse interaction with any of the other ingredients in a typical food (*e.g.* water or salt)
- can be obtained at a reasonable cost
- has no impact on taste and mouthfeel

The last one was not explicitly mentioned in the original list, but is obviously important. Clearly, these criteria can be somewhat relaxed when still in the stage of developing the technology and while learning more about the general principles of oil structuring, but they should always be in the back of the mind of a food structure researcher. It is therefore of

interest to assess the applications that have been proposed in the literature, mainly in the work by Eini and Tamarkin,[4,21] against these criteria.
Their work describes a wide range of potential applications of oil structuring based on fatty acids, fatty alcohols or their mixtures. They do not
indicate a preference for the mixed fatty acid + fatty alcohol structurant, in
spite of the synergistic effects described above. The fatty acids mentioned
explicitly (arachidic, behenic, montanic) seem to be pure and not derived
from any particular source. Different sources of fatty alcohols are considered: chemically pure ones (cetyl, stearyl, behenyl, and in one particular
example, a C_{50}-alcohol), but also some derived from beeswax. Generally,
attention focuses more on fatty alcohols as structurants than on fatty acids,
though both are in scope. It is also of interest that Eini and Tamarkin blend
the fatty acid + fatty alcohol system in some cases with regular melting triglycerides, like palm flakes.[4,21]

The first sub-division is between the completely oil-based systems *versus*
the emulsions. Oil structuring is generally easier in systems that do not
contain both an oil phase and an aqueous phase, as much oil structuring
is based on limited solubility of the structurants. One of the more straightforward ways to achieve this is by having more hydrophilic elements in the
molecular structure of the oleogelling agent, which will act as association
sites in the structured oil phase. It is no surprise, therefore, that the list of
potential applications includes lipid flavour delivery systems, shortening,
oil powder, structured oil, oil-based spreads, fat-based coatings, chocolate
(and Valoppi *et al.* add peanut butter to this list[15]). Since these systems are
essentially water-free, any oil structuring technology is likely to work. The
situation becomes more interesting when other ingredients with oil structuring potential are added: lecithin or crystallising triglycerides. Johansson
and Bergenståhl[22] and Delacharlerie *et al.*[23] demonstrated that lecithins can
influence the interactions between fat crystals, and are likely to influence
the interactions in the fatty acid + fatty alcohol system as well. The effect
crystallizing fats will have on network structure formation in the fatty acid +
fatty alcohol oleogel is unknown at this moment—do the triglycerides form
a separate network or will either the oil structurant or the crystallizing triglycerides form a template for the other component? Such mixed systems are
worthy of further investigation.

The emulsions differ qualitatively from oil-based applications because of
the presence of water. A well-known example of an oleogelling system that
is sensitive to water migration during storage is the sterol + sterol ester system, which can only be kept stable by reducing water activity or water solubility in the organic phase sufficiently.[24] One way to reduce water activity is by
freeze concentration of the aqueous phase, as happens in ice cream during
frozen storage—and therefore ice cream is an example that could circumvent
such problems. In any case, the fatty acid + fatty alcohol system is expected
to be less sensitive to water migration, and therefore several applications
such as margarine and low-fat spreads and dairy alternatives (filled milk
and yoghurt) are proposed. For these products only the proximity of water

matters, so there is no fundamental difference between water-in-oil and oil-in-water emulsions—except that the dispersed phase in an oil-in-water emulsion is less likely to control the texture of the emulsion-based product, making application of oil structurants in water-continuous emulsions more forgiving.

Next to these food applications, Eini and Tamarkin also identified a number of applications in pharmaceutical or cosmetic creams,[21] which we consider out of scope for the present chapter.

As a last step, we should hold the fatty acid + fatty alcohols system against the criteria we mentioned at the start of this application section. The assessment is as follows:

- *Food-grade status*: Neither fatty acids nor fatty alcohols are regular food additives. Fatty acids are allowed as additives in foods, but essentially as lubricants, binders, anti-foaming agents, or as a component in the manufacture of other food-grade additives.[25] On the other hand, free fatty acids do occur at low levels in many fat-based ingredients used in the manufacturing of foods (*e.g.* ≲0.8% for extra virgin olive oil and ≲1.5% for virgin olive oil), and are part of the human metabolism. Stearyl alcohol has been registered only as a self-affirmed GRAS ('generally recognized as safe') ingredient.[26] Thus, both ingredients will probably require further regulatory effort when used in a marketed product.
- *Oil structurant*: The oil structuring capabilities of fatty acids and fatty alcohols and their mixtures have been discussed extensively in this chapter
- *Manipulation of melting temperature/level of structuring*: One of the attractive elements of the fatty acid + fatty alcohol system is that its crystallisation behaviour is rather similar to that of triglycerides. This suggests that more or less the same rules will apply with respect to the effect of concentration, chain length, degree of unsaturation—and possibly the effect of blending several chain lengths. Although not yet demonstrated, it seems likely that it is possible to tailor the melting profile to the requirements of the application.
- *No adverse interactions with other ingredients in a typical food*: Given the similarity in behaviour between fatty acid + fatty alcohol system compared to triglycerides, it is expected that the system will not suffer from storage instabilities in emulsions.
- *Cost*: Both the fatty acid and the fatty alcohol can be produced from triglycerides using more or less industry-standard processing technology. The ingredients will have an on-cost, which is comparable to some of the other ingredients considered in the field of oil structuring.
- *Impact on taste*: Fatty acids are generally considered to be tasteless, although there is some discussion about whether they can be perceived under optimal conditions.[27,28] No studies could be found on the taste of long chain fatty alcohols, but no specific taste contribution is expected either.

5.6 Conclusions

The main hurdle for the application of the fatty acids + fatty alcohols as a structuring system in foods is the regulatory approval, and it will not be a mean feat to achieve this. Most other aspects of the technology look rather favourable. On the other side, the reason why these aspects are so favourable is precisely because these ingredients are so similar to the triglycerides that one might consider replacing. So in the end, the user should also consider whether the nutritional benefit of replacing certain fats with these particular oil structurants is sufficiently big to warrant application of this technology.

References

1. H. M. Schaink and K. F. van Malssen, *Langmuir*, 2007, **23**, 12682.
2. M. Pernetti, K. F. van Malssen, E. Flöter and A. Bot, *Curr. Opin. Colloid Interface Sci.*, 2007, **12**, 221.
3. U. R. Kreutzer, *J. Am. Oil Chem. Soc.*, 1984, **61**, 343.
4. M. Eini and D. Tamarkin, Patent WO 01/50873, 2001.
5. J. Daniel and R. Rajasekharan, *J. Am. Oil Chem. Soc.*, 2003, **80**, 417.
6. F. G. Gandolfo, A. Bot and E. Flöter, *J. Am. Oil Chem. Soc.*, 2004, **81**, 1.
7. H. M. Schaink, K. F. van Malssen, S. Morgado Alves, D. Kalnin and E. van der Linden, *Food Res. Int.*, 2007, **40**, 1185.
8. A. C. T. Teixeira, A. R. Garcia, L. M. Ilharco, A. M. P. S. Goncalves da Silva and A. C. Fernandes, *Chem. Phys. Lipids*, 2010, **163**, 655.
9. C. Blach, A. J. Gravelle, F. Peyronel, J. Weiss, S. Barbut and A. G. Marangoni, *RSC Adv.*, 2016, **6**, 81151.
10. S. S. Sagiri, V. K. Singh, K. Pal, I. Banerjee and P. Basak, *Mater. Sci. Eng., C*, 2015, **48**, 688.
11. N. Garti and K. Sato, *Crystallisation and Polymorphism of Fats and Fatty Acids*, Marcel Dekker, New York, 1988.
12. C. A. Elliger, D. G. Guadagni and C. E. Dunlap, *J. Am. Oil Chem. Soc.*, 1972, **49**, 536.
13. M. A. Rogers and A. G. Maragoni, *Edible Oleogels: Structure and Health Implications*, ed. A. G. Marangoni and N. Garti, AOCS Press, Urbana, Ill, USA, 2011, ch. 5, pp. 101–118.
14. A. J. Wright and A. G. Marangoni, *Edible Oleogels: Structure and Health Implications*, ed. A. G. Marangoni and N. Garti, AOCS Press, Urbana, Ill, USA, 2011, ch. 4, pp. 81–99.
15. F. Valoppi, S. Calligaris and A. G. Marangoni, *Eur. J. Lipid Sci. Technol.*, 2017, **119**, 1600252.
16. F. Valoppi, S. Calligaris and A. G. Marangoni, *Cryst. Growth Des.*, 2016, **16**, 4209.
17. D. G. Kolb and E. S. Lutton, *J. Am. Chem. Soc.*, 1951, **73**, 5593.
18. A. Bot and E. Flöter, *Edible Oil Processing*, ed. W. Hamm, R. J. Hamilton and G. Calliauw, Wiley, Chichester, 2nd edn, 2013, ch. 8, pp. 223–249.
19. F. G. Gandolfo, A. Bot and E. Flöter, *Thermochim. Acta*, 2003, **404**, 9.

20. A. Bot and E. Flöter, *Edible Oleogels: Structure and Health Implications*, ed. A. G. Marangoni and N. Garti, AOCS press, Urbana, Ill, USA, 2011, ch. 3, pp. 49–79.
21. M. Eini and D. Tamarkin, Patent WO 01/51014, 2001.
22. D. Johansson and B. Bergenståhl, *J. Am. Oil Chem. Soc.*, 1992, **69**, 718.
23. S. Delacharlerie, R. Petrut, S. Deckers, E. Flöter, C. Blecker and S. Danthine, *LWT–Food Sci. Technol.*, 2016, **72**, 552.
24. H. Sawalha, R. den Adel, P. Venema, A. Bot, E. Flöter and E. van der Linden, *J. Agric. Food Chem.*, 2012, **60**, 3462.
25. Anonymous, CFR–Code of Federal Regulations Title 21, Volume 3, Chapter I, Subchapter B, Part 172, http://www.accessdata.fda.gov/scripts/cdrh/cfdocs/cfcfr/cfrsearch.cfm?fr=172.860.
26. Anonymous, Self-affirmed GRAS Notice No. GRN 000070 for Stearyl Alcohol, http://www.fda.gov/Food/IngredientsPackagingLabeling/GRAS/NoticeInventory/ucm153987.htm.
27. R. D. Mattes, *Annu. Rev. Nutr.*, 2009, **29**, 305.
28. P. Besnard, P. Passilly-Degrace and N. A. Khan, *Physiol. Rev.*, 2016, **96**, 151.

CHAPTER 6

Gelation Properties of Gelator Molecules Derived from 12-Hydroxystearic Acid

J. F. TORO-VAZQUEZ*, M. A. CHARÓ-ALONSO AND
F. M. ALVAREZ-MITRE

Universidad Autónoma de San Luis Potosí, Facultad de Ciencias Químicas,
Av. Dr Manuel Nava 6, San Luis Potosi, 78210, México
*E-mail: toro@uaslp.mx

6.1   Introduction

The texture, mouthfeel, and flavor of foods, and particularly of lipid-based foods (*i.e.*, butter, margarines, vegetable creams), are a strong function of the nano- and meso-scale organization of the bio-molecules, their phase changes and rheological behavior. Since triacylglycerides are the main component of fats and vegetable oils, in lipid-based foods these functional properties are directly associated with their phase changes, namely crystallization, melting, and polymorphism. In particular, the physical properties associated with saturated triacylglycerides and/or *trans* fatty acids provide the solid fat content, melting and textural properties wanted by the consumers in food products like shortenings, butter, margarines, vegetable creams, and spreads, and also in more complex food systems like processed meat (*i.e.*, comminuted meat products), confectionary and bakery products. Unfortunately, there is now a well-established link between the consumption of *trans* fatty acids with

Food Chemistry, Function and Analysis No. 3
Edible Oil Structuring: Concepts, Methods and Applications
Edited by Ashok R. Patel
© The Royal Society of Chemistry 2018
Published by the Royal Society of Chemistry, www.rsc.org

higher risk of coronary heart disease[1] and with the development of type II diabetes.[2,3] These negative health implications can be reversed by altering the intake of unhealthy lipids and replacing them with healthier alternatives. Within this context, several health organizations worldwide have issued recommendations to limit and even eliminate *trans* fatty acids in processed foods. Thus, in 2003 the World Health Organization recommended that *trans* fats make up no more than 1% of a person's diet. The same year the Danish government introduced laws that limit the content of *trans* fatty acids in vegetable oils and fats intended for human consumption to less than 2%. In contrast, the Food and Drug Administration has issued recommendations stating that by July 18, 2018, partially hydrogenated oils will no longer be recognized as GRAS. On the other hand, although the relationship of saturated fat intake by humans to a higher risk for atherosclerotic cardiovascular disease remains controversial,[4] the response from consumers is to move away from products high in saturated fatty acids. This last situation might be exacerbated by recent reports that claimed that certain components present in refined palm oil might be genotoxic and carcinogenic. Palm oil, rich in saturated fatty acids (*i.e.*, palmitic fatty acids), is the edible oil most used by the food industry worldwide to replace *trans* fatty acids in food formulations. Its functional properties and oxidative stability, make palm oil technologically useful in the production of a wide range of foods, including non-dairy creams, shortenings, margarines, salad dressings, confectionery and baking products (*i.e.*, biscuits, cookies, crackers). The glycidyl fatty acid esters, generated in the edible vegetable oils during the deodorization step at high temperatures, are the center of the cancer concerns. The glycidyl esters are formed from diacylglycerols on heating vegetable oils to temperatures greater than 200 °C during the deodorization stages of refining. Therefore, glycidyl ester formation is a particular problem in palm oil, which usually has a high diacylglycerol content (4–12%) and is processed at high temperatures to remove palm oil's natural red color and neutralize its smell.[5,6]

Within this context, the challenge for the industry is to produce foods containing fewer 'bad fats' while retaining most of the functional properties associated with the saturated triacylglycerides or with *trans* fatty acids. Evidently, this situation means that food formulation, and subsequently our diet, must shift toward the use of highly unsaturated oils, a goal that is not easily accomplished, mainly because unsaturated oils have different functionality.[7] The use of unsaturated oils in food formulations would result in severe modifications to food processing, and also in a deleterious effect in shelf life and overall acceptability of foods by the consumer. Despite health benefits, consumers will not accept a particular food if it does not meet their sensory expectations.

Organogelation is a useful and novel alternative to structure vegetable oils without the use of *trans* and/or saturated fats, providing useful and novel functional properties not only to food systems[8–11] but also to cosmetics[12] and pharmaceutical products.[13] Through this technology, low molecular weight gelators (LMOGs; molecular weight < 3000 Da) self-assemble developing supramolecular structures that physically entrap an apolar solvent (*i.e.*, vegetable or mineral oil) within a three-dimensional, self-assembled fibrillar network (SAFIN) providing the liquid phase with structure and viscoelastic properties. These solid-like materials are referred to as organogels or molecular gels.[14]

The molecular self-assembly, the nature and topology of the supramolecular structure, whether a gel will be formed, and the physical properties of the organogel depend on the molecular structure of the LMOGs, and on a balance of complex gelator–gelator and gelator–solvent interactions.[15–19] Our understanding of the mechanisms involved in organogel formation and the factors involved in engineering SAFIN structures to achieve a particular rheological behavior (*e.g.*, elasticity) have improved greatly during the last three decades.[16,20,21] However, given the wide variety of LMOGs (*i.e.*, cholesteryl anthraquinone derivatives, sterols, *n*-alkanes, lecithin, sorbitan monostearate, monoglycerides, fatty acids, amines, amides) the relationship between the molecular structure of LMOGs, the gelling capability, and the particular physical properties of gels is achieved more by serendipity than by a scientific approach. Although some relevant results have been achieved through systematic studies,[17,19,22–27] it is evident that this is not an easy task.[14–16,20] Within this context, an empirical but practical approach involves starting by using an already known gelator as the basic molecular structure, and from there applying systematic chemical modifications to the parent molecule.[24] Subsequently, the gelling capability of the new gelator is established along with the organogel's microstructural organization and physical properties. 12-hydroxystearic acid (12-HSA) is a well-known gelator molecule that provides a structural platform from which a variety of derivatives can be made relatively easily. Thus, in this chapter we discuss the relationship between the molecular structure of the *R* enantiopure form of 12-HSA, ammonium chloride salts of amines (*i.e.*, "polar" gelators), and primary and secondary amides derived from the *R* enantiopure form of 12-HSA (Figure 6.1) with the thermomechanical properties of the organogels developed using vegetable oils as the liquid apolar phase. As previously noted, both the nature and balance of the gelator–gelator and gelator–solvent interactions are critical in establishing the process of organogelation. This is because organogelation requires a meticulous balance between contrasting parameters that determine solubility and those intermolecular forces that ease the molecular self-assembly.[28] Most systematic studies on organogelation have been done using organic solvents as the liquid phase (*i.e.*, *n*-hexane, silicon oil, mineral oil), for instance for 12-HSA, stearic acid, and some of their derivatives.[17,22,24,29–31] In this chapter we will discuss results obtained using just safflower oil high in triolein as the liquid apolar phase. Given the particular composition of vegetable oils (*i.e.*, 95% to 99% triacylglycerides) and physical properties (*i.e.*, viscosity), its use as solvent results in organogels with different microstructures and physical properties than those obtained with other organic solvents. The role of solvent chemistry on the capacity of LMOG to assemble is as important as the gelator structure.

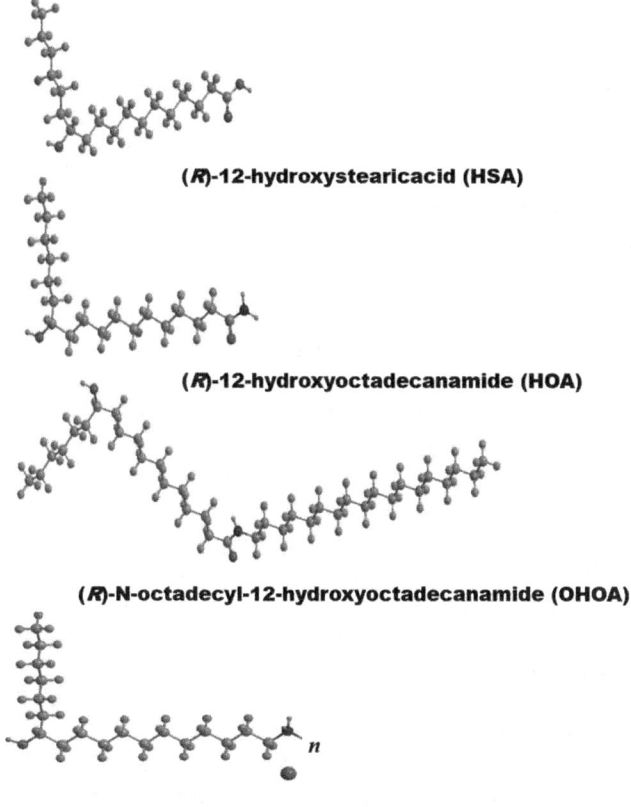

(*R*)-12-hydroxystearicacid (HSA)

(*R*)-12-hydroxyoctadecanamide (HOA)

(*R*)-N-octadecyl-12-hydroxyoctadecanamide (OHOA)

Ammonium chloride salts (*n*-HOA-Cl)

Figure 6.1   Structures of (*R*)-12-HSA, the primary and secondary amides, and ammonium chloride salts of amines synthesized from the *R* enantiopure form of HSA. For the ammonium chloride salts the alkyl chain (*n*) has 3, 4, 6 or 18 carbons. Adapted with permission from J. F. Toro-Vazquez, J. Morales-Rueda, A. Torres-Martínez, M. A. Charó-Alonso, V. A. Mallia, R. G. Weiss, *Langmuir*, 2013, **29**, 7642. Copyright 2013 American Chemical Society.

On the other hand, most of the results discussed involve organogels developed under static conditions. However, it is evident that external thermodynamic (*i.e.*, supercooling)[16,21,32] and mass transfer (*i.e.*, shearing)[33–36] conditions have important implications on the self-assembly of LMOGs, in the organization of SAFIN structures, and subsequently in the physical properties of the organogel. Therefore, in this chapter we will discuss some results obtained with organogels developed under the independent and combined effects of shearing and cooling rate using as gelators (*R*)-12-HSA (HSA) and primary and secondary amides derived from HSA. Tailoring the gelator's capacity to structure the liquid apolar phase and the resulting organogel's physical properties (*i.e.*, melting temperature, viscoelasticity, and thixotropic

behavior) requires the understanding of the effect of external thermodynamic and mass transfer conditions on the gelation of well-defined gelator
chemical structures.

6.2 Molecular Structure and Mechanism for Self-assembly of HSA

Mallia and Weiss pointed out the importance of the 12-HSA backbone for
molecular self-assembly.[24] Racemic 12-HSA and its *R* enantiopure form
(*i.e.*, HSA) are efficient gelators of a variety of low polarity liquids,[23,24,37] including vegetable oil.[19,30,38,39] In contrast, stearic acid, a gelator molecule without a
secondary 12-hydroxyl group, is not an efficient gelator.[40] This highlights the
importance of the hydroxyl group at the 12 position in the molecular self-
assembly of enantiopure and racemic forms of 12-HSA, specifically through
H-bonds. The presence of a carboxylic functional group and a secondary -OH
gives 12-HSA the capability to interact efficiently with other 12-HSA molecules through cyclic carboxylic acid dimer motifs and strong hydrogen bonds
through the -OH groups.[19,22] However, in homochiral interactions among
the hydroxyl groups on the chiral carbons, HSA molecules are connected by
unidirectional hydrogen bonding sequences resulting in pairs of H-bonds
connected in parallel directions within the unit cell[24,37] (Figure 6.2). This
organization leads to an extended secondary crystal growth with ribbon-like
organization resulting from the helical arrangement of the molecules within
the fiber.[37,41] In contrast, in racemic 12-HSA the orientation of the OH groups
results in pairs of H-bonds in sequence between two orthogonal directions
within a unit cell[24,37] (Figure 6.2). Stearic acid observes a similar molecular

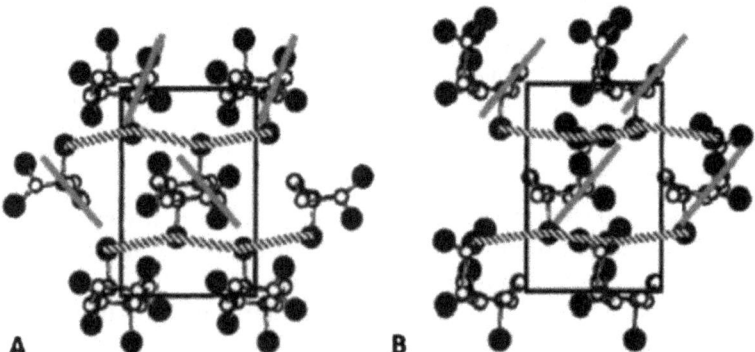

Figure 6.2   Simulated molecular ordering in HSA aggregates (monoclinic symme-
try with *Z* = 4). A: racemic HSA; B: *R* enantiopure form of HSA. The parallel solid lines represent the direction and sequence of the H-bonds
between hydroxyl groups.[37] Reproduced with permission from P. Terech, V. Rodriguez, J. D. Barnes, G. B. McKenna, *Langmuir.*, 1994, **10**,
3406. Copyright 1994 American Chemical Society.

organization but without the H-bonds. Thus, in racemic 12-HSA as in stearic acid, self-assembly occurs mainly through cyclic carboxylic acid dimer formation and lamellar organization stabilized mainly through London forces throughout the hydrocarbon chain. Subsequently, the longitudinal growth is prevented and stearic acid and racemic 12-HSA crystallize as microplatelets or acicular structures.[22,42]

As previously stated, the effect of solvent chemistry on the capacity of LMOG to assemble is as important as the gelator structure, and in this context HSA is not exempt. Thus, using HSA as the gelator, Rogers and Marangoni[42] showed that the nature of the solvent (*i.e.*, hexane, heptane, octane, decane, dodecane, tetradecane, hetanthiol, heptanthiol, octanthiol, decanthiol, nonanenitrile, decanal, dodecanal, octylamine, decylamine, and 5-nonanone) modifies the compatibility of the solvent with the HSA molecule. Subsequently, the solvent–gelator compatibility affects the degree of supercooling required for nucleation of HSA, the kinetics of its gelation, and its crystal morphology. These authors also showed that crystallization parameters are linearly correlated with the hydrogen-bonding solubility parameter of Hansen (δ_H) for the solvents.[42] The δ_H is a parameter associated with the capability of the solvent to develop hydrogen bonds with other solvent or with gelator molecules. Thus, according to these results, as the δ_H of the solvent approaches the δ_H of HSA (*i.e.*, $\delta_H = 6.77\ \text{Mpa}^{0.5}$) higher supercooling is required to nucleate HSA. This shows that the more the incompatibility between the solvent and the gelator molecule (*i.e.*, the solvent–gelator vectorial distance in Hansen space, R_a), the lower the thermodynamic drive to achieve nucleation. For the same token, the authors reported that when R_a decreases (*i.e.*, as chemical differences between the solvent and HSA decreased), higher supercooling is required by the gelator to nucleate.[42] The same study showed that the rate constant for crystallization has a strong negative linear relationship with δ_H and a positive linear relationship with R_a. All these results indicate that the lower the chemical similarities between the solvent and HSA, the lower the supercooling needed to achieve nucleation and the higher the rate constant for crystallization. In the same way, the higher the chemical similarities between the solvent and HSA, the higher the supercooling needed to achieve nucleation and the rate constant for crystallization would be lower.[42] Within this context, we might expect that using a solvent–gelator system that requires high supercooling conditions for nucleation (*i.e.*, the solvent and the gelator have chemical similarities) would result in a low rate constant for crystallization and, subsequently, in the development of lower numbers of crystals with a high dimensionality (*i.e.*, spherulitic crystals). A solvent–gelator system with lower chemical similarities would require lower supercooling, resulting in the formation of fewer crystals with a lower dimensionality (*i.e.*, fibers or microplatelets).[42] Our group did a similar analysis using the concept of the equilibrium melting temperature (T_M°) applied to the crystallization of triacylglycerides. T_M° is a thermodynamic parameter associated with the molecular compatibility between compounds present in a solution (*i.e.*, LMOG + vegetable oil or pure triacylglycerides + vegetable oil).

Within this context, the $T_M°$ of a solution is associated with the molecular compatibility between the major components of the solution.[43] Evidently, each solvent–gelator system would develop a three-dimensional crystal organization with particular rheological behavior. Regarding the use of vegetable oils in the development of organogels, now it is possible to calculate their Hansen solvent solubility parameters by simply knowing their fatty acid composition.[28] Knowledge of the Hansen solvent solubility parameters may be used to rationalize the ability of LMOGs' to develop organogels in vegetable oils and from there tailor their physical properties.

6.3 Shear Rate and Cooling Rate Effect on the Microstructure, Self-assembly, and Rheological Properties of Organogels

6.3.1 Independent Effect of Shearing and Cooling Rate on the Formation of Organogels with Vegetable Oil

Based on studies that investigated the organogelation of 3% candelilla wax (CLW) in safflower oil under static and shearing conditions, we postulated that shearing and the extent of its application as temperature decreases determine the crystal size and crystal–crystal interaction throughout the SAFIN.[33,34] These conditions were studied through the application of a particular shear rate (30 s^{-1} to 600 s^{-1}) until achieving 52 °C, *i.e.* a temperature above the crystallization temperature of CLW (T_{Cr} = 39.0 °C ± 0.08 °C) and even above its melting temperature (T_M = 43.6 °C ± 0.32 °C).[33] The results of this investigation showed that, compared with the CLW gels formed statically, the use of constant shearing while cooling (6 °C min^{-1}) until achieving the gel setting temperature (5 °C) resulted in poorly structured systems with decreasing elasticity as the shear rate increased. In contrast, the use of pre-shearing (*i.e.*, applying the shear rate up to achieving 52 °C) resulted in CLW organogels with larger crystals, a better organized three-dimensional crystal network where the elastic modulus ≫ loss modulus ($G' \gg G''$; *i.e.*, true gel rheological behavior).[33] The organogels' solid content could not explain fully their rheological behavior. In particular, at a given rate (30 s^{-1} to 600 s^{-1}), the statistical analysis showed that the organogels' solid content formed with pre-shearing was the same that in organogels developed using constant shearing. Additionally, the solid content of organogels developed with pre-shearing was mostly independent of the shearing rate applied, showing a higher value just at 300 s^{-1} (Figure 6.3). We explained these results considering that shearing under metastable conditions resulted in flow-induced molecular alignment of CLW components (*i.e.*, mesophase precursors). Upon further cooling under static conditions, these mesophase precursors crystallize, developing a three-dimensional organization with a higher extent of microplatelet–microplatelet interactions and, therefore, higher elasticity than organogels developed statically or under constant shearing.[33]

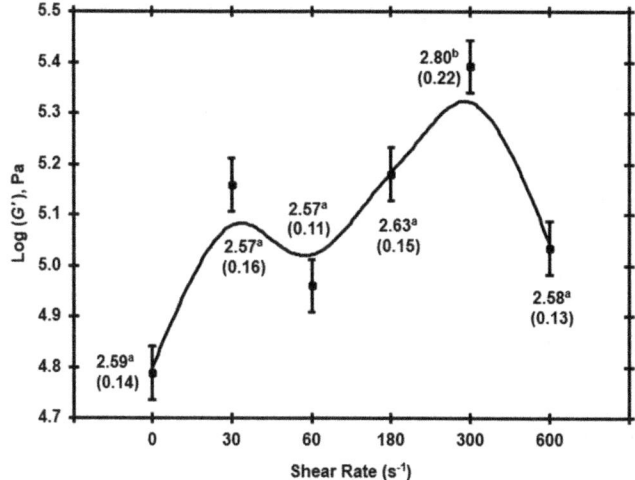

Figure 6.3 Logarithm of the elastic modulus [Log(G')] as a function of shearing rate of 3% CLW organogels developed in safflower oil. Shearing was applied while cooling (6 °C min^{-1}) up to achieving 52 °C. The bars correspond to the standard error of the organogels' elasticity. The values correspond to the percentage of solid content of the organogel developed at each shear rate; values with different superscript indicate a significant shear rate effect on the solid content ($P < 0.10$). Adapted from *Food Research International*, **49**, F. M. Alvarez-Mitre, J. A. Morales-Rueda, E. Dibildox-Alvarado, M. A. Charó-Alonso, J. F. Toro-Vazquez, Shearing as a variable to engineer the rheology of candelilla wax organogels, 580, Copyright 2012, with permission from Elsevier.

Within this framework, Wang *et al.*[32] showed that the proportions of transient (*i.e.*, entanglement of fibers or microplatelets) and permanent (*i.e.*, branching of fibers or microplatelets) junction zones in a SAFIN structure determine the rheological properties of the self-assembled crystal network. Thus, for a given solvent (*i.e.*, vegetable oil), the proportions of transient and permanent junction zones, crystal shape and size are affected by external thermodynamic and mass transfer conditions applied during organogelation.

Additional investigations done by our group addressed the effect of cooling rate under static conditions using as gelator molecules HSA, stearic acid, and primary and secondary amides synthesized from the enantiopure form of HSA. These studies investigated the self-assembly mechanism, microstructure and rheological behavior of organogels developed in safflower oil (high in triolein).[19,44] These studies provided important new insights about the relationship between the gelator's molecular structure, solid content, and the gel's microstructure to determine the rheological properties of organogels developed at two cooling rates (1 and 20 °C min^{-1}). For instance, we showed that the addition to HSA of a primary or a secondary amide bonded with an alkyl group resulted in gelator molecules that crystallized as fibrillar spherulites at both cooling rates, resulting in organogels with thixotropic behavior.

Additionally, we concluded that gel microstructure has a more important role in determining the organogels' rheology than the solid content. This is because the solid content explained just ~25% and ~63% of the total variation in G' of the organogels developed at 1 °C min^{-1} and 20 °C min^{-1}, respectively.[19]

From these studies we selected three gelator molecules to investigate the combined effect of external thermodynamic (*i.e.*, cooling rate) and mass transfer conditions (*i.e.*, shear rate) in the self-assembly process. The gelator molecules selected were HSA, (*R*)-12-hydroxyoctadecanamide (HOA), and (*R*)-*N*-octadecyl-12-hydroxyoctadecanamide (OHOA) (Figure 6.1). The selection was based on its distinctive self-assembly mechanism, and the microstructure and rheological behavior of organogels developed in safflower oil (high in triolein) at two cooling rates (1 °C min^{-1} and 20 °C min^{-1}) under static conditions. Thus, in safflower oil and independent of the cooling rate used, HSA crystallizes as fibers (Figure 6.4A), while HOA as fibrillar spherulites (Figure 6.4B) and OHOA as acicular spherulites (Figure 6.4C). The G' of the organogels (2% wt/wt) developed by these gelators under static conditions using cooling rates of 1 °C min^{-1} or 20 °C min^{-1} to achieve 25 °C is shown in Figure 6.5. The statistical analysis of these results indicated that at both cooling rates G' had the following order: HSA < HOA < OHOA ($P < 0.01$). Despite G' in the gels the solid content showed the following order at 1 °C min^{-1}: OHOA < HOA < HSA, and at 20 °C min^{-1} OHOA = HSA < HOA ($P < 0.01$). Our study indicated that cooling rate did not affect the G' of the HSA organogels. In contrast, gelator molecules with a 12-hydroxyl and a primary or secondary amide group, *i.e.*, HOA and OHOA, showed a significant cooling rate effect on elasticity.[19] This behavior was associated with the cooling rate effect on crystal growth and its implication in microfiber interpenetration between vicinal spherulites developed by HOA and OHOA. Thus, at the higher cooling rate the spherulites were smaller as a consequence of nucleating rates being accelerated at the expense of growth rates.[18,45,46] In contrast, at low cooling rates two phenomena occurred: (1) during cooling gelator molecules have sufficient time to diffuse and to add selectively to a nucleated crystallite surface, and (2) crystal growth of already nucleated species is favored over additional nucleation events and fewer, but larger, crystals are formed.[19,44] An additional factor considered in the microstructural and rheological behavior of HOA and OHOA is that spherulite size decreases with increasing molecular weight, because the growth rate decreases as the molecular weight increases.[47] Within this context, and considering that HOA has a molecular weight of 299.49 Da and OHOA of 551.97 Da, the overall result is that HOA gels developed at 20 °C min^{-1} showed higher fiber interpenetration between vicinal spherulites than the gels formed at 1 °C min^{-1}. Given the smaller size of the spherulites, the opposite occurred with the OHOA gels. Thus, HOA gels formed at 20 °C min^{-1} exhibited higher G' than the gels formed at 1 °C min^{-1}, while OHOA gels formed at 1 °C min^{-1} showed higher G' values than the gels developed at 20 °C min^{-1}. This occurred despite the fact that OHOA gels formed at 1 °C min^{-1} had lower solid content than those formed at 20 °C min^{-1} (Figure 6.5).[19]

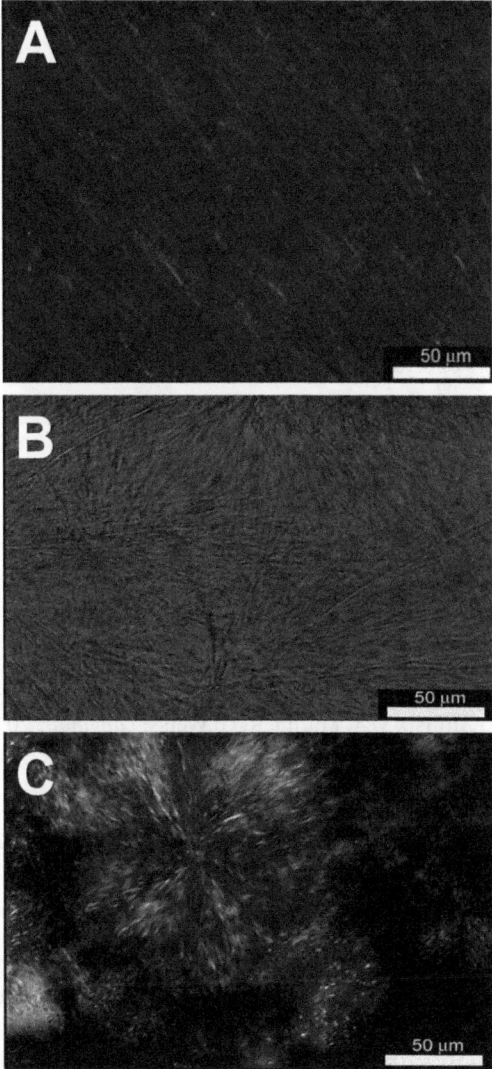

Figure 6.4 Polarized light microphotographs of organogels formed by 2% HSA (A), HOA (B), and OHOA (C) at 25 °C, using a cooling rate of 20 °C min⁻¹. Microphotographs reproduced with permission from J. F. Toro-Vazquez, J. Morales-Rueda, A. Torres-Martínez, M. A. Charó-Alonso, V. A. Mallia, R. G. Weiss, *Langmuir*, 2013, **29**, 7642. Copyright 2013 American Chemical Society.

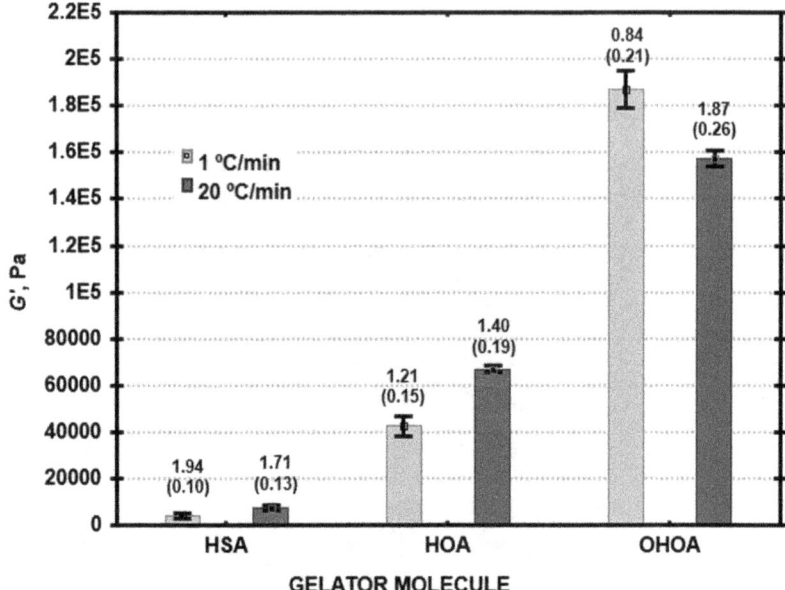

Figure 6.5   Elasticity (G') of organogels (2% wt/wt) of HSA, HOA, and OHOA developed in safflower oil under static conditions. The cooling rate used to
achieve the gel setting temperature (25 °C) was 1 °C min^{-1} or 20 °C min^{-1}.
G' was measured after 75 min at 25 °C (bars correspond to the standard
error for the G' measurements). The organogels' solid content (%) and
corresponding standard deviation are included. Adapted with permission from J. F. Toro-Vazquez, J. Morales-Rueda, A. Torres-Martínez, M. A.
Charó-Alonso, V. A. Mallia, R. G. Weiss, *Langmuir*, 2013, **29**, 7642. Copyright 2013 American Chemical Society.

6.3.2   Combined Effect of Shearing and Cooling Rate on the Formation of Organogels Developed with HSA, HOA, and OHOA in Vegetable Oil

The amides used to investigate the combined effect of cooling rate and
shearing were synthesized through a well-established methodology.[17] Their
characterization and relative purity were determined through melting temperature, infrared and mass spectrometry analysis.[48] The rationale for the
gelator selection was based on having a useful set of molecules synthesized
through systematic modification of the parent molecule, each having a different crystal habit and molecular mechanism for self-assembly. HOA is the
corresponding primary amide of the enantiopure form of HSA, and OHOA
(*i.e.*, a secondary amide) is the *N*-octadecyl derivative of HOA (Figure 6.1).
Additionally, while HSA and HOA have similar molecular weights (300.49
Da and 299.49 Da, respectively), OHOA has a much higher molecular weight
(551.97 Da). Regarding the self-assembly mechanism, as previously indicated

the carboxyl group in HSA develops just intermolecular interactions in "head to head" arrangements through cyclic carboxylic dimers. The secondary 12-hydroxyl group present in 12-HSA and the corresponding primary (HOA) and secondary (OHOA) amide derivatives increases the molecular polarity allowing the occurrence of molecular self-assembly along the secondary axis through hydrogen bonds.[22,41] Furthermore, the introduction of an amide group in HSA molecules increases both the molecule's polarity and, in the primary amide (*i.e.*, HOA), its ability to develop hydrogen bonds on both the oxygen and nitrogen atoms of the amide group. In contrast, in the secondary amide (*i.e.*, OHOA) the introduction of the octadecyl chain (Figure 6.1) increases the involvement of the London dispersion forces in the intermolecular interactions associated with the molecular self-assembly. However, OHOA has lower capacity to develop hydrogen bonds than HOA.[19,44] The importance of hydrogen bonds and van der Waals attractive forces in determining gelator efficiency and rheological properties has been established using mono-, di-, and tri-hydroxymethylated alkylamide gelators in different solvents by Zhang and Weiss.[49]

To develop the organogels, we applied the shear rate (0 s^{-1}, 300 s^{-1}, 600 s^{-1}, and 1200 s^{-1}) during cooling of the gelator solutions (2% wt/wt) from their isotropic phase just until achieving a temperature nearby the melting temperature of the gelator in the vegetable oil at the corresponding cooling rate, *i.e.*, ~55.5 °C for HSA, ~99.3 °C for HOA, and ~86.7 °C for OHOA. Then, cooling was continued under static conditions until achieving 15 °C. We applied these experimental conditions using a rheometer with parallel plate geometry since this set-up allowed the application of particular shear and cooling rates at specific stages during the development of the organogels, while monitoring the rheological properties. The study included four cooling rate protocols: using a constant cooling rate of 1 °C min^{-1} (CR$_1$) or 10 °C min^{-1} (CR$_{10}$) during the shearing and static stages, and using variable cooling rates for the shearing and static stages (VR$_{1\text{-}10}$ and VR$_{10\text{-}1}$).[48] It is important to note that with the use of the CR$_{10}$ and VR$_{10\text{-}1}$ cooling protocols, the shear rate was applied while the supercooling changed at a rate ten times faster than the one used when we formed the organogels using the CR$_1$ and VR$_{1\text{-}10}$ conditions. The G' and G'' were monitored during cooling under static conditions, once 15 °C was reached, and after 60 min at 15 °C. Figure 6.6 shows a diagram for the cooling protocols CR$_{10}$ and VR$_{10\text{-}1}$. Figure 6.7 shows the G' values of the organogels after 60 min at 15 °C formed by HSA (Figure 6.7A), HOA (Figure 6.7B), and OHOA (Figure 6.7C), using the CR$_1$, CR$_{10}$, VR$_{1\text{-}10}$ and VR$_{10\text{-}1}$ cooling protocols. We noted that, independent of the cooling conditions used, the G' of the HSA organogels once 15 °C was reached showed a significant decrease after 60 min of storage at 15 °C (data not shown). In contrast, the HOA and OHOA organogels observed an increment in G' after 60 min of storage at 15 °C, particularly when using the CR$_{10}$ and VR$_{10\text{-}1}$ protocols (*i.e.*, the conditions where shearing was applied while supercooling changed at the faster rate) (data not shown).[48]

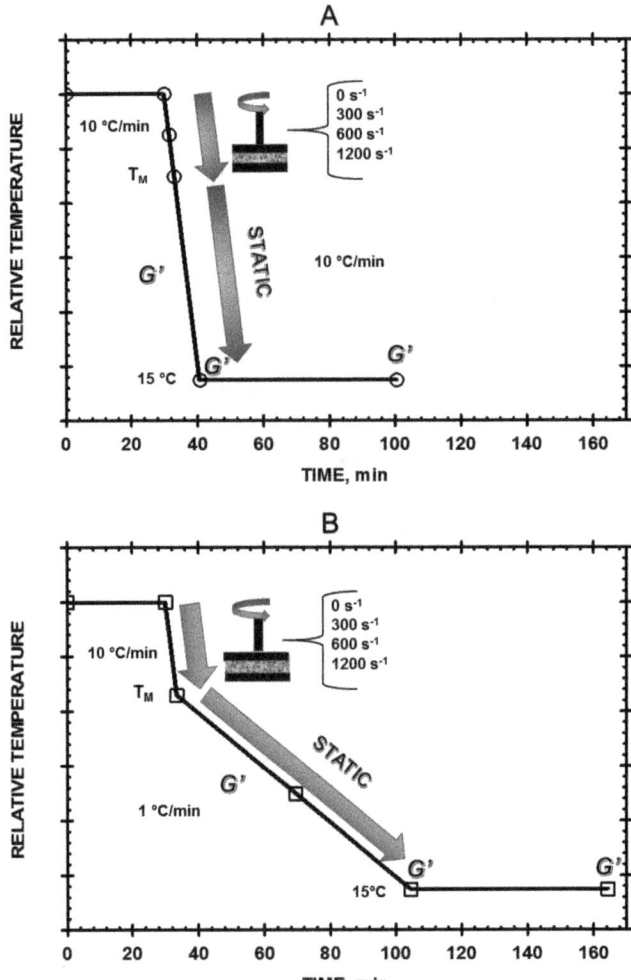

Figure 6.6   Diagrams showing two cooling rate protocols used to investigate the combined effect of shearing and cooling rate on organogelation. A: the same cooling rate (10 °C min⁻¹) is applied during the shearing (0, 300, 600, or 1200 s⁻¹) and static stage (CR$_{10}$). B: a cooling rate of 10 °C min⁻¹ is applied during the shearing stage and 1 °C min⁻¹ during the static stage (VR$_{10-1}$). T_M indicates the meting temperature for the corresponding gelator. The elastic modulus (G') was measured during cooling, once reaching 15 °C, and after 60 min at 15 °C. Diagram adapted from the experimental protocol described in ref. 48.

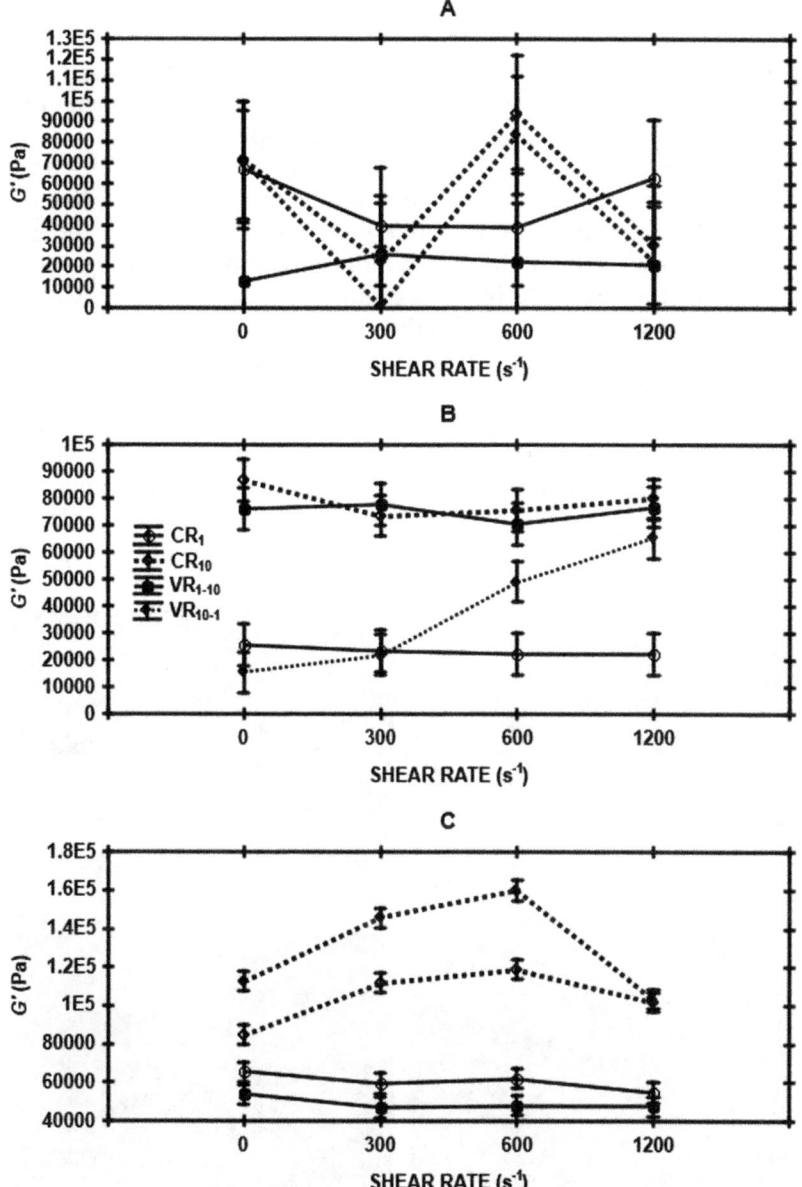

Figure 6.7   Cooling rate effect on the elasticity (G') of the 2% HSA (A), HOA (B) and OHOA (C) organogels as a function of the shear rate applied while cooling at 1 °C min^{-1} (*i.e.*, CR$_1$ and VR$_{1\text{-}10}$) or 10 °C min^{-1} (*i.e.*, CR$_{10}$ and VR$_{10\text{-}1}$). G' was measured after 60 min of storage at 15 °C. Adapted from *Food Research International*, **93**, A. De la Peña-Gil, F. M. Álvarez-Mitre, M. M. González-Chávez, M. A. Charó-Alonso, J. F. Toro-Vazquez, Combined effect of shearing and cooling rate on the rheology of organogels developed by selected gelators, 52, Copyright 2017, with permission from Elsevier.

The main conclusions of this study were the following:

1. Overall, the cooling rate protocols studied for LMOGs without the application of shearing had a differential effect on the fibers' relative diameter, branching extent, and size of the spherulites.

2. The magnification and resolution of our microscope used in this study were not sufficient to establish the cooling rate effect on the length and branching extent of the HSA fibers (Figure 6.8). From all the gelator molecules involved in this study, independent of the cooling rate used, HSA had the lowest T_{Cr} in the vegetable oil (50.38 °C ± 1.36 °C). Consequently, HSA had higher solubility in the vegetable oil than HOA (T_{Cr} = 84.96 °C ± 1.55 °C) and OHOA (T_{Cr} = 76.72 °C ± 0.11 °C). Owing to the partial solubility of HSA in the vegetable oil,[38] independent of the cooling protocol applied, the G' of all HSA organogels had high variability showing statistically the same elasticity (Figure 6.7A). In the same way, the partial solubility of HSA in the vegetable oil was additionally associated with the decrease in G' observed by the HSA organogels during storage at 15 °C. In contrast, the HOA and OHOA had lower solubility

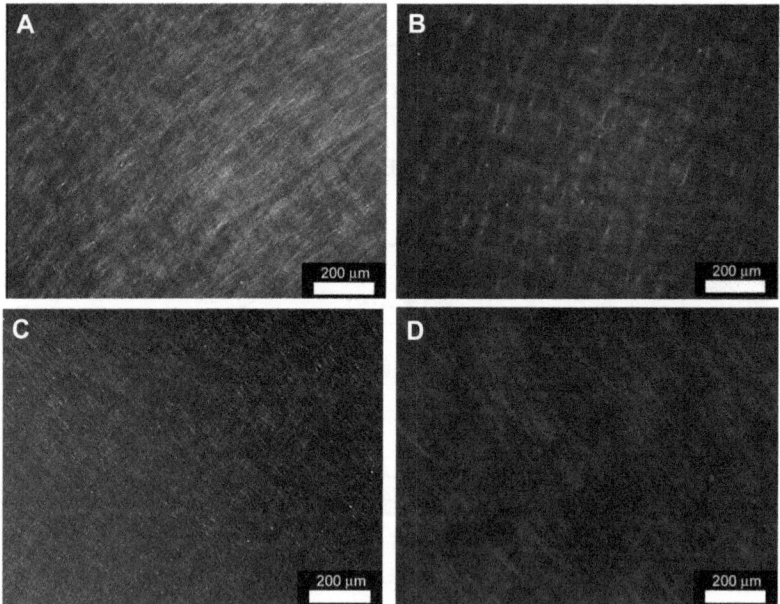

Figure 6.8   Polarized light microphotographs for the 2% HSA organogels after 60 min at 15 °C formed under static conditions (0 s^{-1}) applying the different cooling rate protocols studied. A: CR$_1$; B: VR$_{1-10}$; C: VR$_{10-1}$; D: CR$_{10}$. Reproduced from *Food Research International*, **93**, A. De la Peña-Gil, F. M. Álvarez-Mitre, M. M. González-Chávez, M. A. Charó-Alonso, J. F. Toro-Vazquez, Combined effect of shearing and cooling rate on the rheology of organogels developed by selected gelators, 52, Copyright 2017, with permission from Elsevier.

in the vegetable oil, so the corresponding organogels showed an increment in G' during storage at 15 °C (data not shown).

3. For HOA the spherulites formed by applying a cooling rate of 1 °C min^{-1} after achieving the gelator's T_M (*i.e.*, CR$_1$ and VR$_{10-1}$) resulted in fibers with relative thicker diameter and lower branching extent than the spherulites developed with the CR$_{10}$ and VR$_{1-10}$ protocols. The use of a high rate of supercooling change (*i.e.*, 10 °C min^{-1}) after achieving the gelator's T_M developed thinner fibers with higher extent of branching (Figure 6.9). The permanent junction zones are mainly responsible for the gel's mechanical strength and the transient junction zones are associated with the recovery of gel elasticity after applying stress. Therefore, after 60 min at 15 °C the HOA organogels developed without shearing using the CR$_{10}$ and VR$_{1-10}$ protocols had higher elasticity than the organogels formed with the CR$_1$ and VR$_{10-1}$ conditions (Figure 6.7 B).

4. The inclusion of the *N*-octadecyl chain in the primary amide group of the HOA decreases the H-bonding capacity and therefore limits the branching extent of the gelator molecule and favors the development of needle-like self-assembled structures instead of fibers. Thus, while

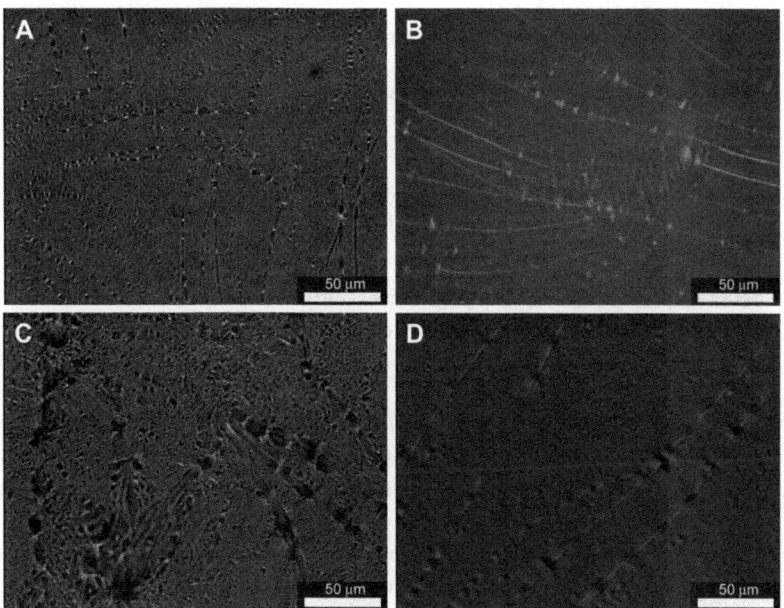

Figure 6.9 Polarized light microphotographs for the 2% HOA organogels after 60 min at 15 °C formed under static conditions (0 s^{-1}) applying the different cooling rate protocols studied. A: CR$_1$; B: VR$_{1-10}$; C: VR$_{10-1}$; D: CR$_{10}$. Reproduced from *Food Research International*, **93**, A. De la Peña-Gil, F. M. Álvarez-Mitre, M. M. González-Chávez, M. A. Charó-Alonso, J. F. Toro-Vazquez, Combined effect of shearing and cooling rate on the rheology of organogels developed by selected gelators, 52, Copyright 2017, with permission from Elsevier.

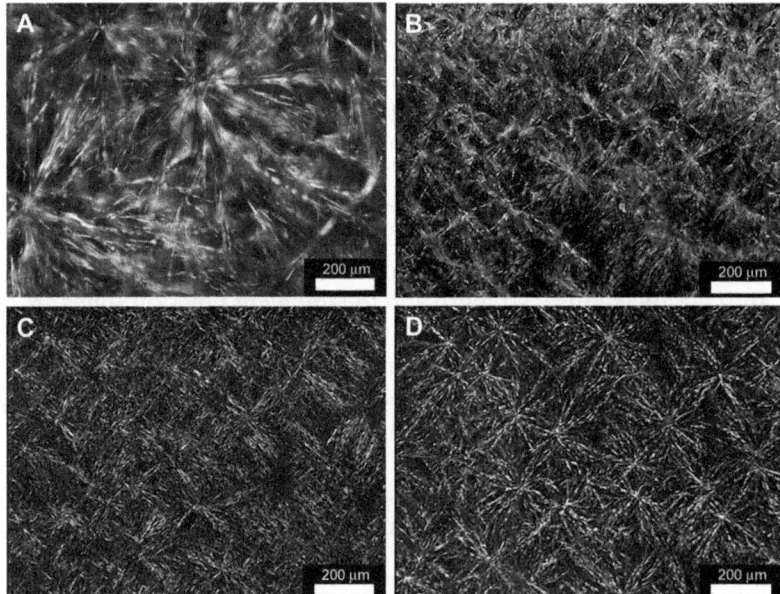

Figure 6.10   Polarized light microphotographs for the 2% OHOA organogels after 60 min at 15 °C formed under static conditions (0 s^{-1}) applying the different cooling rate protocols studied. A: CR_1; B: VR_{1-10}; C: VR_{10-1}; D: CR_{10}. Reproduced from *Food Research International*, **93**, A. De la Peña-Gil, F. M. Álvarez-Mitre, M. M. González-Chávez, M. A. Charó-Alonso, J. F. Toro-Vazquez, Combined effect of shearing and cooling rate on the rheology of organogels developed by selected gelators, 52, Copyright 2017, with permission from Elsevier.

HOA develops fibrillar spherulites (Figure 6.9), OHOA forms acicular spherulites (Figure 6.10). In contrast with the rheological behavior observed in HOA, the OHOA organogels developed statically and applying a cooling rate of 10 °C min^{-1} before achieving the gelators' T_M (*i.e.*, CR_{10} and VR_{10-1}) had higher elasticity than the organogels formed applying a cooling rate of 1 °C min^{-1} before achieving T_M (Figure 6.7 C).

5. For HOA the mechanical properties of organogels obtained under static conditions depend mainly on the spherulitic domain through the permanent junctions (*i.e.*, fiber branching) arranged radially within the spherulite (Figure 6.9). In contrast, the mechanical properties of OHOA organogels are predominantly associated with the network of spherulitic domains resulting from the interpenetration among vicinal acicular spherulites (*i.e.*, transient junctions) (Figure 6.10).

6. The shearing effect on organogels' elasticity was observed particularly with the higher molecular weight gelator (*i.e.*, OHOA), and just when we applied shearing while cooling at 10 °C min^{-1} until achieving T_M (*i.e.*, CR_{10} and/or VR_{10-1} protocols; Figure 6.7 C). Based on the results of Balzano, Kukalyekar, Rastogi, Peters, and Chadwick[50] and Mondello, Grest, Webb, and Peczak,[51] we propose that the molecular relaxation time of

gelator molecules, and its increase as molecular weight increases and as temperature decreases, plays an important role in the gelator's susceptibility to going through a shear-induced crystallization process. Thus, using eqn (5) reported by Mondello, Grest, Webb, and Peczak,[51] assuming a density for the gelator solution of 930 000 g m^{-3} and using the corresponding viscosity, we determined that at the temperature where shearing stopped the rotational relaxation time (τ_R) for OHOA was 2.445×10^{-9} ns (81 °C), for HOA was 0.936×10^{-9} ns (100 °C), and for HSA was 1.171×10^{-9} ns (60 °C) (De la Peña-Gil A. and Toro-Vazquez, J. F., unpublished results). Therefore, when shearing was applied while cooling at 10 °C min^{-1} until T_M was reached (*i.e.*, CR$_{10}$ and VR$_{10-1}$ protocols), OHOA, the gelator with the highest molecular weight, would develop mesophase precursors. Upon further cooling, these shear-induced "liquid structures" crystallize developing organogels with higher elasticity than the ones developed under quiescent conditions. Using the same cooling protocols and shear rates, the stretched low molecular weight molecules (*i.e.*, HSA and HOA) with shorter τ_R, would dissolve back to the isotropic state. In contrast, when we applied shearing while cooling at a rate ten times slower (*i.e.*, CR$_1$ and VR$_{1-10}$ protocols), independent of the gelator's molecular weight, the molecules would have sufficient time to fold down and return to the isotropic phase. Consequently, using these cooling rate protocols we did not observe a shearing effect on the organogels' elasticity with any of the gelator molecules (Figure 6.7).

6.4 Organogels Developed by "Polar" Gelator Molecules Derived from HSA

As already noted, amides have higher capability for developing hydrogen bonds through both the oxygen and nitrogen atoms of the functional group than the amines. Consequently, the amine derivatives of HSA have lower gelling capabilities in organic solvents than do the corresponding amines.[17] Nevertheless, the amine derivatives of HSA and, particularly their corresponding ammonium chloride salts, show interesting gelling behavior.[52] Evidently, the inclusion of positive and negative centers in the *N*-alkyl-(*R*)-12-hydroxyoctadecylammonium chlorides (*n*-HOA-Cl; where *n* is the length of the *N*-alkyl chain, Figure 6.1), increases the role of the electrostatic forces in the molecular self-assembly. This provides the gelator with the potential ability to gel both organic liquids and water.[52] Few known LMOGs are ambidextrous (*i.e.*, capable of forming both hydro- and organo-gels).[53,54]

Within this context, we studied the mechanism associated with the self-assembly behavior of *n*-HOA-Cl molecules with alkyl chains of $n = 3, 4, 6$, and 18.[55] The overall objective was to compare the self-assembly and rheological behavior of SAFIN structures developed by molecules with gelling ability to form both hydro- and organo-gels with molecules capable of gelling just apolar solvents. This was done using safflower oil as the liquid phase.

Previous research showed that *n*-HOA-Cl molecules with $n = 2–5$ can form both hydrogels and organogels, but the salts with longer alkyl chains ($n = 6$ and 18) gel organic liquids only.[52,56]

6.4.1 The Self-assembly Mechanism of Ammonium Chloride Salt Derivatives of HSA

Overall, the *n*-HOA-Cl molecules develop more than one solid phase during crystallization as a result of rotational disorder occurring during cooling, *i.e.*, rotator phases.[57] This behavior is evident in the crystallization thermograms of the 5% *n*-HOA-Cl solution in the vegetable oil, where independent of the cooling rate used (1 °C min^{-1} or 20 °C min^{-1}), each gelator solution showed more than one exotherm (Figure 6.11).[55] The calorimetric behavior of the *n*-HOA-Cl solutions indicated that, given the polar character of the *n*-HOA-Cl molecules, these gelators follow a self-assembly process similar to that of amphiphilic lipids (*i.e.*, monoacylglycerides, phospholipids) in low polarity solvents. Thus, we associated the T_{Cr} for the first exotherm (T_{Cr1}) with the development of lamellar layers achieving electrostatic pairing between linear pairs of *n*-HOA-Cl molecules (*i.e.*, the protonated nitrogen atoms interact with pairs of the corresponding chloride counterions). Figure 6.12 shows the T_{Cr1} of 5% *n*-HOA-Cl solutions in the vegetable oil as a function of the number of carbons of the alkyl chain. Further cooling below T_{Cr1} resulted in a second exotherm, which was associated with the crystallization of the *n*-alkyl chains in layers to maximize London dispersion interactions, developing hydrogen bonds through the 12-hydroxyl groups, and maximizing the efficiency of the molecular packing within the SAFINs. As for other amphiphilic lipids (*i.e.*, monoacylglycerides), the calorimetric results showed that for a given *n*-HOA-Cl molecule, the T_{Cr} for the second exotherm (T_{Cr2}) behaved independent of the gelator concentration in the vegetable oil, varying only as a function of the length of the alkyl chain[55] (data not shown). This self-assembly behavior was particularly evident in *n*-HOA-Cl molecules with $n \leq 6$, where the involvement of electrostatic forces in the molecular self-assembly appears to dominate over the London dispersion forces. However, for 18-HOA-Cl the low-polarity alkyl group makes the molecular self-assembly less dependent on the electrostatic forces and dependent mainly on London dispersion forces. In the same way, the low-polarity alkyl group increases the solubility of 18-HOA-CL in the vegetable oil resulting in the lower T_{Cr1} in comparison with *n*-HOA-Cl molecules with alkyl chains of $n = 3, 4, 6$ (Figure 6.12).

6.4.2 Rheological Behavior of Organogels Developed with Ammonium Chloride Salt Derivatives of HSA

The G' at 10 °C for the *n*-HOA-Cl organogels developed at 1 °C min^{-1} and 20 °C min^{-1} are shown in Table 6.1. Although the organogels showed a high melting temperature for assembly of the alkyl chains layer (60 °C to 125 °C)

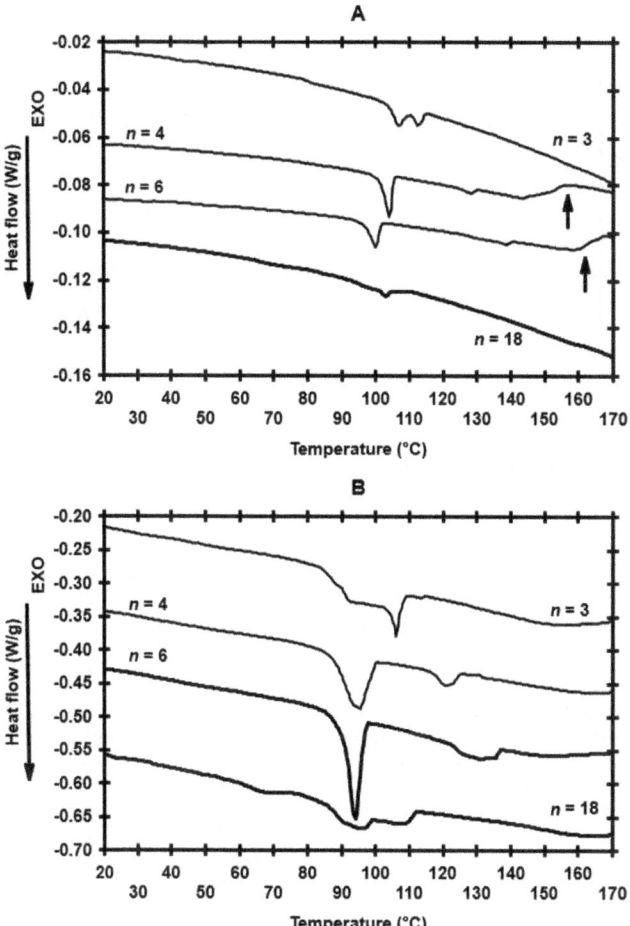

Figure 6.11   Crystallization thermograms for the 5% (wt/wt) *n*-HOA-Cl solutions in vegetable oil at cooling rates of 1 °C min⁻¹ (A) and 20 °C min⁻¹ (B). The arrows in the crystallization thermograms at 1 °C min⁻¹ (A) indicate a second order transition observed just in the 4-HOA-Cl and 6-HOA-Cl solutions. Reproduced from *Food Structure*, F. M. Alvarez-Mitre, V. A. Mallia, R. G. Weiss, M. A. Charó-Alonso, J. F. Toro-Vazquez, Self-assembly in Vegetable Oils of Ionic Gelators Derived from (R)-12-Hydroxystearic Acid, Copyright 2016, with permission from Elsevier.

and the lamellar organization (115 °C to 145 °C), their elasticity at 10 °C was significantly lower than those of the gels developed with only 2% of similar amides [*i.e.*, (R)-N-propyl-12-hydroxyoctadecanamide, and (R)-N-octadecyl-12-hydroxyoctadecanamide] in the same vegetable oil and using similar time–temperature conditions.[55] (Table 6.1). Amines have lower capacity for developing hydrogen bonds than do amides, and this results in the lower gelling capability of the amine derivatives of HSA in comparison with the corresponding amides.[17]

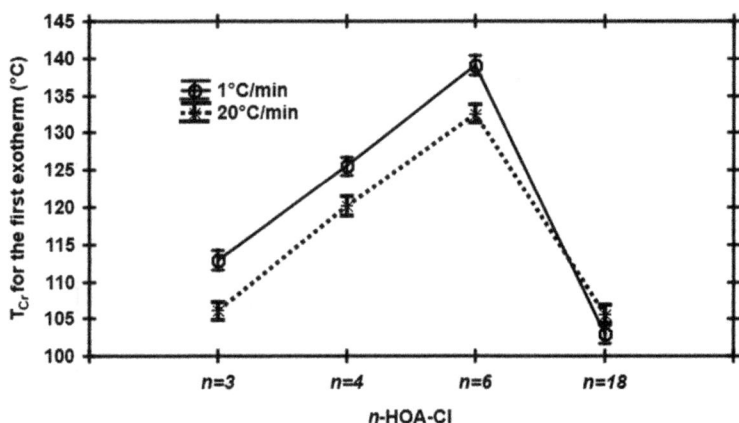

Figure 6.12   Crystallization temperature for the first exotherm (T_{Cr}) as a function of the *n*-alkyl chain of 5% (wt/wt) *n*-HOA-Cl solutions in vegetable oil. T_{Cr} was obtained using a cooling rate of 1 °C min^{-1} or 20 °C min^{-1}. Reproduced from *Food Structure*, F. M. Alvarez-Mitre, V. A. Mallia, R. G. Weiss, M. A. Charó-Alonso, J. F. Toro-Vazquez, Self-assembly in Vegetable Oils of Ionic Gelators Derived from (*R*)-12-Hydroxystearic Acid, Copyright 2016, with permission from Elsevier.

Table 6.1   Elastic modulus (G') at 10 °C for the organogels developed at two cooling rates by the *n*-HOA-Cl molecules in safflower oil high in triolein.

n- HOA-Cl	Cooling rate (°C min^{-1})	G'^{a} (Pa)	$G'^{a,b}$ (Pa)
3- HOA-Cl	1	29.31[b] (1.76)	n.d.[c]
	20	4819.50[c] (1.09)	n.d.[c]
4- HOA-Cl	1	1216.19[b] (1.41)	75 200.0[d] (3000.0)
	20	2449.10[c] (3.04)	15 600.0[d] (3400.0)
6- HOA-Cl	1	5956.60[b] (1.17)	n.d.[c]
	20	5741.16[b] (2.47)	n.d.[c]
18- HOA-Cl	1	1977.00[b] (2.26)	187000.0[d] (11 300.0)
	20	3288.52[c] (1.62)	157500.0[d] (4900.0)

[a]Mean and standard deviation of two independent measurements.
[b]Elasticity values for organogels developed by the corresponding amides using similar time–temperature conditions (values from ref. 19; published with permission of the American Chemical Society).
[c]n.d. = Not determined.
[d]For the same *n*-HOA-Cl a different letter indicates a significant effect of cooling rate ($P < 0.05$) (values from ref. 55; published with permission of Elsevier).

We noted several particularities in the crystallization behavior of the 4-HOA-Cl and 6-HOA-Cl gelators. For instance, from all the *n*-HOA-Cl molecules studied, 4- and 6-HOA-Cl were the only ones showing a differential crystallization behavior during cooling (*i.e.*, two different shapes of crystals were present in the organogels developed at 10 °C; results not shown). Additionally, from all the *n*-HOA-Cl molecules studied, 4-HOA-Cl and 6-HOA-Cl showed the highest gelling capability in the vegetable oil. This is because in comparison with 3-HOA-Cl (4.5%) and 18-HOA-Cl (3.3%), just 2% concentration of 4-HOA-Cl and 6-HOA-Cl was required to achieve gelation.[55] Regarding the rheology of the *n*-HOA-Cl organogels, we also noted a particular behavior of the 4-HOA-Cl and 6-HOA-Cl. This peculiar rheological behavior might be associated with a second-order transition observed by these molecules during cooling. The 4-HOA-Cl and 6-HOA-Cl were the only molecules that showed a second-order transition in vegetable oil when cooled at 1 °C min^{-1} at temperatures of 156.9 °C (± 3.0 °C) and 164.9 °C (± 1.8 °C), respectively. This second-order transition occurred before the T_{Cr1} (transitions indicated with an arrow in Figure 6.11). We could not establish the origin of this second-order transition. However, additional experiments showed its occurrence showed a cooling-rate- and concentration-dependent behavior. Therefore, the mesophase structures tentatively related with this transition must be associated with a kinetic phenomenon followed by the 4- and 6-HOA-Cl molecules during their crystallization. Thus, under isothermal conditions at 113 °C (*i.e.*, a temperature in the interval $T_{Cr1} \leq T \leq T_{Cr2}$, where a gel already existed since $G' > G''$) in the 4-HOA-Cl sol, the elasticity of the system increased exponentially as a function of time. At this crystallization temperature the SAFIN structures of the 4-HOA-Cl organogels changed from microplatelets with needle-like shapes to multidomain spherulite networks (Figure 6.13), resulting in organogels with higher rheological properties ($G' \approx 10\,000$ Pa) and melting temperature than any of the *n*-HOA-Cl organogels obtained at 10 °C (Table 6.1). For the 6-HOA-Cl solution, as cooling proceeded and the temperature approached T_{Cr2} (*i.e.*, 102 °C, where $G' > G''$), we observed a modification in the crystal habit from microplatelets with needle-like shapes (at 121 °C) to small spherulites (at 102 °C) (Figure 6.13), which was followed by a linear increase in the organogel's elasticity as a function of time until achieving a $G' \approx 3500$ Pa. In contrast with the rheological behavior observed by the 4-HOA-Cl organogels, the 6-HOA-Cl organogels developed under isothermal conditions at temperatures in the interval $T_{Cr1} \leq T \leq T_{Cr2}$ had lower elasticity than some of the organogels formed at 10 °C (Table 6.1).

These results support the following conclusions:[55]

1. In 3-HOA-Cl the electrostatic forces appear to control the mode of initial aggregation. However, the lack of a longer alkyl group limited the involvement of the London dispersion forces in the molecular self-assembly. The result is that 3-HOA-Cl has the lowest gelling capability in the vegetable oil providing organogels with the lowest elasticity (Table 6.1). In contrast, with 18-HOA-Cl, and to a lower extent with 6-HOA-Cl,

4-HOA-Cl

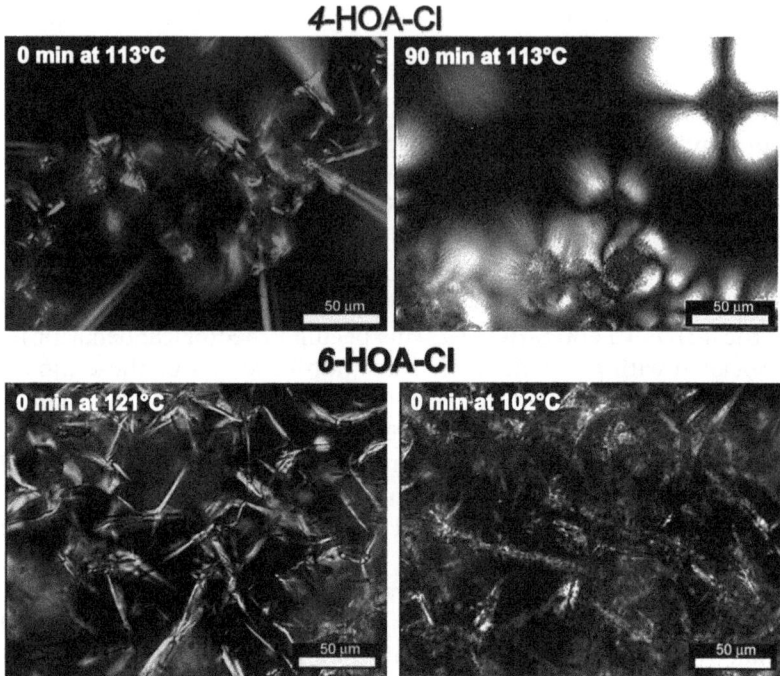

Figure 6.13   Polarized light microphotographs of the 5% 4-HOA-Cl organogels showing the morphological transition from microplatelets (0 min at 113 °C) to spherulites (90 min at 113 °C), and for the 5% 6-HOA-Cl organogels during cooling (1 °C min^{-1}) from 121 °C to 102 °C. Adapted *from Food Structure*, F. M. Alvarez-Mitre, V. A. Mallia, R. G. Weiss, M. A. Charó-Alonso, J. F. Toro-Vazquez, Self-assembly in Vegetable Oils of Ionic Gelators Derived from (*R*)-12-Hydroxystearic Acid, Copyright 2016, with permission from Elsevier.

their longer alkyl group makes the molecular self-assembly less dependent on electrostatic forces relative to London dispersion forces. Thus, after cooling down to 10 °C this resulted in organogels with higher elasticity than 3-HOA-Cl organogels (Table 6.1). Therefore, the gelling behavior of polar gelators derived from HSA might be designed by adjusting the balance between electrostatic and London dispersion forces within the structure of the LMOG.

2. The balance between the electrostatic and London dispersion forces within the 4-HOA-Cl gives this molecule the capability to form both hydrogels and organogels. In contrast, 6-HOA-Cl forms organogels only.[52] Using cooling rates of 1 °C min^{-1} or 20 °C min^{-1} to achieve 10 °C, the 4-HOA-Cl formed organogels with lower G' than 6-HOA-Cl. However, under isothermal conditions (*i.e.*, 113 °C) the 4-HOA-Cl organogels achieved higher elasticity than any of the *n*-HOA-Cl organogels formed at 10 °C. The increase in G' was the result of a structural transition of

the 4-HOA-Cl crystals from microplatelets with needlelike shapes to acicular spherulites. Since at 113 °C a gel already existed, this structural transition occurred in the gel state (Figure 6.13).

3. The combination of viscoelasticy and high melting temperature shown by the 4-HOA-Cl organogels is unique, and opens new alternatives to provide structure to vegetable oil under high temperature conditions.

6.5 Conclusions

HSA is a well-known gelator molecule, and provides a simple structural platform to synthesize a variety of derivatives through systematic chemical modifications. We have summarized the relationship between the molecular structure of HSA and other molecules derived from HSA, with their self-assembly and gelling properties in safflower oil high in triolein (*i.e.*, vegetable oil as the solvent). The long term objective of the results discussed here is to understand the mechanisms involved in molecular self-assembly of gelator molecules and their relationship with organogel formation and the physical properties of organogels (*i.e.*, rheology, melting temperature). In particular, our purpose is to use the organogelation process as a useful and novel alternative to structure vegetable oils without the use of *trans* and/or saturated fats. Unfortunately, most of the basic research done in self-assembly studies and the development of organogels uses pure organic solvents or mineral oil as the liquid phase. Although vegetable oils are mainly composed of triacylglycerides (*i.e.*, 95–99%), they are complex organic systems, making the study of vegetable oils as solvents in the development of organogels a challenge.

Using HSA as the basic molecular structure, followed by systematic chemical modifications, we have established some useful relationships between the gelators' molecular structure, their gelling capability, and the rheological properties or organogels while considering the effect of external thermodynamic (*i.e.*, cooling rate) and mass transfer (*i.e.*, shearing) conditions. However, before using the organogels in the food industry, we need to establish the basic molecular mechanisms associated with the molecular self-assembly (*i.e.*, nucleation) of gelator molecules in vegetable oils, the growth kinetics, and the subsequent development of the three-dimensional network in the organogel.

References

1. A. M. Ascherio, P. L. Katan, M. J. Stampfer and W. C. Willet, *N. Engl. J. Med.*, 1999, 1994.
2. J. Salmerón, F. B. Hu and J. E. Manson, *et al.*, *Am. J. Clin. Nutr.*, 2001, 73, 1019.
3. U. Risérus, W. C. Willett and F. B. Hu, *Prog. Lipid Res.*, 2009, 48, 44.
4. P. W. Siri-Tarino, Q. Sun, F. B. Hu and R. M. Krauss, *Curr. Atheroscler. Rep.*, 2010, 12, 384.

5. F. Destaillats, B. D. Craft, M. Dubois and K. Nagy, *Food Chem.*, 2012, **131**, 1391.
6. EFSA Panel on Contaminants in the Food Chain (CONTAM), *EFSA J.*, 2016, **14**, 4426.
7. S. A. Vieira, D. J. McClements and E. A. Decker, *Adv. Nutr.*, 2015, **6**, 309S.
8. E. D. Co and A. G. Marangoni, *J. Am. Oil Chem. Soc.*, 2012, **89**, 749.
9. L. S. K. Dassanayake, D. R. Kodali and S. Ueno, *Curr. Opin. Colloid Interface Sci.*, 2011, **16**, 432.
10. *Edible Oleogels: Structure and Health Implications*, ed. A. G. Marangoni and N. Garti, AOCS Press, Urbana, IL, 2011, p. 1.
11. M. A. Rogers, T. Strober, A. Bot, J. F. Toro-Vazquez, T. Stortz and A. G. Marangoni, *Int. J. Gastron. Food Sci.*, 2014, **2**, 22.
12. M. E. Morales, V. Gallardo, B. Clarés and M. B. García, *J. Cosmet. Sci.*, 2009, **60**, 627.
13. N. Jibry, T. Sarwar and S. Murdan, *J. Pharm. Pharmacol.*, 2006, **58**, 187.
14. R. G. Weiss, *J. Am. Chem. Soc.*, 2014, **136**, 7519.
15. Y. Lan, M. G. Corradini, R. G. Weiss, S. R. Raghavan and M. A. Rogers, *Chem. Soc. Rev.*, 2015, **44**, 6035.
16. J. L. Li and X.-Y. Liu, *Adv. Funct. Mater.*, 2010, **20**, 3196.
17. V. A. Mallia, M. George, D. L. Blair and R. G. Weiss, *Langmuir*, 2009, **25**, 8615.
18. S. Tang, X. Y. Liu and C. S. Strom, *Adv. Funct. Mater.*, 2009, **19**, 2252.
19. J. F. Toro-Vazquez, J. Morales-Rueda, A. Torres-Martínez, M. A. Charó-Alonso, V. A. Mallia and R. G. Weiss, *Langmuir*, 2013, **29**, 7642.
20. J. H. van Esch, *Langmuir*, 2009, **25**, 8392.
21. B. Yuan, X.-Y. Liu, J.-L. Li and H.-Y. Xu, *Soft Matter*, 2011, **7**, 1708.
22. S. Abraham, Y. Lan, R. S. H. Lam, D. A. S. Grahame, J. J. H. Kim, R. G. Weiss and M. A. Rogers, *Langmuir*, 2012, **28**, 4955.
23. M. Burkhardt, L. Noirez and M. Gradzielski, *J. Colloid Interface Sci.*, 2016, **466**, 369.
24. V. A. Mallia and R. G. Weiss, *J. Phys. Org. Chem.*, 2014, **27**, 310.
25. P. Duan, Y. Li, J. Jiang, T. Wang and M. Liu, *Sci. China: Chem.*, 2011, **54**, 1051.
26. J. H. van Esch and B. Feringa, *Angew. Chem., Int. Ed. Engl.*, 2000, **39**, 2263.
27. N. Zweep and J. H. van Esch, in *Functional Molecular Gels*, ed. B. Escuder and J. Miravet, Royal Society of Chemistry, Cambridge, UK, 2013, ch. 1, pp. 1–29.
28. A. De La Peña-Gil, J. F. Toro-Vazquez and M. A. Rogers, *Food Biophys.*, 2016, **11**, 283.
29. R. S. H. Lam and M. A. Rogers, *Cryst. Growth Des.*, 2011, **11**, 3593.
30. A. Pal, S. Abraham, M. A. Rogers, J. Dey and R. G. Weiss, *Langmuir*, 2013, **29**, 6467.
31. V. A. Mallia and R. G. Weiss, *Soft Matter*, 2015, **11**, 5010.
32. R. Wang, X.-Y. Liu, J. Xiong and J. Li, *J. Phys. Chem. B*, 2006, **110**, 7275.
33. F. M. Alvarez-Mitre, J. A. Morales-Rueda, E. Dibildox-Alvarado, M. A. Charó-Alonso and J. F. Toro-Vazquez, *Food Res. Int.*, 2012, **49**, 580.

34. M. Chopin-Doroteo, J. A. Morales-Rueda, E. Dibildox-Alvarado, M. A. Charó-Alonso, A. de la Peña-Gil and J. F. Toro-Vazquez, *Food Biophys.*, 2011, **6**, 359.
35. A. Jabbarzadeh and R. I. Tanner, *J. Non-Newtonian Fluid Mech.*, 2009, **160**, 11.
36. M. Lescanne, A. Colin, O. Mondain-Monval, K. Heuzé, F. Fages and J. L. Pozzo, *Langmuir*, 2002, **18**, 7151.
37. P. Terech, V. Rodriguez, J. D. Barnes and G. B. McKenna, *Langmuir*, 1994, **10**, 3406.
38. E. Co and A. G. Marangoni, *J. Am. Oil Chem. Soc.*, 2013, **90**, 529.
39. M. A. Rogers, A. J. Wright and A. G. Marangoni, *Soft Matter*, 2008, **4**, 1483.
40. S. S. Sagiri, V. K. Singh, K. Pal, I. Banerjee and P. Basak, *Mater. Sci. Eng., C.*, 2015, **48**, 688.
41. M. A. Rogers, S. Abraham, F. Bodondics and R. G. Weiss, *Cryst. Growth Des.*, 2012, **12**, 5497.
42. M. A. Rogers and A. G. Marangoni, *Langmuir*, 2016, **32**, 12833.
43. D. Pérez-Martínez, C. Alvarez-Salas, J. A. Morales-Rueda, J. F. Toro-Vazquez, M. Charó-Alonso and E. Dibildox-Alvarado, *J. Am. Oil Chem. Soc.*, 2005, **82**, 471.
44. J. F. Toro-Vazquez, J. Morales-Rueda, V. A. Mallia and R. G. Weiss, *Food Biophys.*, 2010, **5**, 193.
45. M. Kobayashi, in *Polymorphism of Fats and Fatty Acids Crystallization*, ed. N. Garti and K. Sato, Marcel Dekker, New York, 1998, Ch. 4.
46. R. Lam, L. Quaroni, T. Pedersen and M. A. Rogers, *Soft Matter*, 2010, **6**, 404.
47. X. Chen, G. Hou, Y. Chen, K. Yang, Y. Dong and H. Zhou, *Polym. Test.*, 2007, **26**, 144.
48. A. De la Peña-Gil, F. M. Álvarez-Mitre, M. M. González-Chávez, M. A. Charó-Alonso and J. F. Toro-Vazquez, *Food Res. Int.*, 2017, **93**, 52.
49. Y. Zhang and R. G. Weiss, *J. Colloid Interface Sci.*, 2017, **486**, 359.
50. L. Balzano, N. Kukalyekar, S. Rastogi, G. W. M. Peters and J. C. Chadwick, *Phys. Rev. Lett.*, 2008, **100**, 48302.
51. M. Mondello, G. S. Grest, E. B. Webb III and P. Peczak, *J. Chem. Phys.*, 1998, **109**, 798.
52. V. A. Mallia, P. Terech and R. G. Weiss, *J. Phys. Chem. B*, 2011, **115**, 12401.
53. F. Delbecq, K. Tsujimoto, Y. Ogue, H. Endo and T. Kawai, *J. Colloid Interface Sci.*, 2013, **390**, 17.
54. G. John and P. Vemula, Patent US 8968784 B2, 2015.
55. F. M. Alvarez-Mitre, V. A. Mallia, R. G. Weiss, M. A. Charó-Alonso and J. F. Toro-Vazquez, *Food Struct.*, 2016, DOI: 10.1016/j.foostr.2016.07.003.
56. V. A. Mallia, H.-I. Seo and R. G. Weiss, *Langmuir*, 2013, **29**, 6476.
57. J. Tsau and D. Gilson, *J. Phys. Chem.*, 1968, **72**, 4082.

Section IV

Structuring Units: Polymeric Strands and Network

CHAPTER 7

Thermo-gelation of Ethyl-cellulose Oleogels

MAYA DAVIDOVICH-PINHAS

Technion, The Israeli Institute of Technology, Haifa, Israel
*E-mail: dmaya@tx.technion.ac.il

7.1 Introduction

Gels are semi-solid materials composed mainly of a liquid phase stabilized by an extensive mesh network composed of self-assembled gelator molecules. The formation of such materials is termed gelation and can be achieved by various mechanisms. Gels can be classified according to the cross-linking type or gelation mechanism responsible for the network formation, *i.e.* chemical or physical interactions. Strong covalent bonds between polymer chains using either a reactive chemical group along the polymer chain or adding a cross-linking agent can be found, in most cases, in synthetic polymers. Physical interactions forming junction zones are usually attributed to weak physical interactions, such as hydrogen bonds, van der Waals interactions, hydrophobic forces, and electrostatic interactions.[1] These types of gels are usually used in food, pharmaceutical and biomedical applications.[2,3] Additional classification of gels can be performed based on the liquid used for their formation: water-based gels, termed *hydrogels*, and organic liquid-based gels, termed *organogels*. The formation of these gels is based on the use of gelator molecules which interact to form a three dimensional network that confines the solvent in it.

Food Chemistry, Function and Analysis No. 3
Edible Oil Structuring: Concepts, Methods and Applications
Edited by Ashok R. Patel
© The Royal Society of Chemistry 2018
Published by the Royal Society of Chemistry, www.rsc.org

The type of gelator molecules used for gel formation can be classified into two major groups: low molecular weight gelators (LMWG), such as peptides and fatty acids, or high molecular weight gelators (HMWG), such as polymers and proteins. In hydrogel systems, examples of LMWG include molecules from classes such as bile salts, functionalized sugars, oligopeptides, surfactant-like molecules and dendrimers.[4,5] While in organogels one can find waxes,[6] fatty acids,[7] mono-glycerides,[8] phytosterols,[9] surfactants,[10] and isotactic and syndiotactic polystyrene.[11] The use of HMWG in hydrogel systems is well known and common. Most hydrogel systems explored in the literature use synthetic biopolymers, such as poly(lactic-*co*-glycolic acid) (PLGA) and polyethylene glycol (PEG)[12] or natural polysaccharides and proteins, such as alginate, pectin, chitosan, gelatin, carrageenans, and whey protein.[12] However, the use of HMWG to form stable organogel systems is less common. There are some synthetic polymers, such as polyesters, poly peptides and others, that have been reported to form organogels. However, their number is small and their organogelation ability is limited.[11] Ethyl-cellulose, which is a semi-synthetic polymer, was reported to gel various organic solvents.[13]

Organogel systems offer a unique spectrum of adjustable physical properties together with additional unique chemical properties, such as hydrophobicity and antibacterial characteristics, resulting from a lack of aquatic media.[14,15] Despite the abundance and variety of studies done with organogel systems, only a few were actually concerned with physiological applications. This is mostly owing to issues of biocompatibility and toxicity relating to the organic solvent and the organogelator used.[14,16] Most pharmaceutical formulations studied to date involve the use of oil as their organic solvent. Oil is a renewable resource with a GRAS status, and is therefore a suitable substance to use in physiological applications. Oil-based systems are typically referred to as *oleogels* rather than organogels due to the specific nature of their solvent. Similar to organogels, oleogel systems are formed using either low or high molecular weight gelators. To date only one polymer, ethyl-cellulose, has been found with the ability to directly gel liquid oil. The current manuscript will focus on ethyl-cellulose characteristics, oleogelation ability and parameters affecting the final gel properties.

7.2   Ethyl-cellulose Characteristics

Ethyl-cellulose (EC) is a linear polysaccharide derived from cellulose—the most abundant polymer in nature. Its commercial production involves exchanging the hydroxyl hydrogen atoms on the cellulose backbone with an ethylene end group thus forming an ether bond.[17] The final polysaccharide chain is composed of β-1,4 poly-glucose backbone with ethylene side chains (Figure 7.1). The polymer is characterized by its degree of substitution (DS), *i.e.* the degree of hydroxyl end groups substitute with ethylene end groups, or ethoxy content, *i.e.* the percent of ethylene addition along the polymer backbone. A DS of approximately 2.4–2.5 (with approximately 47–49.5% ethoxy) leads to the formation of a hydrophobic polymer that is immiscible

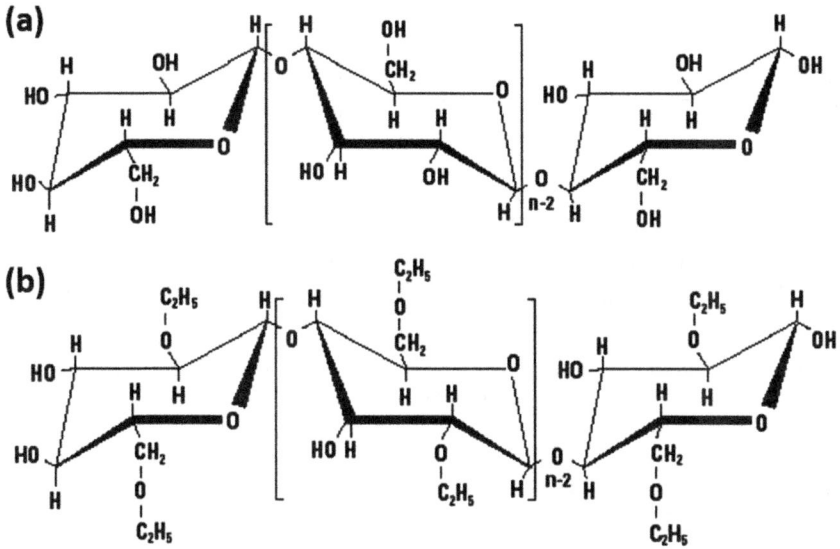

Figure 7.1   Chemical structure of (a) cellulose and (b) ethyl-cellulose.

in water and miscible in various organic solvents. EC with a DS of 1–1.5 forms a hydrophilic polymer soluble in water.[17] EC is commercially marketed by the product viscosity of 5 wt% in 20% ethanol/80% toluene mixture at 25 °C. Previous research has shown a positive correlation between the polymer viscosity and its molecular weight.[18]

Aside from its chemical formulation, which is well known, its structural characteristics are less studied. Thermal analysis of neat EC reveals two characteristic transitions; a glass transition at approximately 130 °C and a melting transition at approximately 180 °C with their reversible transitions, *i.e.* crystallization and vitrification (Figure 7.2a).[18] The presence of a glass transition was reported previously by the manufacturer and other researchers.[13,19,20] Such thermal transition can be attributed to changes in polymer volume leading to an increase in polymer–polymer spaces. The presence of a melting temperature suggests that the polymer consists of crystals or some ordered sections. In order to verify this observation, X-ray diffraction (XRD) analysis in the small and wide angle range (SAXS and WAXS) was performed. The scattering pattern obtained from EC powder revealed two broad characteristic peaks located at $q = 0.5$ and 1.5 Å^{-1} (Figure 7.2b). The broad peak centered at $q = 1.5$ Å^{-1} can be related to the amorphous regions of the polymer, whereas the peak centered at $q = 0.5$ Å^{-1} can be attributed to organized structures with a lattice parameter of 12.6 Å. X-ray scattering analysis at multiple length scales done at elevated temperatures demonstrated the disappearance of the peak centered at $q = 0.5$ Å^{-1} while going above the polymer melting temperature, *i.e.* 180 °C.[18] This result strengthens the assumption that EC consists of ordered or semi-crystalline areas. The EC scattering patterns

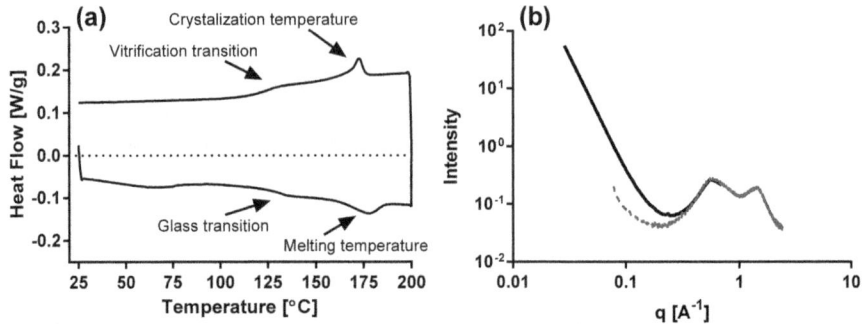

Figure 7.2 The (a) DSC thermogram and (b) X-ray scattering (SAXS: black; WAXS: gray) obtained for EC 10 cP powder.

consisting of two broad peaks at approximately q ~0.5 and 1.5 Å^{-1} were previously observed for native EC. However, the exact nature of these peaks was not discussed.[21–23] The thermal and structural analysis done with DSC and scattering techniques allude to the conclusion that EC is a semi-crystalline polymer.[18]

7.3   Thermo-gelation of Ethyl-cellulose Oleogels

Oleogel systems are gels based on oil as the continuous phase. The production of such gels is based on the mixing of an oil with a structurant agent, *i.e.* a gelator, which is responsible for the gel network formation. Various structurant agent formulations can be found in the literature alluding to various oleogel formulations with a variety of gel properties. As described above, the gelators that assemble the gel network can be classified into two major groups based on their molecular weight; low and high molecular weight oil gelators (LMOG and HMOG, respectively).[10] In the LMOG class, one can find a variety of small molecules, which are able to self-assemble into specific structures that further aggregate into a three-dimensional network. Waxes,[6] fatty acids,[7] fatty alcohols[24] and their mixtures,[25] phytosterols,[9] mono-glycerols,[10] and surfactants[26] are several examples of such gelators. Less common are oleogel systems based on HMOGs such as polymers and proteins. In comparison to LMOG, HMOGs offer different properties for possible applications, which are the consequence of their extended polymeric chains and gelation mechanisms.[11,27] Patel *et al.*[28–29] and Mazzenga *et al.*[30] developed an edible oleogel based on polymers and proteins using an emulsion templated approach. In their work, a two-step approach was used, which included the mixing of a concentrated oil-in-water emulsion together with a water-soluble polymer, followed by selective evaporation of the continuous water phase, which resulted in the formation of the final polymer network. Another approach to form protein-based oleogels has been recently suggested by De Vries *et al.*[31,32] Protein aggregates formed in water medium were able to stabilize oil medium after solvent exchange using an intermediate solvent. The intermediate solvent was based on a mixture of THF or acetone with water at different ratios. Only one polymer gelator, EC, has demonstrated an ability to gel liquid oil without recourse to intermediate processing phases. This direct gelation approach has the great advantage of reducing preparation procedures and environmental interferences and minimizing errors.

The gelation of EC in the presence of liquid oil was first introduced in the early 1990s by Aiache and colleagues.[33] Gelation of EC in liquid oil can be achieved by first increasing the temperature of the polymer–oil mixture above the polymer glass transition (~140 °C), and then cooling it down to room temperature.[34] Lower preparation temperature conditions were reported for various organic solvents. The EC glass transition temperature, *i.e.* ~140 °C, was required in order to dissolve the EC while using vegetable oils such as canola, soybean, flaxseed, and medium chain triacylglycerol (MCT).[35,36] Higher temperatures, *i.e.* 180 °C, were used for the formation of EC gels with di-ester solvents such as di(2-ethylhexyl)phthalate (DOP), dibutyl phthalate (DBP), and diethyl phthalate (DEP).[37] Non-aqueous solvents such as Miglyol 840 (Dow Chemicals) produced gel with EC by continuous stirring at 60 °C,[38] whereas the combination of propylene glycol dicaprylate/dicaprate with EC produced gels by simply mixing the two using 300 rpm at room temperature.[39,40] The different preparation procedures are probably related to the ability of the EC backbone to interact with solvent molecules and thus dissolve in the medium.

A minimum amount of EC is required in order to form a gel network. This amount is strongly affected by the polymer molecular weight and the oil type used. For example, EC 45 cP can gel soybean oil using only 4 wt% polymer while the gelation of canola oil requires at least 6 wt% polymer.[41] The formation of EC-based oleogels using propylene glycol dicaprylate/dicaprate mixture as a solvent requires 11 wt% and 7 wt% with EC 10 and 100 cP FP Premium, respectively.[38] Polymer concentrations as low as 1, 2, and 3 wt% were required while using di(2-ethylhexyl)phthalate (DOP), dibutyl phthalate (DBP), and diethyl phthalate (DEP), respectively.[37]

The sol–gel and gel–sol transitions of 11 wt% EC 10 cP/canola oil samples were observed using temperature sweep experiment at a wide temperature range (Figure 7.3). A clear cross-over between the loss modulus (G″) and the storage modulus (G′) can be detected during cooling and heating, implying the presence of sol–gel and gel–sol transitions, respectively. The sol–gel transition was detected at approximately 110 °C while the gel–sol transition occurred at approximately 125 °C, suggesting a hysteresis behavior between gelation and melting. Similar trends were detected for various EC types and concentrations with a shift in the transition temperature to higher values with an increase in EC molecular weight. Gel point temperature analysis demonstrated a decrease in the gel point temperature while reducing the EC molecular weight. No significant differences in gel point temperature were observed using polymer concentration in the range of 7–13 wt%.[34] Lizaso *et al.*[37] worked with EC oleogels made with different di-ester solvents and observed significantly different

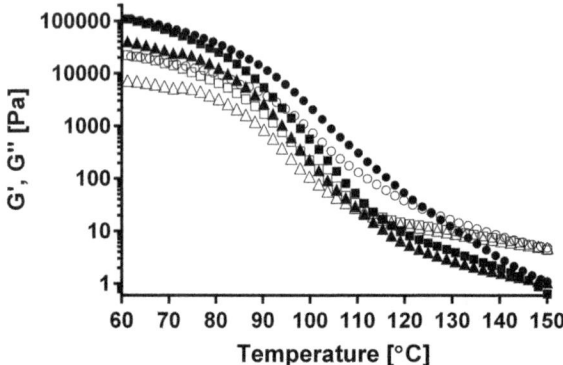

Figure 7.3 Typical temperature sweep experiment obtained for 11 wt% EC 10 cP/
canola oil sample at 1 °C min⁻¹ cooling/heating rate during first cooling
(■), heating (●) and second cooling (▲). G′: filled symbols; G″: open
symbols.

transition temperatures in the range of 40–120 °C using different polymer concentration and solvents. EC oleogels also demonstrated reversible sol–gel and gel–sol transitions using several cooling–heating cycles, as shown in Figure 7.3. Such gelation scheme is referred to as cold set thermo-reversible gelation process, *i.e.* gelation upon cooling.

7.4 Ethyl-cellulose Gelation Mechanism

Induction of gel formation can be achieved using various environmental conditions, such as pH and temperature. The transformation of sol–gel or gel–sol due to temperature stimulation is referred to as thermo-reversible gelation. The ability of a three-dimensional network to form and disassemble due to temperature change is typical for physical gels. Such gels demonstrate unique behavior arising from the thermodynamic nature of the junction zones, which can fluctuate in size and position over time, thus never approaching equilibrium. This is owing to secondary chain extension, formation or breaking of junction zones, and re-arrangement or stacking of the already formed junction zones.[42]

The formation of thermo-reversible networks can be found in aqueous and organic solvents. Natural polymers such as polysaccharides and proteins are able to form gel structures in aqueous phase upon thermal stimuli. This process is governed by conformation changes leading to crystallization, coil-to-helix transition, or denaturation, which can further aggregate to form three-dimensional networks. Physical bonds such as hydrogen bonds, hydrophobic interactions, van der Waals forces, *etc.* govern the formation of such structures. These interactions are strongly affected by environmental conditions such as temperature. Due to the different effects of temperature on each type of bond, "hot" and "cold" setting gelation systems can be found. "Hot" setting gelation is referred to as systems which go through sol–gel transition upon heating.

Such systems are usually formed by hydrophobic interactions, which are more pronounced at higher temperatures, for example in methylcellulose[43] and ethyl(hydroxyethyl)cellulose,[44] or by denaturation of proteins, for example whey protein and myosin.[45] "Cold" setting gels, on the other hand, are formed upon cooling. In this group, one can find various polysaccharides, such as agar and pectin, as well as proteins such as gelatin.[45] Thermo-reversible oleogel systems are usually formed by "cold" setting gelation procedure. Such systems are usually characterized by the formation of crystal structures,which further aggregate to form the three-dimensional network.[10] EC-based oleogels are unique in the sense of their gelation mechanism compared to other oleogel systems.

EC oleogels are formed due to the formation of hydrogen bonds between polymer chains. The role of hydrogen bonds in the EC network formation was first suggested by Laredo *et al.*[46] In this study, Raman and infra-red (IR) spectroscopy showed that hydrogen bonds between polymer chains are responsible for the network formation. This observation was further confirmed using a rheometer. The effect of temperature on EC oleogels' viscoelastic properties was examined using small deformation rheology. Frequency sweep experiments at elevated temperatures up to 90 °C using a low frequency range were performed on EC oleogels. A decrease in gel strength owing to temperature increase was detected, suggesting temperature-induced destabilization of the polymer network as a result of hydrogen bonds weakening and breaking during the temperature increase.[34] Another confirmation was obtained using attenuated transmission reflection infrared (ATR-IR) analysis. EC oleogels exhibit an O–H stretching band at ~3440 cm^{-1}, which implies the involvement of these hydrogen atoms in hydrogen bonding.[47] This observation strengthens the hypothesis that EC gels are formed by hydrogen bond polymer network. The involvement of hydrogen bonds between polymer chains and dipole–dipole interactions between the polymer's monomer units and the solvent molecules were also suggested by Heng *et al.*,[38] Bruno *et al.*[39,40] and Lizaso *et al.*[37] Additional investigation of EC oleogel properties using small amplitude and thermal analysis demonstrated a continuous sol–gel transition in the rheometer (Figure 7.3), without any thermal event in the DSC. These results suggested that EC network formation does not include any kind of secondary structures that will produce a distinct change in the rheological properties or a change in the thermal heat flow.[34] This observation suggests that EC oleogels are formed *via* unique structural architecture in comparison to other known thermo-reversible gels discussed above. Further in-depth studies on the gel's nano- and micro-structure are required in order to understand the characteristic building blocks assembling the gel network.

7.5   Ethyl-cellulose Gel Properties

The polymeric nature of EC and the physical bonds forming its gel network produce unique gel properties, which can be altered and tuned for a specific application. Various strategies have been suggested to control gel properties, such as solvent quality, addition of small surface-active molecules, and thermal treatment.

7.5.1 The Effect of Polymer Concentration and Molecular Weight on the Gel Properties

According to the rubber elasticity theory, polymer molecular weight and concentration have a significant impact on gel properties.[1] Positive relation between the gel hardness and the EC molecular weight and concentration has been detected by several researchers.[35,37,38]

A predictive model was developed in order to evaluate the mechanical properties of EC oleogels based on several parameters, *i.e.* polymer molecular weight, polymer concentration, oil type, and the addition of surfactant.[48] The sole effect of polymer concentration was successfully fitted using a power law model, regardless of the other parameters, *i.e.* oil type, polymer molecular weight and surfactant type. Response surfaces were developed using the EC mass fraction and the EC/surfactant ratio as variables while the remaining parameters, *i.e.* oil type, polymer molecular weight and surfactant type, remained fixed. The EC gel mechanical strength was fitted to a single universal equation:[48]

$$Z = X^{\mu}(a + bY + cY^2) \tag{7.1}$$

Whereas Z represents the normalized gel strength, and X and Y represent the EC and surfactant concentrations, respectively. The scaling factor μ and the coefficients a, b, and c were obtained from the fitting of each independent response surface using the fixed parameters: oil type, polymer molecular weight and surfactant type.

7.5.2 The Effect of Oil Type on the Gel Properties

The effect of oil type was previously reported by Zetzl *et al.*[41] and Laredo *et al.*[46]. Different oils, *i.e.* canola, soybean, and flaxseed, produced oleogel formulations with different mechanical properties in the order of canola < soybean < flaxseed (Figure 7.4). The fatty acid composition of these oils consists of more than 50% oleic (18:1), linoleic (18:2) and linolenic (18:3) acids for canola, soybean, and flaxseed oil, respectively. Therefore, it was concluded that the higher unsaturation levels of fatty acids in the oil produce stronger gels.[41] Further analysis done using Fourier transform infrared spectroscopy (FTIR) and Raman spectroscopy suggested that higher unsaturation levels produce higher molar volume of the triacylglycerides (TAGs) molecules in the solvent. This conformational volume expansion promotes polymer–polymer interactions, *i.e.* an increase in the number of junction zones, thus producing stronger gels.[46]

More recently, it was suggested that the oil polarity is the source of this solvent effect on gel properties.[36] The polarity of the oil phase can be altered using various approaches. Oxidation of lipids in the oil phase was suggested as a way to increase oil polarity.[49] Oil samples were thermally treated using various holding times at 140 °C in order to obtain various oxidation levels. Oil oxidation was measured using peroxide and 2-thiobarbituric acid (TBA)

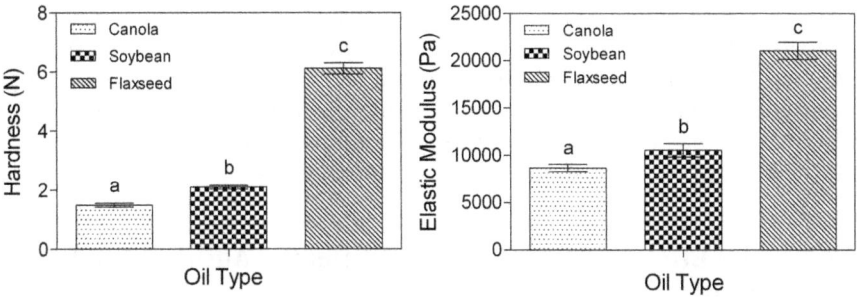

Figure 7.4   First compression hardness and elastic modulus from texture profile analysis of 10 wt% EC 10 cP made with different vegetable oils. Values followed by the same letter are not significantly different ($p > 0.05$). Reproduced from ref. 41 with permission from The Royal Society of Chemistry.

values while the total polar components (TPC) were monitored using a commercially available unit for deep fryer analysis. An increase in TPC values with the holding time at 140 °C were observed suggesting a relation between the oxidation level and the oil polarity. In addition, a direct correlation between the holding time and oxidation level with the gel strength was detected. The effect of oil polarity on EC gel properties was further explored by altering solvent properties using two strategies: mixing different oil compositions and addition of small surface-active molecules.[36] Soybean oil polarity was altered by supplementation of either castor oil or mineral oil at different ratios. Castor oil consists of high levels of ricinoleic acid, an 18-carbon monounsaturated fatty acid with a hydroxyl functional group at C12, alluding to a very high polar oil. Mineral oil, on the other hand, consists of high levels of *n*-alkanes, which do not possess any polar functional groups, thus forming an apolar oil. Another approach was taken with the addition of free fatty acids and fatty alcohol in order to increase oil polarity. Both strategies produced a dose–response behavior where an increase in gel mechanical properties was observed in response to the increase in oil polarity. However, the extent of the effect varied using the different strategies.[36]

The effect of solvent polarity on EC gel mechanical properties was related to the role of hydrogen bonds in the gel network formation and possible polymer–solvent interactions with free hydroxyl end groups. It was suggested that during polymer dissolution higher solvent polarity provides an improved environment for polymer extension thus allowing better polymer–polymer interactions during gelation. The role of solvent–polymer interactions was further examined using Hansen solubility parameters (HSPs).[36,50] This analysis relies on the evaluation of the key energies involved during molecule dissolution, *i.e.* dispersive (δ_d), polar (δ_p), and hydrogen bonding (δ_h). The total HSP value (δ_t) is defined as:[50]

$$\delta_t = \sqrt{\left(\delta_d\right)^2 + \left(\delta_p\right)^2 + \left(\delta_h\right)^2} \tag{7.2}$$

The closer the HSP values of the solute and solvent are, the greater their compatibility and thus solubility. The HSP values of the EC and the oil were matched by adding castor or mineral oil. Gel formulations with matched HSP values produced significantly stronger gels. It was assumed that at this composition the polymer is in its most extended conformation, providing an improved platform for polymer–polymer interactions.[36]

7.5.3 The Effect of Surface-active Molecule Addition on Gel Properties

The addition of surface-active molecules that can crystalize and form a crystal network was studied using surfactants,[41,48,51] monoglycerides (MAG),[52] free fatty acids, and fatty alcohol[36] and their mixtures.[53] The effect of surfactant addition on EC-based oleogels was examined using four different surfactants: sorbitan monostearate (SMS), sorbitan monooleate (SMO), glycerol monostearate (GMS), and glycerol monooleate (GMO).[51] It was demonstrated that the addition of surfactant significantly increases the oleogel strength (Figure 7.5). This behavior was attributed to possible surfactant–polymer physical interactions, which can stiffen the polymer network. Different levels of strength enhancement were demonstrated by the different surfactants owing to the surfactant melting temperature, which produced either liquid or crystal form at room temperature. Recent studies that utilized EC gels as fat replacers in meat products demonstrated the ability of surfactant addition, SMS, to improve textural attributes of meat products using tensile and sensory analysis.[54–56]

Oleogel systems formed by a MAG network were successfully stabilized using EC. According to this study, synergistic interactions between EC and MAG allude to stronger gels with lower oil loss.[52] Free fatty acids (FA), fatty alcohols (FO) and their mixtures produced a dose–response oleogel system with improved mechanical properties. This behavior was attributed to

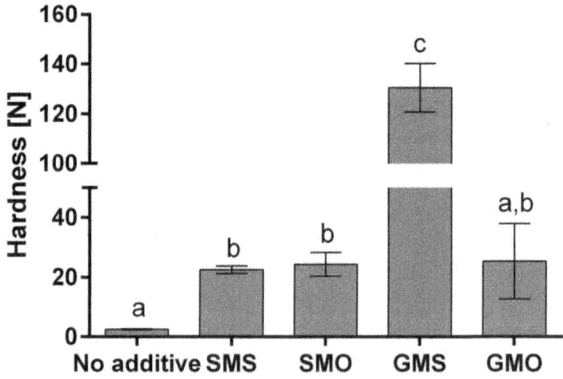

Figure 7.5 Gel hardness measured with texture analyzer instrument using 11 wt% EC 45 cP with the addition of 3.67 wt% surfactant. Values followed by the same letter are not significantly different ($p > 0.05$).

possible molecular interactions between FA and FO molecules with the EC backbone, thus enhancing the network stiffness.[36,53] Previous studies on EC gel systems demonstrated significantly higher gel strength while using flaxseed oil.[41] This behavior was attributed to the unrefined, cold-pressed processing of the flaxseed oil, which maintained a high content of free fatty acids in the oil. In order to validate this hypothesis, free fatty acids were removed from flaxseed oil using a dry neutralization process. EC gels prepared using neutralized and neat flaxseed oil were prepared and their mechanical properties were evaluated. Significantly softer gels were obtained with the neutralized oil, implying a direct relation between the free fatty acid content and the final gel strength.[36] According to these studies, the combination of the viscoelastic EC network with the crystal network can potentially provide a way to control the gel properties. Moreover, the gel properties can be optimized more accurately to mimic the functional attributes of traditional edible fats.[53]

7.5.4 The Effect of Thermal Treatment on Gel Properties

The effect of temperature on the mechanical properties of EC gels can be related to the role of hydrogen bonds in the gel network formation. Hydrogen bonds are strongly affected by temperature, *i.e.* they tend to break at elevated temperatures. The effect of setting temperature on the mechanical properties of EC gels was examined by incubating samples at a wide temperature range between −20 °C and 80 °C for 24 h, followed by equilibration at 25 °C for an additional 24 h. The texture profile analysis curves obtained demonstrated a decrease in the gel hardness as well as a decrease and eventual disappearance of the fracture peak, with a decrease in the gelation setting temperature.[57] This setting temperature effect was detected using two different EC molecular weights (Figure 7.6a). These results suggest that

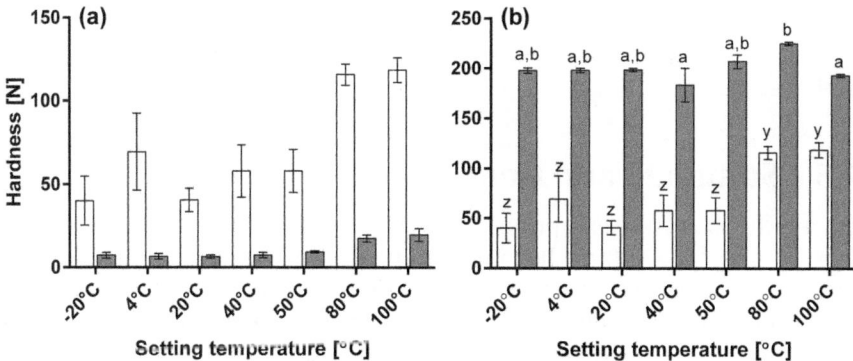

Figure 7.6 Effect of setting temperature on the hardness of EC gels using texture analyzer. (a) 15 wt% of EC 20 cP (white bars) and EC 45 cP (gray bars) in canola oil. (b) EC 20 cP in canola oil, 15 wt% (white bars) and 17 wt% (gray bars). Values followed by the same letter are not significantly different ($p > 0.05$).

setting at higher temperatures produces stronger and more brittle gels, possibly owing to an enhanced ability to form junction zones between polymer chains. It was concluded that setting at lower temperatures leads to fast cooling, which produces rapid and disordered association of hydrogen bonds between the polymer chains. On the other hand, setting at higher temperatures provides a slower association process, which results in a more efficient hydrogen bonding network.[57] In addition, it appears that the effect of setting temperature was significantly minimized with higher EC concentrations (Figure 7.6b) or while dissolving the EC using higher temperatures, *i.e.* above its melting temperature.[57] This observation suggests that increasing the amount of hydrogen bonding sites, while using higher concentration or by exposing more hydrogen bonding sites by melting the polymer, leads to the disappearance of the setting temperature effect.[57]

7.5.5   Ethyl-cellulose Oleogel Fractionation

Specific formulations of EC oleogels tend to separate into two distinct phases; a soft interior core surrounded by a firm exterior sheath. This behavior is referred to as fractionation, which produces inhomogeneity in the mechanical properties of the material. The control of this behavior is achieved by the polymer concentration, polymer type, *i.e.* polymer molecular weight, oil type, the addition of surfactants, and thermal conditions. For each EC type, fractionation was not detected while using concentrations above a typical critical concentration. This concentration seems to depend on the polymer molecular weight, type of surfactant, and type of oil. It was found that SMS and SMO induce the fractionation process while GMO does not, suggesting that the surfactant head group plays a role in this process. In addition, the cooling procedure was found to strongly affect the fractionation process; rapid setting of the gel increased the fractionation effect while slow cooling produced more homogenous samples. The source of this effect is not fully understood. However, it was shown that both regions are chemically equivalent suggesting that the role of temperature gradient in the sample is critical during setting.[58]

7.6   Summary and Conclusion

Ethyl-cellulose is an exceptional polysaccharide demonstrating a unique array of properties, thus allowing it to be used in various applications. In order to maximize its properties wisely it is important to understand its native properties as well as its properties in the oleogel form. The current manuscript reviews the latest studies and research published on EC in its different forms and formulations.

Ethyl-cellulose's semi-crystalline nature was demonstrated by the presence of melting temperature in addition to glass transition measured by DSC and the typical *d*-spacing length analyzed by an XRD instrument. Such characteristics suggest that the derivatization of cellulose maintains some of its original crystalline features.

The ability of EC to dissolve in oil and form a gel network was demonstrated using a temperature sweep experiment in a rheometer. The formation of the gel network was related to the constitution of physical polymer–polymer interactions, *i.e.* hydrogen bonds between EC chains. Polymer-solvent interactions, such as dipole–dipole and hydrogen bonds, were found to affect polymer dissolution in oil, thus affecting the final gel properties. Additional parameters such as temperature, solvent polarity, and addition of small surface-active molecules were found to significantly affect the EC gels' mechanical properties. In conclusion, EC gels' properties can be altered and tuned for a specific application using a variety of different approaches.

References

1. M. Rubinstein and R. H. Colby, *Polymer Physics*, Oxford University Press Inc., 2003.
2. A. H. Clark, in *Physical Chemistry of Foods*, ed. H. G. Schwartzberg and R. W. Hartel, Marcel Dekker Inc., New York, 1992, pp. 263–306.
3. J. M. Aguilera and D. V. Stanley, *Microstructural Principles of Food Processing and Engineering*, Aspen publisher Inc., 1999.
4. J. Raeburn, A. Z. Cardoso and D. J. Adams, *Chem. Soc. Rev.*, 2013, **42**, 5143–5156.
5. L. E. Buerkle and S. J. Rowan, *Chem. Soc. Rev.*, 2012, **41**, 6089–6102.
6. A. I. Blake, E. D. Co and A. G. Marangoni, *J. Am. Oil Chem. Soc.*, 2014, **91**, 885–903.
7. S. S. Sagiri, V. K. Singh, K. Pal, I. Banerjee and P. Basak, *Mater. Sci. Eng., C*, 2015, **48**, 688–699.
8. A. López-Martínez, J. A. Morales-Rueda, E. Dibildox-Alvarado, M. A. Charó-Alonso, A. G. Marangoni and J. F. Toro-Vazquez, *Food Res. Int.*, 2014, **64**, 946–957.
9. A. Bot and W. G. M. Agterof, *J. Am. Chem. Soc.*, 2006, **83**, 513–521.
10. E. D. Co and A. G. Marangoni, *J. Am. Chem. Soc.*, 2012, **89**, 749–780.
11. M. Suzuki and K. Hanabusa, *Chem. Soc. Rev.*, 2010, **39**, 455–463.
12. A. S. Hoffman, *Adv. Drug Delivery Rev.*, 2012, **64**, 18–23.
13. *Ethocel: Ethylcellulose polymers technical handbook*, Dow Cellulosics, Dow Chemical Company, 2005.
14. A. Vintiloiu and J.-C. Leroux, *J. Controlled Release*, 2008, **125**, 179–192.
15. K. J. Skilling, F. Citossi, T. D. Bradshaw, M. Ashford, B. Kellama and M. Marlow, *Soft Matter*, 2014, **10**, 237–256.
16. S. Murdan, *Expert Opin. Drug Delivery*, 2005, **2**, 489–505.
17. W. Koch, *Ind. Eng. Chem.*, 1937, **29**, 687–690.
18. M. Davidovich-Pinhas, E. Co, S. Barbut and A. G. Marangoni, *Cellulose*, 2014, **21**, 3243–3255.
19. M. M. Crowley, B. Schroeder, A. Fredersdorf, S. Obara, M. Talarico, S. Kucera and J. W. McGinity, *Int. J. Pharm.*, 2004, **269**, 509–522.
20. M. Tarvainen, R. Sutinen, S. Peltonen, H. Mikkonen, J. Maunus, K. Vähä-Heikkilä, V.-P. Lehto and P. Paronen, *Eur. J. Pharm. Sci.*, 2003, **19**, 363–371.

21. J. Desai, K. Alexander and A. Riga, *Int. J. Pharm.*, 2006, **308**, 115–123.
22. J.-X. Guo and D. G. Gray, *Macromolecules*, 1989, **22**, 2082–2086.
23. P. Shi, Y. Li and L. Zhang, *Carbohydr. Polym.*, 2008, **72**, 490–499.
24. F. R. Lupi, D. Gabriele, V. Greco, N. Baldino, L. Seta and B. d. Cindio, *Food Res. Int.*, 2013, **51**, 510–517.
25. H. M. Schaink, K. F. v. Malssen, S. Morgado-Alves, D. Kalnin and E. v. d. Linden, *Food Res. Int.*, 2007, **40**, 1185–1193.
26. D. K. Shah, S. S. Sagiri, B. Behera, K. Pal and K. Pramanik, *J. Appl. Polym. Sci.*, 2013, 793–805.
27. S. S. Sagiri, B. Behera, R. R. Rafanan, C. Bhattacharya, K. Pal, I. Banerjee and D. Rousseau, *Soft Mater.*, 2014, **12**, 47–72.
28. A. R. Patel, N. Cludts, M. D. B. Sintang, A. Lesafferb and K. Dewettincka, *Food Funct.*, 2014, **5**, 2833–2841.
29. A. R. Patel, P. S. Rajarethinem, N. Cludts, B. Lewille, W. H. D. Vos, A. Lesaffer and K. Dewettinck, *Langmuir*, 2015, **31**, 2065–2073.
30. A. I. Romoscanu and R. Mezzenga, *Langmuir*, 2006, **22**, 7812–7818.
31. A. d. Vries, J. Hendriks, E. v. d. Linden and E. Scholten, *Langmuir*, 2015, **31**, 13850–13859.
32. A. d. Vries, A. Wesseling, E. v. d. Linden and E. Scholten, *J. Colloid Interface Sci.*, 2017, **486**, 75–83.
33. J. M. Aiache, P. Gauthier and S. Aiache, *Int. J. Cosmet. Sci.*, 1992, **14**, 228–234.
34. M. Davidovich-Pinhas, S. Barbut and A. G. Marangoni, *Carbohydr. Polym.*, 2015, **117**, 869–878.
35. M. R. Ceballos, V. Brailovsky, K. L. Bierbrauer, S. L. Cuffini, D. M. Beltramo and I. D. Bianco, *Food Res. Int.*, 2014, **62**, 416–423.
36. A. G. Gravelle, M. Davidovich-Pinhas, A. K. Zetzl, S. Barbut and A. G. Marangoni, *Carbohydr. Polym.*, 2016, **135**, 169–179.
37. E. Lizaso, M. E. Munoz and A. Santamaría, *Macromolecules*, 1999, **32**, 1883–1889.
38. P. W. S. Heng, L. W. Chan and K. T. Chow, *Pharm. Res.*, 2005, **22**, 676–684.
39. L. Bruno, t. Kasapis, V. Chaudhary, K. T. Chow, P. W. S. Heng and L. P. Leong, *Carbohydr. Polym.*, 2011, 644–651.
40. L. Bruno, S. Kasapis and P. W. S. Heng, *Carbohydr. Polym.*, 2012, **88**, 382–388.
41. A. K. Zetzl, A. G. Marangoni and S. Barbut, *Food Funct.*, 2012, **3**, 327–337.
42. J. Lefebvre and J.-L. Doublier, in *Polysaccharides: Structural Diversity and Functional Versatility*, ed. S. Dumitriu, Marcel Dekker, New York, 1998, pp. 357–409.
43. K. Kobayashi, C.-i. Huang and T. P. Lodge, *Macromolecules*, 1999, **32**, 7070–7077.
44. A.-L. Kjøniksen, B. Nystrom and B. Lindman, *Macromolecules*, 1998, **31**, 1852–1858.
45. P. Harris, *Food Gels*, Elsevier Science Publishers Ltd, Essex, 1990.
46. T. Laredo, S. Barbut and A. G. Marangoni, *Soft Matter*, 2011, **7**, 2734–2743.

47. M. Davidovich-Pinhas and A. G. Marangoni, *Annu. Rev. Food Sci. Technol.*, 2016, 7, 4.1–4.27.

48. A. J. Gravelle, S. Barbut, M. Quinton and A. G. Marangoni, *J. Food Eng.*, 2014, **143**, 114–122.

49. A. J. Gravelle, S. Barbut and A. G. Marangoni, *Food Res. Int.*, 2012, **48**, 578–583.

50. T. A. Stortz and A. G. Marangoni, *Green Chem.*, 2014, **16**, 3064–3070.

51. M. Davidovich-Pinhas, S. Barbut and A. G. Marangoni, *Carbohydr. Polym.*, 2015, **127**, 355–362.

52. A. Lopez-Martínez, M. A. Charó-Alonso, A. G. Marangoni and J. F. Toro-Vazquez, *Food Res. Int.*, 2015, **72**, 37–46.

53. A. J. Gravelle, M. Davidovich-Pinhas, S. Barbut and A. G. Marangoni, *Food Res. Int.*, 2017, **91**, 1–10.

54. S. Barbut, J. Wood and A. G. Marangoni, *Meat Sci.*, 2016, **122**, 155–162.

55. S. Barbut, J. Wood and A. G. Marangoni, *J. Food Sci.*, 2016, **81**, 2183–2188.

56. S. Barbut, J. Wood and A. G. Marangoni, *Meat Sci.*, 2016, **122**, 84–89.

57. M. Davidovich-Pinhas, A. J. Gravelle, S. Barbut and A. G. Marangoni, *Food Hydrocolloids*, 2015, **46**, 76–83.

58. A. J. Gravelle, S. Barbut and A. G. Marangoni, *Food Funct.*, 2013, **4**, 153–161.

CHAPTER 8

Proteins as Building Blocks for Oil Structuring

E. SCHOLTEN*[a,b] AND A. DE VRIES[a,b]

[a]Wageningen University, Physics and Physical Chemistry of Foods, Bornse Weilanden 9, 6708 WG, Wageningen, The Netherlands; [b]Top Institute Food and Nutrition, Nieuwe Kanaal 9A,6709 PA, Wageningen, The Netherlands
*E-mail: elke.scholten@wur.nl

8.1 Introduction

In recent years, research performed in the field of oil structuring has increased substantially. More and more molecules have been identified that can structure oil, and the list of potential organogelators or oil gelators is increasing. Besides many chemically synthesized molecules, more focus has been directed towards the identification of biocompatible gelators. Two main classes of oil gelators can be identified: (i) the low molecular weight oleogelators (LMWOG) and (ii) the biopolymeric gelators. Gelators of low molecular weight are often of amphiphilic nature, and their self-assembly behaviour in hydrophobic environments leads to the formation of three-dimensional structures, such as inverse micelles, cylindrical micelles or bilayers. When connected at larger length scales, these structures provide sufficient network formation to solidify the oil. Known examples of such LMWOG are waxes,[1] phytosterols,[2] fatty acids,[3] oligopeptides,[4] lecithin,[5] and combinations of LMWOG, such as fatty acids/fatty alcohols,[6,7] and lecithin/tocopherol.[8] Gelators of biopolymeric original are less common as,

Food Chemistry, Function and Analysis No. 3
Edible Oil Structuring: Concepts, Methods and Applications
Edited by Ashok R. Patel
© The Royal Society of Chemistry 2018
Published by the Royal Society of Chemistry, www.rsc.org

due to their predominant hydrophilic nature, their dispersibility in hydrophobic solvents is limited. One exception is the cellulose derivative ethylcellulose, which contains a sufficient amount of hydrophobic groups to be dispersible in oil, and network formation can be obtained upon heating above the glass transition temperature.[9,10] In other research, it has been shown that the polysaccharide chitin can be used in its native form, or as modified chitin nanocrystals.[11] Surfactants may also be used to aid the dispersibility of hydrophilic biopolymers in hydrophobic media. A more indirect approach that has been used to introduce polysaccharides in oil is the emulsion-templated approach.[12,13] In this case, surface-active ingredients are used to provide interfacial stabilization, and after evaporation of water, a gel with a very high oil content is obtained, also known as high internal phase emulsions (HIPE).[14] Although oil volume fractions of 99% can be obtained, the continuous phase is still of hydrophilic nature, and upon hydration the system returns to an oil-in-water emulsion. Polysaccharides may be added to the system to change the interfacial characteristics and thereby the network properties of the dried emulsions.[15] Although a long list of organogelators are known, they often fail to be incorporated in foods, and the commercialization of oleogels is hampered.

The requirements for a good organogelator depend on the envisioned application. For food applications, the gelator should of course be food-grade, but also cheap, widely available and accepted by ever more critical consumers. Oleogels based on food-grade biopolymers are scarce, owing to their limited dispersibility in oil. Next to polysaccharides, proteins would be an ideal candidate as an oil gelator. Proteins are widely available, can be obtained from different sources, are relatively cheap, and do not have any detrimental effects regarding health. In aqueous systems, proteins give a large diversity in terms of gelling mechanisms and the type of networks formed. Gelation can either be induced by heat treatment or by removing electrostatic repulsion between the proteins by the addition of salt or changing the pH towards the isoelectric point of the proteins. Depending on the balance between repulsive electrostatic and attractive hydrophobic interactions, a dense or more open network can be formed. A high degree of attractive interactions leads to the formation of random aggregates, leading to more brittle and opaque gels. In case of increased repulsive interactions, the formation of fibrillary structures is induced, leading to more elastic and transparent gels.[16] The gel characteristics can be controlled very easily by external environmental conditions, such as pH, ionic strength, and temperature. Even though the use of proteins as water gelators is very common, their potential as a direct gelator for oil has barely been investigated. Direct addition of a protein powder to oil does not lead to the formation of an oleogel as the proteins will not be distributed finely throughout the continuous oil phase. To improve the applicability for proteins to be used as a gelator for oil or other hydrophobic solvents, new approaches are required. Recently, it has been shown that proteins can be dispersed in oil when using a solvent exchange approach.[17,18]

8.2 Solvent Exchange Route

Direct addition of proteins to oil leads to a large degree of agglomeration and precipitation of protein flocs. To increase the dispersibility in oil, two steps need to be taken into account:

- Heat treatment to create protein building blocks with increased hydrophobicity
- Use of a solvent exchange procedure to prevent agglomeration of the proteins.

Owing to the tertiary structure of proteins, their hydrophobic sites are buried within their core and the outside of the proteins has more hydrophilic groups, hence, the good dispersibility in water. When heated, the proteins denature, after which the more hydrophobic groups become exposed to the outside, increasing the hydrophobicity of the proteins. This exposure leads to increased hydrophobic interactions between the proteins and, depending on the protein concentration, either small protein aggregates or a space continuous network can be formed. The increased hydrophobicity of the protein aggregates contributes to more protein–solvent (oil) interactions, and therefore higher dispersibility in oil or other hydrophobic solvents is to be expected. To increase the dispersibility even further, a solvent exchange procedure can be used. In such procedures, the solvent is changed stepwise from water to oil, using an intermediate solvent that is miscible both with water and oil. This procedure changes the polarity of the solvent gradually, allowing the proteins to adjust slowly to their environment. Additionally, the solvent should be rather volatile in order to remove the intermediate solvent from the final oleogel by simple evaporation. The choice of the intermediate solvent is therefore limited, and acetone and tetrahydrofuran (THF) have been shown to work well.[17] During this procedure, the solvents are removed from the sample either by decanting or centrifugation, and the solvent is then replaced by another solvent with a slightly lower degree of polarity. The solvents are changed until the samples are surrounded by oil only. The initial state of the protein building blocks is determined in the aqueous phase. For high concentrations of proteins, the main building block is a macroscopically large hydrogel (mm–cm scale), whereas at low concentrations, small proteins aggregates can be made (submicrometer–micrometer scale). The properties of the protein hydrogels and the protein aggregates (size, network type, *etc.*) depend a lot on the environmental conditions used (pH, ionic strength, *etc.*). The properties of the initial protein building block determine to a large extent the macroscopic properties of the final oleogel. As the properties of the protein building blocks can be easily tuned in water and there is extensive knowledge on how to control the network formation, it is possible to also control the properties of the final protein oleogels.

8.3 Protein Oleogels Prepared from Protein Hydrogels

Protein hydrogels can be made using different approaches, including acid-induced and heat-induced gelation. To create hydrogels suitable for the development of protein oleogels, heat-induced aggregation is preferred. During heating, the hydrophobic interactions lead to strong interactions between the proteins as a result of the exposure of hydrophobic groups. Additionally, heat catalyses the formation of disulfide bonds between the proteins. In this case, the protein network is hold together by both physical (hydrophobic) interactions and covalent (disulfide) bonds. The disulfide bonds provide more permanent stabilization to the protein network. The type of network formed is dependent on the environmental conditions. Figure 8.1 shows examples of whey protein hydrogels prepared at pH 7 and different ionic strengths. At this pH, the proteins exhibit a negative charge as the isoelectric point of whey proteins is approximately 5.2. The accompanying electrostatic

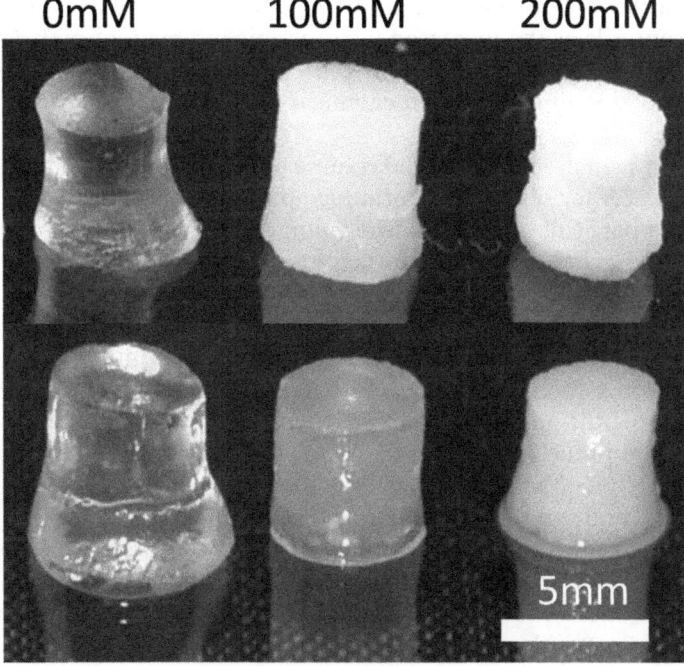

Figure 8.1 Appearance of WPI hydrogels (upper pictures) and oleogels (lower pictures) after the solvent exchange. The composition of the hydrogels were 15% WPI with 0 mM, 100 mM and 200 mM NaCl.[17] Reprinted with permission from de Vries, A.; Hendriks, J.; van der Linden, E.; Scholten, E. Protein Oleogels from Protein Hydrogels *via* a Stepwise Solvent Exchange Route. *Langmuir*, 2015, **31** (51), 13850–13859. Copyright 2015 American Chemical Society.

repulsion between the proteins limits direct mutual contact, but the stronger hydrophobic interactions lead to the aggregation of the proteins. The proteins aggregate in the form of a fine-stranded network with a limited degree of contact between the proteins. As the structural features within the gel are smaller than the wavelength of light, the gels are transparent. When salt is added to decrease the electrostatic repulsion, the increase in contact area between the proteins leads to the formation of random dense aggregates of larger sizes, and therefore an opaque gel is formed.

These gels are then subjected to the solvent exchange procedure, and the appearance of the final gels is similar to that of the initial hydrogels. Colourless gels turn into yellow gels owing to the presence of the oil, but the transparency or the opaqueness remains. This indicates that the structure of the protein network remains unchanged when transformed into an oleogel. The interactions in the protein network are apparently strong enough that changes in the solvent quality do not lead to rearrangements in the protein network. This suggests that the disulfide bridges may play an important role in the stability of the protein network. Hydrophobic interactions commonly responsible for the network formation in aqueous environments do not play a prominent role as the increased hydrophobic interactions with the hydrophobic solvent actually limit protein–protein interactions. This suggests that suitable proteins should contain chemical groups capable of forming disulfide bridges.

As can be seen in Figure 8.1, the volume of the gels slightly changed when replacing water for oil as the continuous phase. In the case of the gel with the opaque particulate network, the volume change is limited, but for the fine-stranded gels, the change in volume is more pronounced. The total amount of oil that can be incorporated in the protein network depends on the swelling capacity of the gel. Swelling is a result of the balance between the tendency of polymer strands to dissolve in the continuous phase and an elastic retractive force of the network.[19] The oil holding capacity (or ability) therefore depends on the flexibility of the protein network.[17] Fine stranded gels are more prone to swelling than networks composed of more random aggregates. The oil binding capacity is therefore related to the strength of the network, and can therefore be linked to the modulus of the gel network. Research has shown that for protein gels with 15 wt% protein, the gels swell up by a factor of 2 when the Young's modulus of the gels is around 10 kPa, but this is reduced to a swelling capacity of 1.3 when the Young's modulus exceeds 50 kPa. Besides the swelling ability of the gels, the quality of the different intermediate solvents also has an effect on the total amount of oil that can be incorporated in the protein gels. As the protein network is still more hydrophilic than hydrophobic, the swelling ability in water is still larger than the swelling ability in oil. When initially the hydrogels are placed in mixtures containing a large amount of water, the hydrogels initially swell and the volume increases. When the polarity is then decreased, the solvent quality decreases and as a result the hydrogels shrink again. The changes in the polarity of the solvent during the different steps in the process also have an effect on the final change in volume. For small changes in the polarity, the oil binding capacity was found to be larger than in cases where larger differences in polarity for proceeding solvents were used.[17] This indicates that the total oil binding capacity is also related to a kinetic component. When the polarity of the solvent is lowered too quickly, the proteins do not get enough time to "adjust" to the new environment, and the gel shrinks and collapses.

Owing to the change in the polarity of the continuous phase, the balance between the protein–protein and protein–solvent interactions changes. As the protein–solvent interactions are less favourable, the protein–protein interactions become stronger, which is reflected in the mechanical properties of such gels. Figure 8.2 gives an overview of the mechanical properties

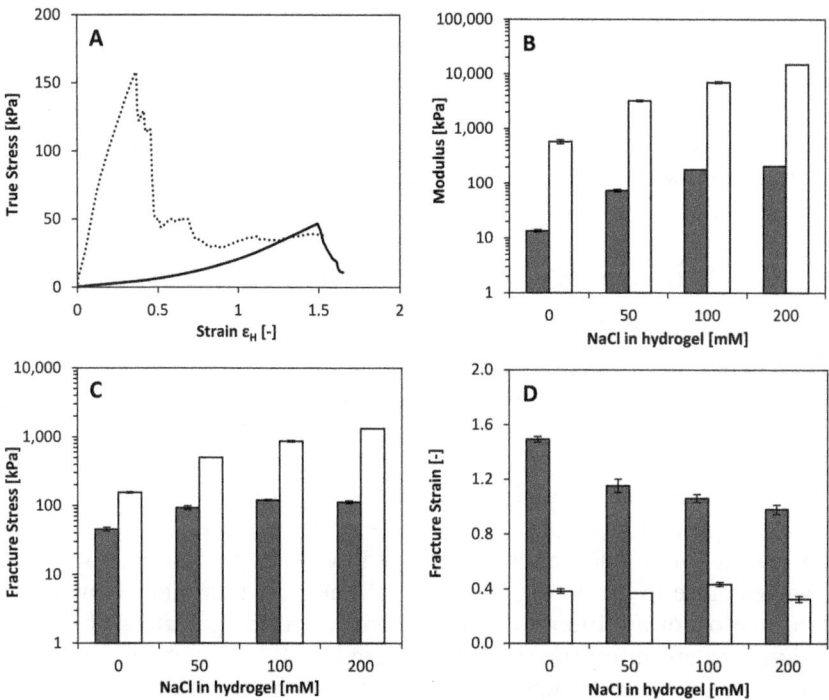

Figure 8.2 Large deformation properties of WPI hydrogels and oleogels after the solvent exchange procedure using acetone as the intermediate solvent. A: Stress–strain curve for a representative 15% 0 mM NaCl hydrogel (solid line) and the resulting oleogel after the solvent exchange (dotted line). B–D: Modulus, fracture stress and fracture strain of hydrogels (grey bars) prepared at 15% WPI with different NaCl concentrations and the resulting oleogels (white bars) after the solvent exchange. Error bars represent the standard deviation of duplicate measurements.[17] Reprinted with permission from de Vries, A.; Hendriks, J.; van der Linden, E.; Scholten, E. Protein Oleogels from Protein Hydrogels *via* a Stepwise Solvent Exchange Route. *Langmuir*, 2015, **31** (51), 13850–13859. Copyright 2015 American Chemical Society.

of both the protein hydrogels and the protein oleogels. Figure 8.2a gives an example of the specific stress–strain curve, from which the modulus (slope in the initial linear region, Figure 8.2b), the fracture stress (Figure 8.2c) and the fracture strain (Figure 8.2d) are obtained. As can be seen from Figure 8.2a, the shape of the curve changes to a large extent. The hydrogels can be deformed easily with a low amount of stress, indicating an elastic network with a high fracture strain. The protein oleogels, on the other hand, show a much faster increase in the stress for smaller deformation, reflecting a more brittle behaviour. Although the fracture stress does not change much (Figure 8.2c), the fracture strain of the oleogels decreases a lot compared to the hydrogels (Figure 8.2d). The gel stiffness, represented by the modulus, is roughly two orders of magnitude larger for the protein oleogels, which highlights the large change within the interactions in the protein network.

These results show that the solvent exchange route is an effective route to create protein oleogels. The structure and the flexibility of the protein network can be altered in the aqueous phase by changing the concentration and the charge density of the proteins. The resulting protein oleogels are found to be much stiffer than the protein hydrogels, and resemble the hardness of fats structured with fat crystals (saturated fats).

8.4   Protein Oleogels Prepared from Protein Aggregates

As seen in the previous paragraph, the initial protein building blocks remain the same after the solvent exchange route. The building blocks can be prepared at any length scale, and these can be used to create different types of networks. For example, the building blocks can be prepared at submicrometer length scale, for which heat-induced gelation can be used to create such small protein aggregates. In the case of whey proteins, protein aggregates with a size of 200 nm can be formed by heating a protein solution at pH values close to the isoelectric point.[20] The heat treatment may lead to the formation of additional disulfide bonds, providing extra stabilization through covalent interactions in the aggregates. When such aggregates are present in a hydrophilic (aqueous) environment, they are finely dispersed in water, as is visualized by the microscopy image in Figure 8.3A. Furthermore, in this case, the solvent exchange procedure is required for gel formation. When a conventional freeze-drying method is used and the dried aggregates are then dispersed in oil, no gel is formed and a very liquid-like system is obtained (Figure 8.3B and E). Apparently, a large degree of clustering leads to limited dispersibility and collapse of the network. When a solvent exchange procedure was applied, the network appears to consist of strong flocs of the protein aggregates, and therefore a gel is obtained through sufficient protein–protein interactions (Figure 8.3C and F).

The difference in dispersibility is related to the degree of aggregation that the particles display under these conditions. Upon drying, clustering of the

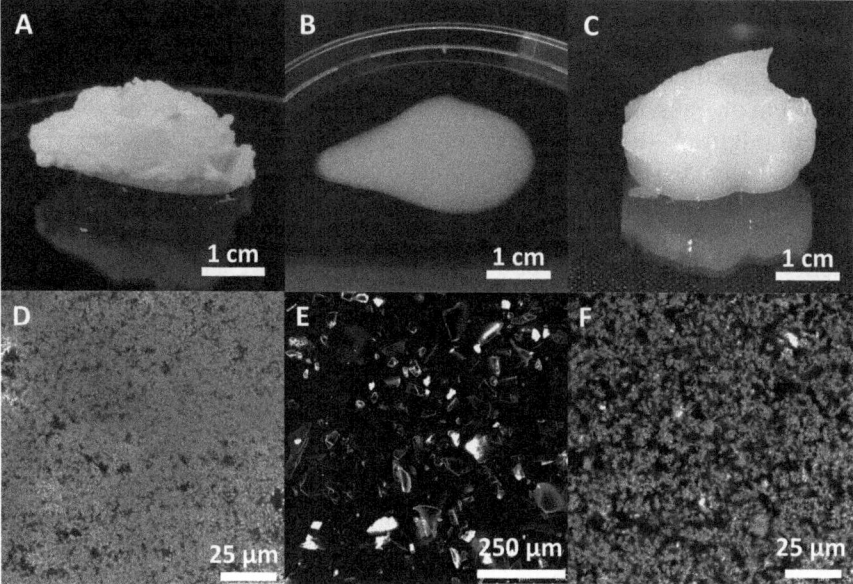

Figure 8.3 (A) Appearance of heat-set WPI aggregates after centrifugation, (B) dispersion of freeze-dried WPI aggregates in sunflower oil and (C) WPI aggregates in sunflower oil *via* solvent exchange with corresponding CLSM micrographs (D–F). To visualize the protein aggregates for CLSM, samples were stained with rhodamine B.[18] Reprinted from *Journal of Colloid and Interface Science*, **486**, de Vries, A.; Wesseling, A.; van der Linden, E.; Scholten, E., Protein oleogels from heat-set whey protein aggregates. 75–83. Copyright 2017, with permission from Elsevier.

particles is very prominent, and the particle clusters present in the oil show a size of roughly 10–100 micrometers.[18,21] When the solvent exchange method was applied, the aggregate size was found to be 200 nm, the same as the initial size of the protein aggregates prepared in water.[18] This shows that during the exchange procedure, the clustering of the protein aggregates seems to be completely prevented, and therefore effective network formation can be accomplished. Compared to the aqueous system (Figure 8.3A and D), the oleogel (Figure 8.3C and F) seems to have a more clustered network, which is also visible in the appearance of the gels.

Apparently, the networks of protein in a hydrogel and an oleogel are quite similar, which was also observed for the macroscopic networks discussed in the previous paragraph. Although the networks appear to be similar, the interactions involved in the protein network formation must be of different origin. In hydrogels, the protein–protein interactions are mostly of hydrophobic nature, and the interactions between the proteins and the aqueous solvent are of hydrophilic nature. In the oleogels, however, we have the opposite situation. In this case, protein–solvent interactions are of hydrophobic nature, which leaves the protein–protein interactions to be of a hydrophilic nature. The network formation and the strength of the interactions in the network are a result of these hydrophobic and hydrophilic interactions. The rheological properties of the gels depend on the strength of the interactions within the protein network. Figure 8.4 shows the storage and loss modulus of protein oleogels at two concentrations: 4.1 wt% (Figure 8.4A), and 6.1 wt% protein (Figure 8.4B). Increasing the protein concentration from 4.1 wt% to 6.1 wt% increased the storage modulus, G', with an order of magnitude, indicating that the gel becomes stronger.

The gels show a predominant elastic behaviour as $G' > G''$ over a large frequency (ω) range.[18] Already at a concentration of 3 wt% proteins, an elastic behaviour is found, which shows that the protein aggregates are very effective at creating a protein network. A measure of the elasticity can be taken from the relation $G' \sim \omega^n$, in which the exponent n can be related to the type of gel formed. For $n = 0$, the gel is considered a pure elastic covalent gel, whereas for $n > 0$, the gels are known as physical gels.[22] For increasing protein concentrations, the value for the exponent has been shown to decrease from 0.15 to 0.04, indicating that the gels are classified as physical gels, although at higher concentrations, they start to resemble more covalent elastic gels.[18] The addition of more proteins effectively contributes to a more elastic behaviour of the gel.

In protein hydrogels, the structure of the network is often evaluated with a fractal dimension. The fractal dimension comes from scaling theories, in which the protein network is seen as an assembly of fractal flocs, which consists of smaller primary particles. To be considered a fractal structure, the network has to show self-similarity at multiple length scales. Although this is difficult to prove, and may even be impossible from a practical point of view, the fractal dimension can be used to obtain information about the "openness" of the network. In literature, different models can be found that

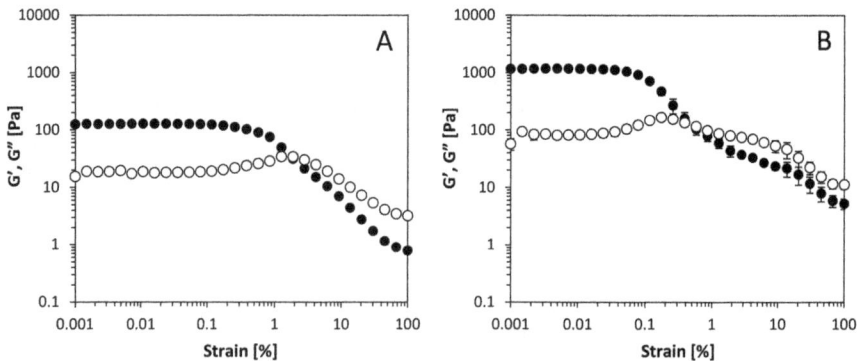

Figure 8.4   Oscillatory strain sweeps of a (A) 4.1 wt% and (B) 6.1 wt% WPI aggregate oleogel. G': filled symbols, G'': open symbols. Error bars indicate standard deviation of duplicate measurements.[18] Reprinted from *Journal of Colloid and Interface Science*, **486**, de Vries, A.; Wesseling, A.; van der Linden, E.; Scholten, E., Protein oleogels from heat-set whey protein aggregates. 75–83. Copyright 2017, with permission from Elsevier.

describe the behaviour of elastic networks and the accompanying interactions between the particles within the network. These models use the elastic modulus and the critical strain to describe the behaviour, as they show a power law relationship with the particle volume fraction. To describe protein networks, often Shih's model is used, which uses two regimes to describe the interactions within the protein network[23]:

- Strong-link regime. In this regime, the interactions between the larger flocs are stronger than those between the particles within the flocs. In this regime, the storage modulus and the critical strain are related to the volume fraction as:

$$G' \sim \varphi^{(d+x)/(d-d_f)} \tag{8.1}$$
$$\gamma_0' \sim \varphi^{-(1+x)/(d-d_f)} \tag{8.2}$$

- Weak-link regime. In this regime, the interactions between the particles within the flocs are stronger than the interactions between the flocs. In this regime, the following relations can be found:

$$G' \sim \varphi^{(d-2)/(d-d_f)} \tag{8.3}$$
$$\gamma_0' \sim \varphi^{1/(d-d_f)} \tag{8.4}$$

where d denotes the Euclidean dimension, d_f is the fractal dimension of the flocs and x is the backbone fractal dimension of the flocs of the primary particles. The type of regime can best be distinguished by the relationship between the critical strain and the volume fraction. In the case of a strong-link regime, the critical strain, γ_0, decreases with volume fraction, whereas in the weak-link regime, the critical strain increases. This difference can be used to experimentally identify the most relevant regime in the gels. From the results in Figure 8.4, we can conclude that the critical strain decreases for increasing protein concentrations, which implies that the protein networks in the oleogel represent a strong link regime. The critical strain for different concentrations is given in Figure 8.5, in which the storage modulus, G', is also given as a function of the volume fraction (Figure 8.5b). From these data, a slope of 5.3 can be extracted, which can be used to estimate a fractal dimension. Considering the strong-link regime, this gives a fractal dimension of 2.2.

Similar fractal dimensions can also be found for protein networks in hydrogels. For example, acid-induced aggregation of WPI leads to a network with a fractal dimension, d_f, of 2.2–2.3, and for salt-induced aggregation, values in the range of 2.3–2.7 can be found. Although the values for the fractal dimensions are similar, the networks found in protein hydrogels are often in the weak-link regime, whereas for the protein oleogels, the network represents a more strong-link regime. This differences shows that the interactions within the proteins and the protein flocs are different when comparing

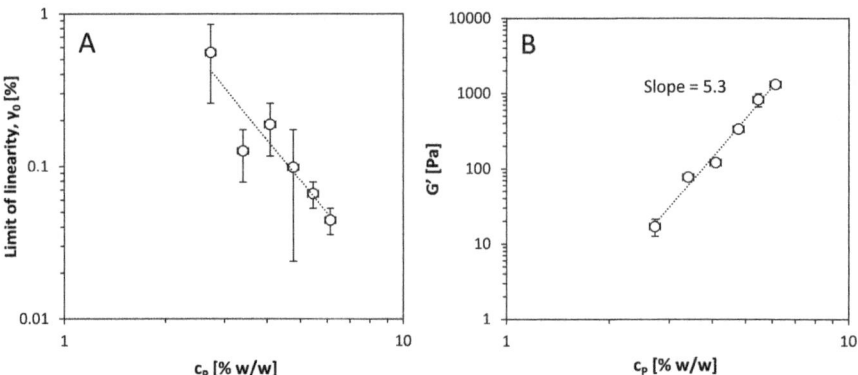

Figure 8.5 A: Log–log plots of (A) limit of linearity (γ_0) and (B) G' *versus* protein concentration (c_P). The dotted lines are the best fit through the experimental data points.[18] Reprinted from *Journal of Colloid and Interface Science*, **486**, de Vries, A.; Wesseling, A.; van der Linden, E.; Scholten, E., Protein oleogels from heat-set whey protein aggregates. 75–83. Copyright 2017, with permission from Elsevier.

networks in hydrophilic or hydrophobic environments. In hydrogels, the interactions between the particles within the flocs seem to be stronger than between the flocs, whereas for protein oleogels, the interactions between the flocs is stronger than that within the flocs. The difference may arise from the differences in protein concentration used to form the network,[18] or a different balance between the interactions related to the protein–protein or protein–solvent interactions. Nevertheless, independent of the details of the network, the network formation of proteins in oil is very similar to the network formation in water, and the precise behavior may be altered by the interactions present in the system.

8.5 Effect of Oil Type

The network formation of the proteins in oil is dependent on the interactions present in the system. As previously discussed, the most relevant difference between hydrogels and oleogels is the balance between the types of interactions. For hydrogels, the protein–solvent interactions are of hydrophilic nature (hydrogen bonds) and the protein–protein interactions are of hydrophobic nature, whereas in oleogels, the protein–solvent interactions are of hydrophobic nature and the protein–protein interactions are of hydrophilic nature. To change the interactions within the network, one approach is to change the protein–solvent interactions. These interactions are dependent on the solvent quality and are therefore related to the polarity (or hydrophobicity) of the surrounding hydrophobic oil phase. The effect of solvent polarity has already been shown before for other gelators, such as a combination of γ-oryzanol and β-sitosterol,[24] monoglycerides,[25] and ricinelaidic acid.[26] In addition, for protein oleogels, the effect of solvent type has a large influence

Table 8.1 Viscosity and oil–water interfacial tension of the oils used to form oleogels[27].[a]

Oil type	Viscosity at 20 °C [mPa s]	Oil–water interfacial tension at 20 °C [mN m^{-1}]
MCT oil	29.8 (±0.1)	26.47 (±0.39)
Sunflower oil	63.3 (±0.1)	24.41 (±0.04)
EVO	80.1 (±0.1)	14.72 (±0.02)
Castor oil	1045.0 (±5.0)	11.66 (±0.01)

[a]EVO: extra virgin olive oil; MCT: medium chain triglycerides.

on the rheological properties of the gels.[27] The type of oil can be altered based on their composition, their polarity, and their affinity to lower the surface tension. Table 8.1 gives an overview of four different oils that differ in composition and polarity. Here, the interfacial tension is taken as a measure for the polarity of the oil.[27]

MCT (medium chain triglycerides) have the shortest fatty acids and therefore the lowest viscosity and highest interfacial tension, indicating the lowest polarity. On the other hand, castor oil has the highest viscosity and the lowest interfacial tension, indicating the highest polarity. Both sunflower oil and extra virgin oil (EVO) have a similar viscosity and composition. The difference lies in the impurities present in the EVO, which may be surface active, thereby lowering the interfacial tension. In oil, the protein aggregates form a fractal-like structure, and the interactions between the protein aggregates determine the density of the network. As proteins are of hydrophilic nature, the protein aggregates will have both hydrophobic protein–solvent interactions and hydrophilic protein–protein particle interactions. The strength of the network depends on the balance between these two contributions. To gain insight into the separate contributions of the interactions, the protein aggregates can be compared to other particles that are either predominantly hydrophilic or hydrophobic. As silica particles can be obtained in a wide variety of hydrophobicities, they can give insights into the effect of these type of interactions. Figure 8.6 shows the large deformation rheology of samples with particles of different hydrophobicity.

Hydrophobic silica particles (Figure 8.6b) and hydrophilic silica particles (Figure 8.6c) are compared to protein aggregates with a more amphiphilic character (Figure 8.6a). The modulus of the protein oleogels increases up to values of 7000 kPa. The moduli of the oleogels containing hydrophilic particles go up to 30 000 kPa, whereas the hydrophobic particles only provide a network with a modulus of 200 kPa. The protein aggregates therefore resemble the hydrophilic particles more than the hydrophobic particles. It is clear from these measurements that the hydrophilic nature of particles leads to more particle–particle interactions, and therefore a larger network strength. A hydrophobic nature leads to large particle–solvent interactions, and therefore particle–particle interactions are not favourable. The type of oil has an influence on the interactions. For example, when olive oil is used

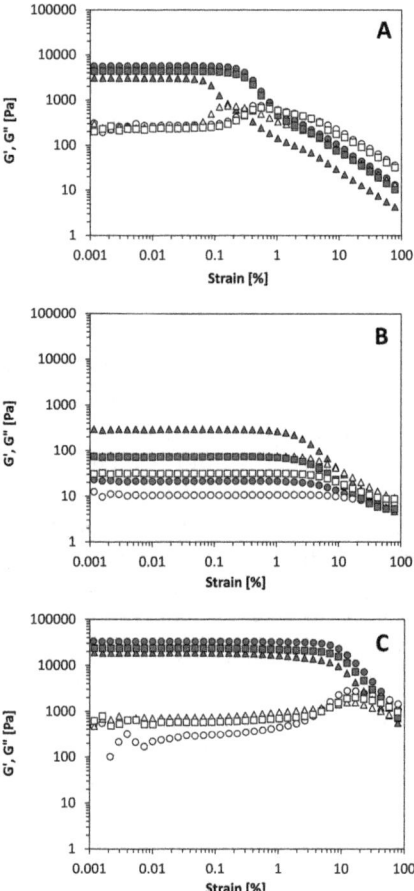

Figure 8.6 Large deformation properties of 10% WPI aggregate oleogels (A), hydrophobic silica oleogels (B) and hydrophilic silica oleogels (C) in MCT oil (o), sunflower oil (□) and olive oil (△). Filled symbols: *G'*, open symbols: *G"*. Reprinted from ref. 27 with permission from the Royal Society of Chemistry.

(triangles), the lower hydrophobicity increases the interactions between the protein aggregates and the surrounding oil, and therefore a decrease in particle–particle network formation is obtained (Figure 8.6a). A similar trend can be seen for the hydrophilic particles (Figure 8.6b). Increased particle–solvent interactions lead to a decrease in the storage modulus, and an increase in the loss modulus. As the MCT oil has the lowest polarity, the particle network is stronger, which is reflected in the increase of the storage modulus. For the hydrophobic particles, we see the opposite trend, and the differences between the oils become more pronounced. For the low polarity MCT oil (circles), we see that increased particle–solvent interactions lead to a lower degree of network formation. For the more polar olive oil (triangles), network formation of hydrophobic particles is enhanced (Figure 8.6b).

The storage modulus of this network is at least one order of magnitude larger than the network of hydrophobic particles in the less polar MCT oil.

Besides the polarity of the oil, also the chemical structure of the oils can influence the network formation. When castor oil is used, no gel formation is observed. Castor oil is highly polar due to the presence of ricinoleic acid. This acid contains a hydroxyl group, which is able to form hydrogen bonds with the particles. The ability to form hydrogen bonds with the surrounding solvent "solvates" the particles, and therefore they are finely dispersed in the oil without extensive network formation. This phenomenon is observed for both silica particles, and no gel formation is obtained. In the case of the hydrophilic silica particles, it is to be expected that the increased polarity would lead to less network formation as more particle–solvent interactions are present. However, for the hydrophobic particles, an increase in network formation is expected based on a decrease in particle–solvent interactions. As also in this case no gel formation was obtained, the particle network formation is not driven solely by the polarity of the solvent, but also by the presence of specific chemical groups. The network formation of protein aggregates in oil can therefore not only be controlled by the polarity of the oil, but also by its composition.

The balance between the particle–particle and the particle–solvent interactions also has an effect on the structural organization of the network, and how the structure recovers after deformation. This may be of relevance when such oleogels are used in different applications. Structure breakdown will occur during processing conditions, and fast recovery of the network will be beneficial for the solid-like behaviour of the product. In the case of silica particles, the storage modulus (Figure 8.6b and c) shows a long linear viscoelastic region, indicating that the network is not very sensitive to yielding. The protein particles, on the other hand, show a much smaller linear viscoelastic region, and a lower value for the critical strain, γ_c. The structure breakdown is mostly dominated by the microstructure of the gels. In the case of protein aggregates, the aggregates form larger flocs, in which the degree of connection between the flocs determines the breakdown of the strands within the network. For the silica particles, their fine distribution leads to the formation of smaller flocs with more connection points, and therefore they are less prone to yielding. The flocculated structure of the protein oleogels is also obvious from the small strain overshoot at strains close to the critical strain. As the structure breakdown and the structure creation occur simultaneously as a result of the rupture and reformation of connection points within the network, the behaviour of the gel depends on the balance between these contributions. When structure build-up occurs at a rate similar to the structure breakdown, a small increase in the loss modulus can be observed. Since such a small overshoot can be seen for the protein aggregates and the hydrophilic silica particles only, rapid structure formation is expected to happen owing to fast hydrogen bond formation between the particles. Owing to the formation of these hydrogen bonds, the structure recovers after deformation, and after recovery the storage modulus, G', is larger than the loss modulus, indicating that the solid-like behaviour of the gels is maintained. In these types of protein aggregate oleogels, the recovery occurs within several minutes. After applying high strains, the network almost instantaneously recovers again owing to strong interactions between the particles. The degree of the recovery depends on the oil type used. In the case of MCT oil, the structure recovered to roughly 40%, whereas for olive oil, the recovery reached 20% only in the case of the protein aggregates. In the case of the hydrophilic and hydrophobic silica particles, the recovery is slightly higher, and reaches values of 50–60%.[27]

8.6 Suitable Protein Building Blocks

Although protein interactions in hydrophilic environments are well investigated, the interactions acting on proteins in hydrophobic environments are largely unexplored. In aqueous conditions, protein aggregation and network formation are commonly induced by different methods, and different types of interactions are involved in the gelation mechanism. For example, in the case of flexible coil proteins, such as gelatin, network formation is induced by the formation of hydrogen bonds. For globular proteins, network formation is often obtained by hydrophobic interactions (heat-induced aggregation) or the reduction of the electrostatic repulsion (pH or salt-induced aggregation). Heat-induced aggregation may also lead to the formation of disulfide bonds, providing additional stabilization through covalent bonds. These disulfide bonds may be important in the preparation of the initial building blocks as the hydrophobic interactions within the aggregates may not be present anymore when the proteins are transferred to the oil *via* intermediate solvents. In oil, the hydrophobic interactions lead to increased dispersibility of the proteins as the protein–solvent interactions are of hydrophobic nature. In the case of whey proteins, such disulfide bonds are mainly formed between the proteins bovine serum albumin and β-lactoglobulin.[21] The protein α-lactalbumin is less prone to form disulfide bonds, and therefore the content of these proteins within the aggregates is lower than in the native proteins. During the solvent exchange procedure, these proteins are washed out, and are incorporated in the aggregates to a limited extent. The protein composition of the aggregates therefore depends on the reactivity of the native proteins towards disulfide bridge formation.

In protein oleogels, the attractive protein–protein interactions must be of hydrophilic nature, and these interactions are most likely owing to hydrogen bond formation, as discussed in the previous paragraphs. The polarity (or hydrophilicity) of the proteins will therefore have an influence on the gel strength, and specific groups on the proteins capable of forming hydrogen bonds will promote network formation. The choice of the proteins therefore depends on the protein's tertiary structure. During the solvent exchange procedure, the proteins are subjected to different solvent types. These solvents may change the tertiary structure of the proteins. However, since the proteins have already been denatured during the heating step, no additional conformational changes in the proteins are observed in the case of whey proteins.[21] Therefore, the final properties of the oleogels are already determined by the formation of the protein aggregates in the aqueous phase. In this phase, the properties of the aggregates can easily be altered, as the size, density, *etc.*, of the protein aggregates depends largely on the environmental conditions used during heating. For example, at pH values close to the isoelectric point of the proteins, the charge of the proteins is minimal, and in the absence of repulsive interactions, the proteins are allowed to come into close proximity. The increased degree of contact between the proteins leads to the formation of more dense, coarse protein aggregates owing to the attractive van der Waals forces. Such aggregates have a higher density and a lower porosity. Charge neutralization can also be accomplished at pH values further from the isoelectric point and with the addition of salt. At pH values further from the isoelectric point and no added salt, the charges present on the proteins provide sufficient electrostatic repulsion to prevent direct aggregation of the proteins. This limits close proximity between the proteins, and sufficient hydrophobic interactions are required to form aggregates, which are more open and fibrillar-like, leading to aggregates with higher porosity and low density. By controlling the balance between the interactions, the density or the porosity of the particles can therefore easily be adjusted in the aqueous phase. The size can also be controlled using different concentrations of the proteins, time of the heat treatment, and temperatures.

As many types of proteins can be used for the formation of small aggregates, a large variety of protein building blocks ranging in size, hydrophilicity, surface roughness, porosity and density can be made. The properties of the protein aggregates will determine to a large extent the final network formation, and the physical properties of the gel, such as the gel strength, the yielding behaviour and the recovery of the gel after large deformation.

8.7 Role of Capillary Interactions

To introduce the protein aggregates in oil, the solvent exchange procedure has been shown to be an effective method. This procedure is necessary as direct dispersion of proteins in oil leads to a large degree of agglomeration and sedimentation of the proteins. One reason is the large hydrophilic nature of the proteins. When the proteins are heated above the denaturation temperature, the exposed hydrophobic groups increase the hydrophobicity of the proteins and the resulting protein aggregates. Fluorescence intensity measurements have indeed shown that the hydrophobicity of the protein aggregates is higher than that of native proteins.[21] However, when these protein aggregates are dried and then dispersed in oil in a dry state, no network formation is observed and a large degree of agglomeration can still be observed. Therefore, the dispersibility of the proteins and the subsequent network formation are not governed by the hydrophobicity of the protein aggregates only, but other phenomena also play a role. Changes in the proteins may arise from contact with solvents other than water and oil, such as acetone, which is used as the intermediate solvent in the exchange process. As acetone contains an oxygen atom, acetone is capable of hydrogen bond formation with the proteins. This may disrupt hydrogen bond formation between proteins and water (solvation) molecules or hydrogen bond formation between different chemical groups within the proteins. This may lead to additional changes in the conformation of the proteins in the aggregates, affecting the hydrophobicity even more. Such potential changes were investigated using FTIR, which allows us to examine the secondary structure and the hydration level of proteins.[28] The protein conformation is related to changes in the amide I region (1600–1680 cm^{-1}), which can be identified as C=O stretching. The amide II region (1480–1560 cm^{-1}) is caused by NH bending and CN stretching, and is supposed to be closely related to the degree of hydration of the proteins.[29] In the case of protein aggregates that were dried from different solvents (water, acetone and hexane) increasing in hydrophobicity, no obvious changes in the protein conformation or hydration level were found.[21] Only minor changes in hydration level were found, where aggregates dried from organic solvents contain more water than the freeze-dried aggregates. In any case, no relation was found between the hydration level and the dispersibility of the dried powder. These results show that the enhanced dispersibility of the protein aggregates is not related to any changes in the conformation or hydrophobicity of the proteins.

However, even though the protein conformation does not change, the properties of the aggregates themselves do change when dried from different solvents. When dried from water, large agglomeration of proteins into dense and compact agglomerates is observed. When added to the oil, the agglomerates cannot be dispersed to particles of the initial protein aggregate size. The attractive interactions between the aggregates within the agglomerates are sufficiently large to prevent re-dispersion. However, when the protein aggregates are dried from solvents with low polarity, the degree of agglomeration decreases, and more open agglomerates are obtained. If hexane is used, the agglomeration is reduced to such an extent that the density of the protein aggregate powder decreases by a factor of 4 compared to protein aggregates dried from water. The drying process therefore has the largest effect on the differences in the properties of the protein aggregates. The differences can be explained by the interactions between the aggregates during the drying process. When the solvent slowly evaporates in the drying step, the remaining solvent will become concentrated between the protein aggregates at sufficiently low concentrations. This leads to the formation of a solvent bridge, which is depicted schematically in Figure 8.7.

A liquid bridge provides additional attractive interactions between particles, also known as capillary interactions. The extent of the particle attraction depends on the properties of the solvent, and the capillary force, F_c, which is calculated as[30]:

$$F_c = 2\pi R \gamma \cos \theta \qquad (8.5)$$

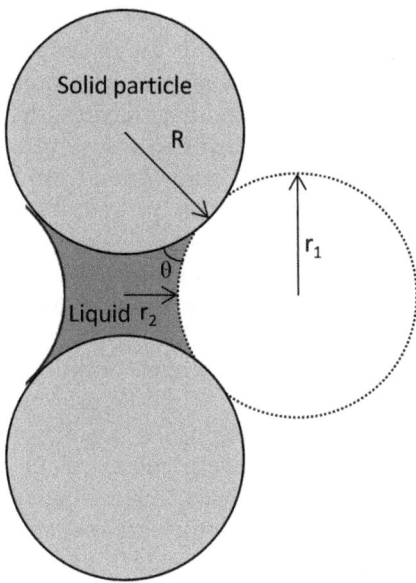

Figure 8.7 Schematic drawing of a liquid bridge between two spherical particles.[21] Modified from de Vries, A.; Lopez Gomez, Y.; Jansen, B.; van der Linden, E.; Scholten, E. Controlling Agglomeration of Protein Aggregates for Structure Formation in Liquid Oil: A Sticky Business. *ACS Applied Materials & Interfaces*, 2017, **9** (11), 10136–10147 (http://pubs.acs.org/doi/full/10.1021/acsami.7b00443) with permission from the American Chemical Society.

when the particles approach each other at close proximity. Here, R is the particle radius, γ is the interfacial tension, and θ is the contact angle between the liquid and the solid. As can be seen from this equation, the force depends largely on the interfacial tension and the contact angle, which are both very much dependent on the hydrophobicity of the solvent. For a solvent of high polarity (water), the large interfacial tension (73 mN m^{-1}) and the low contact angle (large degree of wetting) results in a larger capillary force, which explains the large degree of agglomeration of the protein aggregates dried from water. On the other hand, when a solvent of low polarity is used, such as hexane with a low interfacial tension (19 mN m^{-1}), the wetting ability is much lower (high contact angle) and lower capillary forces are created. This explains why drying from hexane leads to a much lower degree of agglomeration. During drying from hexane, the forces obtained through capillary bridge formation are not sufficiently large to provide permanent attractive interactions between the protein aggregates. This also explains why the solvent exchange procedure is effective to prevent such agglomeration and leads to a finely dispersed suspension. During the solvent exchange procedure, the protein aggregates are kept in a wet state, preventing capillary bridge formation, and therefore subsequent capillary forces are not present.

Agglomeration of the protein aggregates is therefore limited, and the initial size of the protein aggregates created in the aqueous phase remains when transferred to oil. As the total surface area of the protein aggregates remains available for contact with other protein aggregates, the large degree of contact between the small aggregates leads to sufficient attractive interactions to provide network formation. Aggregates dried from hydrophobic solvents may therefore also be effective as an oil structurant without the use of an intermediate solvent.

From an application and practical point of view, it is desirable to circumvent the solvent exchange procedure. As these results show that the role of the intermediate solvent is to prevent the agglomeration of the aggregates, alternative methods may also provide the conditions required for retaining small aggregates. Agglomeration might be prevented by optimizing the freeze-drying conditions. During freezing, the formation of ice crystals drives the particles closer together, effectively leading to an increased concentration and contact between the particles. When the freezing process is fast enough, faster nucleation leads to the formation of smaller ice crystals. This fast freezing prevents the dispersed particles from approaching one another, and remain more evenly distributed throughout the samples. When drying is obtained from this state, particle agglomeration might be prevented. When liquid nitrogen is used to freeze the samples, particle agglomeration has been shown to be partly prevented.[21] The results show that small aggregates are present, but also that a certain degree of agglomeration is still obtained.[21] However, a larger fraction of protein aggregates retains the original size. Centrifugation of such samples allows concentration of the smaller particles and allows for removal of the larger agglomerates. Rheological measurements show that oleogels prepared with such fast freeze-dried protein aggregates show a similar behaviour as the ones prepared using the solvent exchange procedure.[21] Both samples have a similar viscoelastic region and the same overshoot in the loss modulus (G'') is present, indicating that the network formation and network rupture occurs in a similar way, and that rearrangements within the network occur quickly.

Although the current investigations have not resulted in complete limitation of the agglomeration of the particles, it should be possible to obtain small protein aggregates by optimization of the drying process by changing conditions such as the freezing rate, the drying rate, and the protein aggregate concentration. This optimization process may lead to the formation of a powder consisting of small protein aggregates, which should be directly dispersible in liquid oil. This process therefore allows the easy tuning of the protein content, and therefore a large variety of protein oleogels can be obtained. Such protein aggregate powders can be made with different properties, in which the protein type, protein aggregate size, aggregate porosity and hydrophobicity can be controlled. These factors determine the interactions between the protein aggregates, and together with the choice of the oil, protein oleogels with different rheological characteristics can be obtained.

8.8 Potential Applications: First Trials

As proteins are highly available, cheap, have a high nutritional value, and are accepted by consumers as natural ingredients, the potential to use protein oleogels as an alternative for solid fats is high. As a proof of principle, cookies and sausages were made that contain protein oleogels as a replacement for margarine and pork fat.

No extensive study was performed, and the results of this study should be taken with care. For the cookies, a standard recipe for shortbread cookies was used, including flour (42%), salt (0.5%), vanilla sugar (3.5%), sugar (21%) and margarine (33%). This was used as a reference cookie containing solid, saturated fats. For the recipes with the protein oleogel, we replaced the margarine with the same amount of protein oleogel. As a control, a recipe was included in which the margarine was replaced completely by oil. As we noticed that the sugar did not always completely dissolve, we replaced 50% of the sugar with glucose syrup in the case of the samples with protein oleogels and oil. The cookie dough was baked for 15 minutes at 170 °C. Figure 8.8 shows the results after baking. As can be seen, the cookies with margarine (Figure 8.8a) formed a regular cookie. When the margarine was replaced by oil, the dough was not consistent enough, and no cookie could be made (Figure 8.8c). The consistency, hardness and the stickiness of the dough with protein oleogel was very similar to the consistency of the dough with margarine, and therefore the dough resisted the rheological changes during baking. When the protein oleogel was used as a replacement, the appearance of the oleogel cookie (Figure 8.8b) was quite similar to that of the reference cookie with margarine (Figure 8.8a).

Figure 8.9 shows a cross section of the cookies, and also here it is visible that the structure of the cookie with the protein oleogel (Figure 8.9b) is quite similar to that of the reference cookie prepared with margarine (Figure 8.9a).

The baked cookies were tested for their consistency and tasted by a small sensory panel. Mechanical tests show that the cookies with protein oleogels break easier, which was also observed from the sensory test, because the protein oleogel cookies scored lower on the sensory attribute of crunchiness.

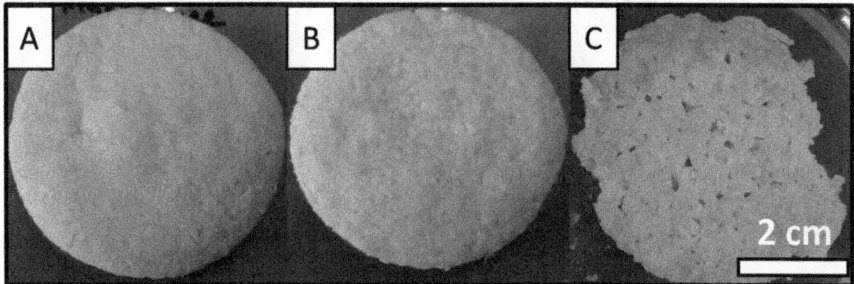

Figure 8.8 Appearance of shortbread cookies prepared with (A) margarine, (B) protein oleogel and (C) sunflower oil.

This leads to slightly lower scores for liking, although some panelists also preferred the more crumbly structure of the cookies prepared with the protein oleogel, compared to the cookies prepared with margarine. However, as this was just a proof of principle study, optimization of the recipe may lead to closer resemblance in the mechanical and sensory properties, and improved liking.

The protein oleogels were also used to prepare frankfurter sausages. The sausages were prepared from low fat pork meat, ice water, nitrite salt brine (2%) and spices. As a reference sample, 15% back fat was added as a representative of a full fat sausage. For the low-fat samples, the added back fat was replaced with protein oleogels or oil. The total fat percentage of the sausages was 25%. The fat reduction level was therefore 60%. Figure 8.10 shows the results after cooking the sausages for 15 minutes at a temperature of 78 °C.

From a mechanical point of view, the sausages with the protein oleogels (Figure 8.10b) showed the same breaking/fracture force as the reference sausages with pork fat (Figure 8.10a). The sausages with only oil (Figure 8.10c) showed a much lower breaking force, indicating that the protein provides stability to

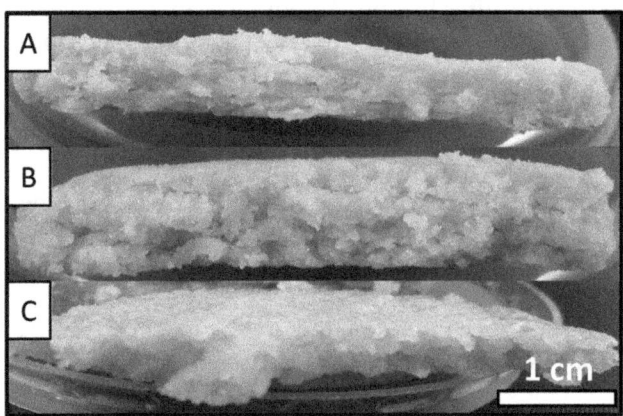

Figure 8.9   Cross section of shortbread cookies prepared with (A) margarine, (B) protein oleogel, and (C) sunflower oil.

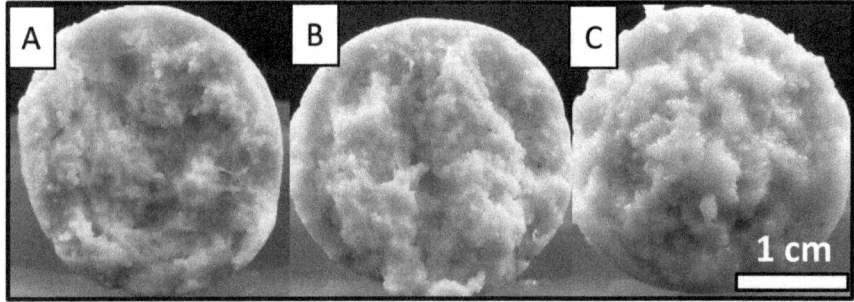

Figure 8.10   Appearance of cross section of frankfurter sausages prepared with (A) pork fat, (B) protein oleogel, and (C) sunflower oil.

the oil phase and gives the sausages a more solid-like behavior. Moreover, the liquid that remained in the sausage was higher for the sausage with protein oleogel than for the sausage with oil, which can be related to the juiciness of the sausage. Overall, no large differences in the sensory profiles of the sausages prepared with protein oleogels and pork fat were found, but the sausages prepared with oil were considered lower in firmness and more fatty.

Although these results were obtained with a small set of samples and no recipe optimization was performed, they show that the protein oleogels have potential as a replacement for solid fat. As many factors relate to the sensory perception of attributes such as crunchy, fatty, mouth coating, crumbly and firm, the optimization of the recipe will lead to a different outcome. As a first proof of principle, at least these results show that evaluation of oleogels based on protein as an alternative for solid fat is highly interesting and has potential.

8.9 Conclusion

The development of structure formation in hydrophobic material, especially edible liquid lipid phases, has gained considerable interest in recent years. With respect to food, the development of structured oils would allow the reduction of saturated fats, and the implementation of more healthy poly-unsaturated fats can have considerable health benefits. It also provides more flexibility towards food formulation and the choice of the lipid phase. Next to the food industry, such oleogels could also be very interesting for pharmaceutical purposes or in the field of cosmetics. Currently, there are many approaches available to impart more solid-like behaviour to liquid oils, but many different factors still hamper the commercialization of products with such structured oils. In this chapter, we have discussed how proteins can be used to provide network formation in liquid oils. Protein is considered a healthy ingredient in the modern diet, and therefore protein would have great potential in many applications once the properties of the protein oleogels can be controlled.

The results in this chapter show that the type of protein networks formed in oil are similar to those in protein hydrogels. The networks may range from fine-stranded to coarse, or a more fractal type of network can be obtained. As the type of network does not change by replacing the initial water phase for the oil phase, the properties of the final oleogels can be altered by controlling the properties of the initial protein building blocks in the aqueous phase. When macroscopic firm hydrogels are used, a firm protein oleogel is also formed. When smaller submicrometer protein aggregates are used as the initial building blocks for network formation, a more spreadable thick oil continuous viscoelastic material is obtained. As the size of the initial building block can easily be varied (from submicrometer to cm), the mechanical properties of the protein oleogels can easily be altered. Firm oleogels with distinct fracture properties can be obtained, or spreadable oleogels with plastic (yielding) behaviour. Owing to the large variation in mechanical properties, various applications can be envisioned.

Although network formation of proteins in aqueous environments is well known and has been studied extensively, there is currently little known about the interactions between the proteins in hydrophobic environments. The results discussed in this chapter show that the interactions involved in the network formation are driven by those of hydrophilic nature. The hydrophilic nature of the protein aggregates and the hydrophobicity of the surrounding solvent will determine the strength of the interactions between the proteins in the network. Starting from a hydrogel, it was already shown that the large increase in the interactions between the proteins in the fine-stranded and coarse networks even led to an increase in the storage modulus by 2 orders of magnitude. An understanding of the relevant interactions within different types of network will contribute greatly to the development of protein oleogels. Parameters such as particle size, hardness, porosity, and hydrophobicity will have a large effect on the network formation, the mutual interactions between the proteins or protein aggregates, and the accompanying mechanical properties of the final protein oleogel. This knowledge may provide guidelines to alter specific desired properties of the protein oleogels for certain commercial applications.

Although the solvent exchange procedure allows the dispersal of proteins in hydrophobic environments, this technique is not very efficient and realistic from an industrial point of view. For commercialization, it would be beneficial to circumvent the solvent exchange technique and be able to use a dried powder directly to create liquid oil suspensions. The results in this chapter indicate that, with a proper drying technique, the agglomeration of the proteins may be prevented during the drying step, and small protein aggregates can be obtained in dry form as a very low density powder. This protein powder may be used directly as an oil structurant. However, owing to the hygroscopic nature of the powder, the powder may have to be stored under dry conditions. The development of a suitable drying technique would make the use of such protein oleogels in industrial environments easier.

References

1. J. F. Toro-Vazquez, R. Mauricio-Pérez, M. M. González-Chávez, M. Sánchez-Becerril, J. d. J. Ornelas-Paz and J. D. Pérez-Martínez, Physical properties of organogels and water in oil emulsions structured by mixtures of candelilla wax and monoglycerides, *Food Res. Int.*, 2013, **54**(2), 1360–1368.
2. A. Bot and W. G. M. Agterof, Structuring of edible oils by mixtures of γ-oryzanol with β-sitosterol or related phytosterols, *J. Am. Oil Chem. Soc.*, 2006, **83**(6), 513–521.
3. M. A. Rogers, A. J. Wright and A. G. Marangoni, Nanostructuring fiber morphology and solvent inclusions in 12-hydroxystearic acid/canola oil organogels, *Curr. Opin. Colloid Interface Sci.*, 2009, **14**(1), 33–42.

4. R. Gong, Y. Song, Z. Guo, M. Li, Y. Jiang and X. Wan, A clickable, highly soluble oligopeptide that easily forms organogels, *Supramol. Chem.*, 2013, **25**(5), 269–275.

5. R. Kumar and O. P. Katare, Lecithin organogels as a potential phospholipid-structured system for topical drug delivery: A review, *AAPS PharmSciTech*, 2005, **6**(2), E298–E310.

6. H. M. Schaink, K. F. van Malssen, S. Morgado-Alves, D. Kalnin and E. van der Linden, Crystal network for edible oil organogels: Possibilities and limitations of the fatty acid and fatty alcohol systems, *Food Res. Int.*, 2007, **40**(9), 1185–1193.

7. F. G. Gandolfo, A. Bot and E. Flöter, Structuring of edible oils by long-chain FA, fatty alcohols, and their mixtures, *J. Am. Oil Chem. Soc.*, 2004, **81**(1), 1–6.

8. C. V. Nikiforidis and E. Scholten, Self-assemblies of lecithin and [small alpha]-tocopherol as gelators of lipid material, *RSC Adv.*, 2014, **4**(5), 2466–2473.

9. T. Laredo, S. Barbut and A. G. Marangoni, Molecular interactions of polymer oleogelation, *Soft Matter*, 2011, **7**(6), 2734–2743.

10. M. Davidovich-Pinhas, S. Barbut and A. G. Marangoni, The gelation of oil using ethyl cellulose, *Carbohydr. Polym.*, 2015, **117**, 869–878.

11. C. V. Nikiforidis and E. Scholten, Polymer organogelation with chitin and chitin nanocrystals, *RSC Adv.*, 2015, **5**(47), 37789–37799.

12. A. R. Patel, P. S. Rajarethinem, N. Cludts, B. Lewille, W. H. De Vos, A. Lesaffer and K. Dewettinck, Biopolymer-Based Structuring of Liquid Oil into Soft Solids and Oleogels Using Water-Continuous Emulsions as Templates, *Langmuir*, 2015, **31**(7), 2065–2073.

13. A. I. Romoscanu and R. Mezzenga, Emulsion-templated fully reversible protein-in-oil gels, *Langmuir*, 2006, **22**(18), 7812–7818.

14. C. V. Nikiforidis and E. Scholten, High internal phase emulsion gels (HIPE-gels) created through assembly of natural oil bodies, *Food Hydrocolloids*, 2015, **43**, 283–289.

15. I. Tavernier, A. R. Patel, P. Van der Meeren and K. Dewettinck, Emulsion-templated liquid oil structuring with soy protein and soy protein: kappa-carrageenan complexes, *Food Hydrocolloids*, 2017, **65**, 107–120.

16. E. A. Foegeding, E. L. Bowland and C. C. Hardin, Factors that determine the fracture properties and microstructure of globular protein gels, *Food Hydrocolloids*, 1995, **9**(4), 237–249.

17. A. de Vries, J. Hendriks, E. van der Linden and E. Scholten, Protein Oleogels from Protein Hydrogels *via* a Stepwise Solvent Exchange Route, *Langmuir*, 2015, **31**(51), 13850–13859.

18. A. de Vries, A. Wesseling, E. van der Linden and E. Scholten, Protein oleogels from heat-set whey protein aggregates, *J. Colloid Interface Sci.*, 2017, **486**, 75–83.

19. P. J. Flory and J. Rehner Jr, Statistical Mechanics of Cross-Linked Polymer Networks II. Swelling, *J. Chem. Phys.*, 1943, **11**(11), 521–526.

20. C. Schmitt, C. Bovay, A.-M. Vuilliomenet, M. Rouvet and L. Bovetto, Influence of protein and mineral composition on the formation of whey protein heat-induced microgels, *Food Hydrocolloids*, 2011, **25**(4), 558–567.

21. A. de Vries, Y. Lopez Gomez, B. Jansen, E. van der Linden and E. Scholten, Controlling Agglomeration of Protein Aggregates for Structure Formation in Liquid Oil: A Sticky Business, *ACS Appl. Mater. Interfaces*, 2017, **9**(11), 10136–10147.

22. A. H. Clark and S. B. Ross-Murphy, Structural and mechanical properties of biopolymer gels, *Biopolymers*, Springer Berlin Heidelberg, Berlin, Heidelberg, 1987, pp. 57–192.

23. W.-H. Shih, W. Y. Shih, S.-I. Kim, J. Liu and I. A. Aksay, Scaling behavior of the elastic properties of colloidal gels, *Phys. Rev. A*, 1990, **42**(8), 4772–4779.

24. H. Sawalha, G. Margry, R. den Adel, P. Venema, A. Bot, E. Floter and E. van der Linden, The influence of the type of oil phase on the self-assembly process of gamma-oryzanol plus beta-sitosterol tubules in organogel systems, *Eur. J. Lipid Sci. Technol.*, 2013, **115**(3), 295–300.

25. F. Valoppi, S. Calligaris, L. Barba, N. Šegatin, N. Poklar Ulrih and M. C. Nicoli, Influence of oil type on formation, structure, thermal, and physical properties of monoglyceride-based organogel, *Eur. J. Lipid Sci. Technol.*, 2017, **119**(2), 1500549.

26. A. J. Wright and A. G. Marangoni, Formation, structure, and rheological properties of ricinelaidic acid-vegetable oil organogels, *J. Am. Oil Chem. Soc.*, 2006, **83**(6), 497–503.

27. A. de Vries, Y. L. Gomez, E. van der Linden and E. Scholten, The effect of oil type on network formation by protein aggregates into oleogels, *RSC Adv.*, 2017, **7**(19), 11803–11812.

28. A. Barth, Infrared spectroscopy of proteins, *Biochim. Biophys. Acta, Bioenerg.*, 2007, **1767**(9), 1073–1101.

29. N. Wellner, P. S. Belton and A. S. Tatham, Fourier transform IR spectroscopic study of hydration-induced structure changes in the solid state of omega-gliadins, *Biochem. J.*, 1996, **319**(Pt 3), 741–747.

30. S. Strauch and S. Herminghaus, Wet granular matter: a truly complex fluid, *Soft Matter*, 2012, **8**(32), 8271–8280.

CHAPTER 9

Oleogels from Emulsion (HIPE) Templates Stabilized by Protein–Polysaccharide Complexes

WAHYU WIJAYA*[a], PAUL VAN DER MEEREN[a] AND ASHOK R. PATEL*[b]

[a]Ghent University, Particle and Interfacial Technology Group, Department of Applied Analytical and Physical Chemistry, Coupure Links 653, B-9000 Gent, Belgium; [b]Sci Five Consulting Services, Coupure 164F, 9000 Gent, Belgium
*E-mail: wahyu.wijaya@UGent.be, ashok2510@gmail.com

9.1 Introduction

Oil structuring using food-grade polymers is an emerging strategy and holds significant promise in the area of food and nutrition because a variety of polymers are approved for use in foods. However, since most of the food polymers are inherently hydrophilic in nature, they are ineffective in structuring oils due to their limited dispersibility in oil. To date, the only known food polymer to gel edible oil through direct dispersion is a hydrophobic cellulose derivative, i.e. ethyl cellulose (EC). However, the process of obtaining EC oleogels has limitations owing to its requirement for high temperature

Food Chemistry, Function and Analysis No. 3
Edible Oil Structuring: Concepts, Methods and Applications
Edited by Ashok R. Patel

processing (dispersing temperature above the glass transition temperature of EC, 130–145 °C) to induce oleogelation.[1] This high processing temperature may affect the quality of the oleogels by oxidative deterioration of the oil medium.[2]

Proteins and polysaccharides play an important role in providing a structural framework to water-based gels, but they cannot be dispersed into liquid oil. Thus, an indirect approach is needed to arrest the structure of the polymers in the water phase and to use it for physical entrapment of oils after evaporation of water from the bulk phase.[3] Practically, stable emulsions are obtained by stabilization of oil–water interfaces using protein particles, after which the water is removed to obtain a 3D interpolymeric matrix entrapping the liquid oil. This approach was first introduced by Romoscanu and Mezzenga,[4] who thermally or chemically cross-linked β-lactoglobulin (β-lg) at the oil–water interface and subsequently dried the emulsion at room temperature under continuous ventilation of the sample.

To the best of our knowledge, only a few reports on liquid oil structuring using unmodified protein–polysaccharide complexes have been published.[5,6] However, these studies used a relatively high amount of water phase ($\geq 40\%$) to disperse the oil phase. There are disadvantages of using a high amount of water in the bulk system, *i.e.* the evaporation time is long using conventional or modern drying methods, and the drying temperature should be high enough to efficiently remove the water. The high drying temperature in conventional drying can deteriorate the quality of the oil and more modern drying methods (lyophilization) are not economically feasible for scale-up. Therefore, we propose a strategy to first formulate high internal phase emulsions (HIPEs), of which the oil fraction is above 0.74 (more than the close packing of spherical oil droplets), in order to have a low amount of water in the bulk phase so the emulsions can be dried at relatively low temperature (≈ 35 °C). Stabilizing HIPEs in itself (with the use of small molecular weight surfactants) is quite challenging, and doing so with the use of only biopolymers is even more problematic. Complexation of a surface-active biopolymer (proteins) with another biopolymer (polysaccharides) is necessary to improve the stability of HIPEs, since the surface-active hydrophilic polymer is not known to stabilize such emulsions on its own.

In the past few years, HIPEs have become an attractive subject of investigation owing to their wide range of applications, particularly in structuring of edible formulations.[7,8] A HIPE is a complex colloidal system with an unusual phase distribution. HIPEs are emulsified systems with an internal phase volume fraction of at least 0.74 ($\varphi_{\mathrm{disp}} \geq 0.74$), which is the maximum packing density of monodispersed hard spheres.[9] Using HIPEs as a template to obtain oleogels has significant advantages, *i.e.* a relatively short conventional drying time because of the low water content, yielding high quality oleogel products and a high amount of hydrophobic bioactives that can be dissolved in the high fraction of hydrophobic phase.

The process of obtaining an oleogel using the HIPE-templated approach consists of three general steps: preparation of protein–polysaccharide complexes, preparation of HIPE, and drying of HIPE followed by shearing of the dried product. Pre-formation of complexes is preferable in the HIPE formation, as the layer-by-layer (LbL) technique is less feasible owing to the foreseen structure disruption of HIPEs when subjected to a step-wise emulsification process. Furthermore, emulsions prepared by the LbL approach have the tendency to become extensively flocculated during preparation, while stable emulsions are obtained with pre-formed protein–polysaccharide complexes.[10]

HIPEs stabilized by protein–polysaccharide complexes are obtained *via* homogenization of a complex containing aqueous phase and a high fraction of oil phase. An appropriate concentration of biopolymers in the aqueous phase is critical to ensure the assembly of conformational networks that also determine the stability of the dried HIPE. Subsequently, an oil-in-gel continuous phase is obtained after the HIPE is subjected to drying followed by shearing at a fixed force for a certain duration.

Electrostatic complexation between proteins and polysaccharides in a mixture is mainly driven by pH.[11] This formation of pre-complexes in the mixture greatly affects the stabilization of HIPEs and oleogels. At a pH where the two biopolymers are oppositely charged, proteins and polysaccharides may form attractive electrostatic complexes. This interaction may be weak or strong leading to the formation of soluble (dispersible) and insoluble complexes, respectively. Soluble complexes are formed by binding of anionic polysaccharides to certain cationic reactive sites of proteins (at a pH close to the iso-electric point (pI) of the protein), which causes the colloidal complexes to be relatively stable as the number of opposite charges carried by the two macro-ions is not equal in number. The resulting net charge allows the solubilization of the complexes by interaction with solvent molecules. However, when the number of opposite charges carried by both biopolymers is equal, the resulting complex charge is zero and the complexes become insoluble, leading to a phase separation between the complexes and solvent.[12] At pH conditions where the two biopolymers carry similar charges (pH > pI of protein), proteins and polysaccharides may form co-soluble complexes by interacting with each other *via* non-covalent interactions other than electrostatic (such as hydrogen bonding). Hydrogen bonds are formed *via* interaction between a hydrogen atom attached to an electronegative atom (*i.e.*, nitrogen, oxygen, or sulfur) and another electronegative atom (*i.e.*, oxygen of carbonyl or carboxyl group).[13] These bondings could be obtained over a wide range of pH values, including values above the pI of the protein. For instance, gelatin–pectin complexes were formed by hydrogen bonding due to the compatibility of the two biopolymers at a pH above the pI of gelatin.[14] Moreover, the occurrence of hydrogen bonding between proteins and polysaccharides is favoured by low temperatures at the critical point of miscibility of gelatin and pectin (*i.e.*, 20 °C).[14,15]

In this chapter, we discuss the properties of whey protein isolate (WPI)–low methoxyl pectin (LMP), sodium caseinate (SC)–LMP and SC–alginate (ALG) mixtures for stabilization of HIPEs and oleogels. WPI is commonly used as a natural emulsifier in foods. It consists of a mixture of different globular proteins, with β-lg and α-lactalbumin as the most important components.[16,17] Unlike gelatin, which has gelling properties similar to those of other hydrocolloids,[18] native WPI is typically not known to form an aqueous gel without heat[19] or enzymatic treatment (*e.g.* transglutaminase)[20] or addition of divalent salts (*e.g.* calcium chloride).[21] SC is also commonly used as a dairy protein emulsifier consisting of a soluble mixture of four principal caseins (αs1-casein, αs2-casein, β-casein, and κ-casein).[22] Compared to other food proteins, SC can form a thick (steric) stabilizing layer on the emulsion droplet surface that protects newly formed droplets against flocculation and coalescence. SC demonstrates a satisfactory thermal stability due to its random coil nature and ability to form extensive intermolecular hydrogen, electrostatic and hydrophobic bonds.[23] On the other hand, LMP with a degree of esterification (DE) of 50% is obtained by de-esterification of high methoxyl pectin with acid or ammonia in an alcoholic medium.[24] Gelation of LMP can be induced by the addition of divalent calcium ions, which hold together two or more polyuronate chains to form side by side aggregates that have been likened to egg boxes.[25,26] ALG is a naturally occurring anionic and hydrophilic polysaccharide extracted with dilute alkali from various species of brown seaweed (*Phaeophyceae*) used as thickening, stabilizing, suspending, film forming, gel producing, and emulsifying agent. In molecular terms, ALG is a linear water-soluble polysaccharide comprising of (1-4)-linked units of *R*-D-mannuronate (M) and α-L-guluronate (G) at different proportions and different distributions in the chain.[27]

The aim of the current chapter is to specifically discuss the preparation and characterization of protein–polysaccharide complexes, HIPEs stabilized by the complexes, and oleogels obtained by drying of the HIPEs. The effect of pH, total concentration, and type of proteins and polysaccharides on the formation of the complexes, HIPEs and oleogels is discussed. Lastly, we also reveal the potential applications of structured liquid oil in food or bio-related fields.

9.2 Formation of Biopolymer Complexes

9.2.1 Preparation of Complexes: Effect of pH and Concentration

A mixture of proteins and polysaccharides in dispersion may behave differently as either a segregative phase or a single phase, depending on pH, ionic strength, charge density of the biopolymers, molecular nature of biopolymers, biopolymer ratios and concentrations.[11] The segregative phase of a protein and polysaccharide mixture is a thermodynamically unstable condition, which may resemble the instability if it is applied for stabilisation of bulk systems (*e.g.* emulsions). In a single phase system, biopolymers may have a strong or weak interaction to form soluble complexes or co-soluble biopolymers (where two separate entities distribute uniformly throughout the medium), respectively. This single phase mixture is thermodynamically stable. Thus, it can be used for hydrophobic–hydrophilic interface stabilization of bulk phases either *via* formation of pre-complexes or co-soluble complexes in the mixture.

Attractive electrostatic complexation is mainly driven by pH, which controls the magnitude of the interaction. The electrostatic interaction can be strong or weak depending on the overall number of surface charges of the mixed biopolymers. The net overall or unequal number of opposite charges will lead to the formation of insoluble complexes (larger particulation) or soluble complexes, respectively. The effect of pH in the mixed biopolymer systems has been extensively studied with different types of protein–polysaccharide mixtures: WPI-(κ-, ι-, and λ-type) carrageenan electrostatic complexes,[28] WPI–ALG,[29] β-lg–pectin,[30] β-lg–acacia gum,[31] lactoferrin–pectin,[32] SC–gum Arabic (GA),[33] and pea protein isolate–GA.[34]

In our recent research, the electrostatic complexes of WPI–LMP formed at different pH values (*i.e.* pH < pI and pH > pI of WPI) were demonstrated to stabilize HIPEs.[35] WPI–LMP electrostatic complexes were pre-formed by pH adjustment of aqueous mixtures at pH ≤ pI of WPI (4.5–5.5), *i.e.* 3.5, 4.5, and 5.5. These aqueous dispersions containing complexes of WPI–LMP were then used as the aqueous phase and homogenized together with oil to obtain HIPEs.

On the other hand, co-soluble biopolymers of WPI–LMP and SC–LMP obtained by mixing of two components at pH far above pI (pH > pI of WPI) could be used to prepare HIPEs with high physical stability against coalescence. It is speculated that the molecular layer formed by biopolymer molecules is more stable as compared to a layer of biopolymer particulates formed at pH ≤ pI of WPI. Co-solubility of the two biopolymers was obtained by using a relatively high ratio (1 : 50 of protein : polysaccharide) and a low concentration of biopolymers (<1%). Homogenization of the mixtures using equipment such as a magnetic stirrer or a sonication device might be required to uniformly distribute the two biopolymers because of strong electrostatic repulsion and high steric exclusion between the two biopolymers.[36]

Mixed co-soluble biopolymers could form complexes at the oil–water interface creating stable HIPEs. Complexation of co-soluble biopolymers occurs during the emulsification at the hydrophobic–hydrophilic interfaces, leading to a cooperative biopolymer adsorption such that proteins tend to be easily adsorbed at the interface (owing to surface activity) covered by the layer and networks of polysaccharide between the interfaces. In addition, the HIPE preparation using a dispersion of co-soluble biopolymers (prepared at pH ≈ 7.0) has advantages: it is not only a straightforward method, but it also enables a wide range of applications in food products as the pH of the mixture is neutral.

9.2.2 Properties of Complexes

General measurement techniques such as dynamic light scattering, viscosity, contact angle, and surface tension measurements were used to characterize the properties of WPI–LMP complexes. Dynamic light scattering was used to measure the Z-average diameter of the complexes to investigate if the particle size could determine the performance of the complexes for HIPE stabilization.

Table 9.1 presents the effect of pH on the *Z*-average diameter of WPI–LMP complexes and on the volume weighted mean diameter $(D[4,3])$ of HIPEs.[35] For comparison, the data of WPI are also given. WPI at pH 5.5 was slightly aggregated ($\approx$1 μm) as the pH was in the range of the pI of WPI, which was also shown by the slightly white appearance of the WPI solution, indicating an increase of turbidity in the WPI solution. The effect was even greater at pH 4.5 as this value is very close to the pI of WPI ($\approx$pH 4.8), in which the scattered intensity of the WPI solution showed a maximum with a milky appearance and some precipitation.

Besides the effect of pH, increasing the concentration of WPI–LMP in the complexes led to an increase in the size of the particles (Table 9.2). However, although the particle size of complexes increased with the increase of biopolymer concentration, the dispersions were stable without showing any precipitation, probably attributed to the increase in the viscosity of the dispersion medium. The increasing viscosity provides better HIPE stabilization, as shown by a decrease in the size of the oil droplets (Table 9.2).

The performance of WPI–LMP pre-complexes in the stabilization of high internal phase oil-in-water emulsions also depends on the wettability of the WPI–LMP particles by the liquid phases at each side of the interface. In this case, it appears that optimal stabilization is obtained if the continuous phase wets the particles somewhat better than the dispersed phase does.[37] The wetting properties of WPI–LMP particles were investigated by measuring the contact angle θ_w of a drop of water on a dry film of WPI–LMP (0.5:0.25%)

Table 9.1 *Z*-average diameter of WPI (1.0 wt%) and complexes (1.0:0.5 wt%), and volume-weighted average diameter $D[4,3]$ of HIPEs stabilized by WPI and complexes at different pH values. Reproduced from ref. 35 with permission from the Royal Society of Chemistry.

	Z-average (nm)		$D[4,3]$ (μm)	
pH	WPI	Complexes	WPI	Complexes
3.5	153 ± 3	9171 ± 2340	16.1 ± 0.3	19.0 ± 0.8
4.5	35 413 ± 5000	222 ± 12	44.1 ± 2.0	14.8 ± 0.3
5.5	1049 ± 101	440 ± 4	16.0 ± 0.6	12.1 ± 0.6

Table 9.2 Apparent viscosity (at a shear rate of 10 1 s⁻¹) and *Z*-average diameter of complexes, as well as volume-weighted average diameter $D[4,3]$ of HIPEs stabilized by complexes prepared at different total concentrations of WPI–LMP at pH 4.5. Reproduced from ref. 35 with permission from the Royal Society of Chemistry.

WPI–LMP (wt%)	Apparent viscosity (mPa s)	*Z*-average diameter (nm)	$D[4,3]$ (μm)
1.0:0.5	6 ± 1	255 ± 4	14.4 ± 0.6
1.5:0.75	16 ± 2	267 ± 5	12.3 ± 0.5
2.0:1.0	31 ± 4	302 ± 5	11.8 ± 0.5
2.5:1.25	45 ± 4	342 ± 6	10.3 ± 0.5

formed at pH 4.5.[35] The value of θ_w (= 59.8 ± 1.0) revealed that the surface of the WPI–LMP complexes was partially wetted by the hydrophilic and hydrophobic phases, which is a requirement for Pickering stabilization.[37] Furthermore, the role of surface tension reduction might also be associated with the surface active properties of the particles in the dispersion. The surface tension value for WPI–LMP dispersion (2.5 : 1.25 wt%) was found to be 47.7 ± 0.2 mN m^{-1}, which is substantially lower than that of pure water, 72 mN m^{-1}.[35]

9.3 HIPEs Stabilized by Biopolymer Complexes

9.3.1 Preparation and Microstructure of HIPE

Pre-complexes of WPI–LMP formed by pH adjustment were first reported to stabilize HIPEs (φ_{oil} = 0.82).[35] In the previous work, HIPEs stabilized by WPI–LMP electrostatic complexes were simply prepared by homogenizing together aqueous dispersions containing complexes of WPI–LMP with sunflower oil (oil : aqueous phase mass ratio = 4 : 1). Homogenization was done under high shear (13 500 rpm) mixing to force adsorption of complexes at the surface of the oil droplets for efficient coverage of the oil–water interfaces. Owing to the high volume of dispersed phase, the closely packed oil droplets provided a structural framework that held the emulsion together with resultant formation of a self-standing elastic gel-like structure. In comparison, HIPEs stabilized by uncomplexed WPI did not show the presence of a complex network, resulting in poor stability properties of the HIPEs (Figure 9.1).

The effect of pH on the preparation of WPI–LMP complexes strongly influences the stabilization of HIPEs.[35] All HIPEs stabilized by uncomplexed WPI

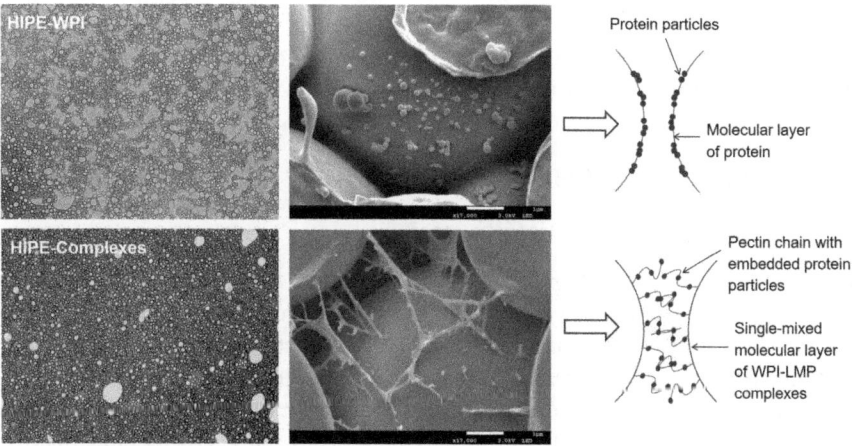

Figure 9.1 Optical microscopy, cryo-SEM images and schematic representation of HIPEs stabilized by WPI particles (top) and complexes (below) formed at pH 5.5. Bars are 1 μm. Adapted from ref. 35 with permission from the Royal Society of Chemistry.

showed poor stability and the instability was even greater at pH 4.5 owing to severe aggregation of WPI particles at its pI (≈pH 4.8) (Table 9.1). Likewise, the HIPEs stabilized by pre-formed aggregated complexes at pH 3.5 (pI of complexes ≈ 3.4) had very poor stability (Table 9.1). On the other hand, HIPEs prepared using complexes at pH 4.5 and 5.5 showed the finest appearance with the absence of oiling-off at macroscale level owing to the complexation with pectin, which prevented protein aggregation at its pI. In addition, the viscoelasticity of the HIPEs also contributed to their easy spreadability without any signs of emulsion breakage and oil expulsion under shear encountered during spreading.

A comparative microstructure of HIPEs stabilized by uncomplexed WPI and WPI–LMP complexes and their schematic mechanism of the stabilization at pH 5.5 is shown in Figure 9.1. HIPEs stabilized by uncomplexed WPI (at pH 5.5) showed evidence of coalescence, while emulsions stabilized by the complexes showed close packing of oil droplets. The coalescence occurred owing to the presence of a few large particulates of β-lg, which were formed around the pI range of WPI (4.5–5.5); the β-lg molecules formed a condensed film around the oil droplets, resulting in stabilization of HIPEs. On the other hand, stable HIPEs formed by the complexes at pH 5.5. were obtained by a combination of an interpolyelectrolyte network (comprising pectin chains linked together with clustered protein particulates) between the droplets and an intermixed layer of WPI–LMP complexes.[35] Similar types of structures have been reported by Wijaya *et al.*,[38] where they suggest that intramolecular interactions between polysaccharide chains and protein particles can occur at pH ≤ pI or pH ≥ pI of the protein for WPI–LMP combinations.

An increase of the total biopolymer concentration from 1.0 : 0.5 to 2.5 : 1.25 (wt%) of WPI–LMP resulted in an increased viscosity of the mixed dispersion and the bulk phase. The consistency of stable emulsions prepared at pH 4.5 and 5.5 could be tuned to smooth (mayonnaise-like) and firm (fat spread-like) spreadable textures depending on the concentration of the complexes used in emulsion preparation.

9.3.2   Properties of HIPEs

9.3.2.1   Particle Size Distribution

Tables 9.1 and 9.2 and Figure 9.2 show the effect of the pH and concentration of WPI–LMP on the volume-weighted average diameter $D[4,3]$ and the oil droplet size distributions of the HIPEs, respectively. The oil droplet size distributions of HIPEs were represented by bimodal and trimodal curves. The minor peaks on the left that were observed at all pH values represent the unadsorbed particles in the bulk phase. Generally, HIPEs stabilized by either aggregated particles of WPI at pH 4.5 or WPI–LMP at pH 3.5 showed a larger droplet size with the major peak position shifted towards the left. Both HIPEs stabilized by WPI or complexes at pH 5.5 showed a narrow distribution range with HIPEs stabilized by complexes showing comparatively smaller

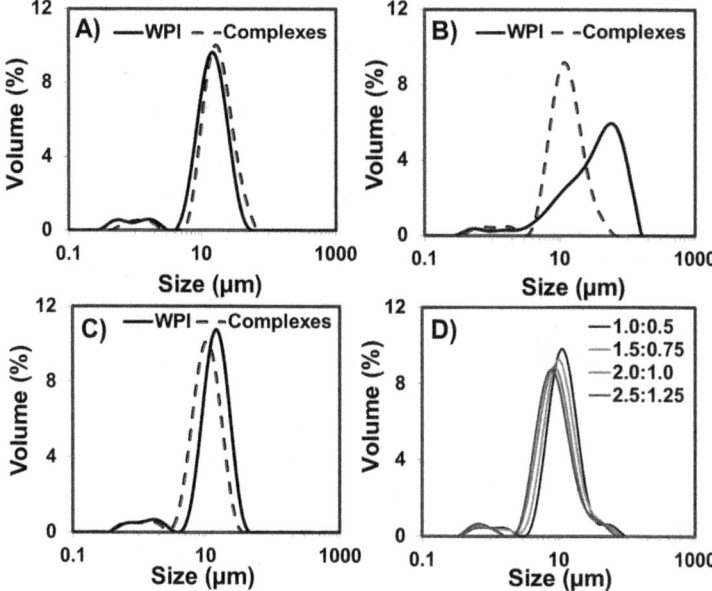

Figure 9.2 Droplet size distribution of oil droplets in fresh HIPEs samples stabilized by WPI (1.0 wt%) and complexes (1.0 : 0.5 wt%) formed at pH 3.5 (A), 4.5 (B), and 5.5 (C), and as a function of WPI–LMP concentration (wt%) in the complexes (formed at pH 4.5) (D). Reproduced from ref. 35 with permission from the Royal Society of Chemistry.

droplet sizes. On the other hand, the increase in the total biopolymer concentration led to a decrease in the droplet size, although the particle size of the complexes increased. This suggests that the size of the particles alone is not the deciding factor for emulsion stabilization; it is rather the total concentration of particles, which can provide an efficient coverage of the interfacial boundaries[39] and increase the viscous drag on the droplets.[40] Moreover, the protein coverage onto the interface was also affected by the total protein concentration, with a higher concentration resulting in a smaller droplet size. Increasing the total biopolymer concentration reduced the volume-weighted mean diameter of the oil droplets resulting in a larger surface area per unit volume of emulsion.[41]

9.3.2.2 Rheology

The rheological properties of HIPEs were evaluated to study the influence of pH and concentration on the emulsion stability and textural properties to establish a relationship between the microstructure and food textural properties. The rheological properties of viscoelastic materials are best determined through small-amplitude oscillatory shear (SAOS) rheometry in which parameters such as the viscoelastic limit (critical oscillatory stress), crossover point, and elastic (G') and viscous (G'') moduli are identified.[42]

   The effect of the pH of complex formation on amplitude sweeps, frequency sweeps and flow behaviour of HIPEs stabilized by WPI and complexes is shown in Figure 9.3. In the amplitude sweeps (Figure 9.3A), all emulsions showed an elastic or solid-like behaviour. The measured G' values for HIPEs stabilized by complexes were always higher than those of HIPEs stabilized by WPI for all the pH values, but at higher amplitudes, a distinct crossover point ($G'' > G'$) was observed, suggesting the structural rearrangement or flow of the emulsion droplets at high applied stress. As shown by the particle size distribution data, the aggregation of WPI and WPI-LMP particles directly affected the textural consistency of the emulsion samples. At pH 3.5, where the complexes were aggregated, the HIPEs displayed a higher firmness than the ones stabilized by complexes formed at pH 4.5 and 5.5. In the case of emulsions stabilized by WPI particles, soft gel-like textures were displayed at pH 3.5 and 5.5 in comparison to pH 4.5, where a lumpy structure with poor stabilization was obtained. Overall, the firmest gel was produced by HIPEs stabilized by complexes at pH 3.5, a pH closer to the pI of the complexes. The controlled aggregation of complexes at pH 3.5 was associated with the addition of pectin, which may cause the increase of particulate mass, thus leading to HIPEs with a firmer structure. However, this was not found in the HIPEs stabilized by WPI at pH 4.5 in which the emulsion showed a lower G' value caused by larger aggregation and poor stability (oiling-off) in the absence of LMP.

   In the frequency sweeps tests (Figure 9.3B), all samples showed slightly positive slopes of G' in the range of applied frequency, indicating the characteristics of a strong gel.[43,44] This is likely attributed to the higher viscoelasticity of protein–polysaccharide layers formed at the oil–water interface[45] in addition to benefiting from the high viscosity owing to the close packing of oil droplets. Furthermore, all HIPEs showed a strong shear-thinning behaviour, as clearly seen from the log–log plot (Figure 9.3C). The viscosity of the sample with the lowest concentration (1.0:0.5 wt% of complexes) is significantly higher than the viscosity of sunflower oil (~0.05 Pa s)[46] even at the end of the flow measurement (at a shear rate of 20 1 s^{-1}), hence confirming that no gel–sol transformation occurs under the applied stress, which is attributed to the well-structured coverage of the complexes at the oil–water interfaces.

   Figure 9.4 shows the effect of total biopolymer concentration on the amplitude and frequency sweeps of HIPEs stabilized by WPI-LMP complexes. An increase in the total concentration of complexes resulted in a progressive increase in the gel strength, G' (elastic modulus), attributed to the stronger network formation in the continuous water phase of the emulsion, as expected. The HIPEs showed the characteristics of a strong gel (curves with slightly positive slopes), as shown in Figure 9.4B, with higher values of G' than G'' throughout the entire frequency range. The inverse proportional relationship of complex viscosity (η^*) with increasing frequency (Hz) as a function of total biopolymer concentration also confirmed the stable gel structure of the samples.

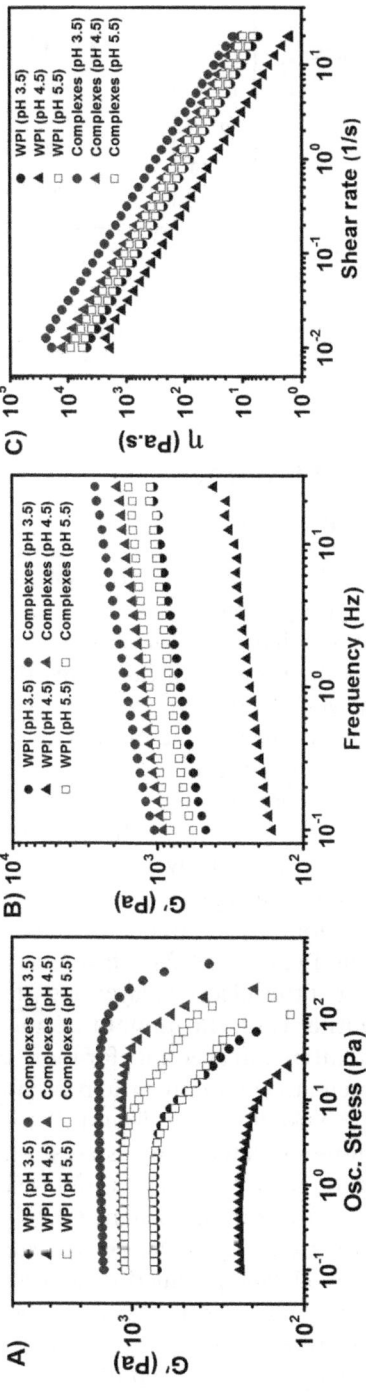

Figure 9.3 Amplitude stress sweeps (A), frequency sweeps (B), and flow measurements (C) (at 5 °C) carried out on HIPEs stabilized by WPI (1.0 wt%) and complexes (1.0 : 0.5 wt%). Reproduced from ref. 35 with permission from the Royal Society of Chemistry.

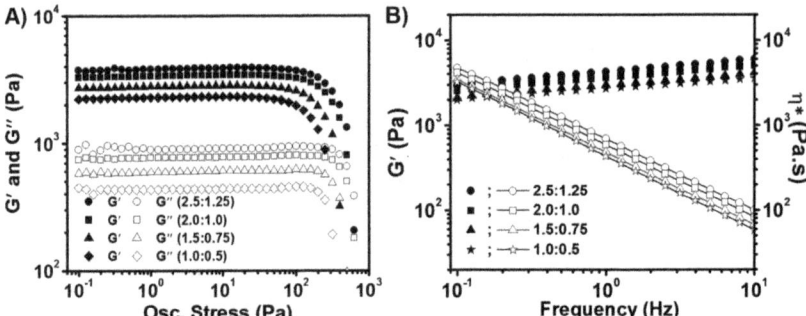

Figure 9.4 Amplitude stress sweeps (A) and frequency sweeps (B) (at 5 °C) carried out on HIPEs stabilized by complexes (formed at pH 4.5) as a function of concentration. Reproduced from ref. 35 with permission from the Royal Society of Chemistry.

9.4 HIPE-templated Oleogels

9.4.1 Preparation of Oleogels

Hydrophilic biopolymers such as proteins and polysaccharides are quite ineffective in structuring hydrophobic oils. Thus, indirect methods such as using the functionality of the polymers to form a structural framework at the oil–water interface followed by removal of the aqueous phase is employed. The conformational framework formed in the water continuous emulsion remains even after drying and entraps the oil phase, resulting in an oil-in continuous polymeric network gel phase.

Emulsions as templates for creating dried structures containing liquid oils for food applications are typically accomplished by spray drying of oil-in-water emulsions to quickly evaporate the water phase and the coalescence of oil droplets into a macrophase during drying is usually prevented by either introducing a substantial amount of carrier/filler (such as maltodextrin) in the water phase[47] or by stiffening the interface through chemical, ionic, enzymatic, or thermal annealing of the adsorbed layer of surface-active polymers, such as proteins.[4,48,49] In either case, the amount of non-oil components in the dried products is substantial and accounts for more than 30 wt%. To get a gel-like consistency, we need to ensure that the percentage of non-oil components in the dried sample is kept to a minimum (*i.e.* oil > 90 wt%).

Most research on emulsion-templated oleogels uses either dilute O/W emulsions (10 wt%)[50] or concentrated O/W emulsions (60 wt%)[51,52] to obtain firm gels. Our approach is a significant improvement over the above-mentioned work as we use a high oil fraction ($\varphi_{oil} > 0.74$, exceeding the maximum close packing value of solid spheres). The production process is straightforward and economically feasible, and requires no cross-linking or tedious colloidal particle formation steps. Moreover, owing to the relatively lower content of water in the bulk phase, the drying of these emulsions is more efficient.

The molecular nature of the biopolymers, such as the structure of the protein and the polysaccharide, becomes the main factor in obtaining a stable oil gel. Gelling proteins provide better stabilization of oleogels after drying of HIPEs than non-gelling proteins (at low concentrations), as illustrated with a case study of oleogels stabilized by WPI–LMP and SC–LMP in Section 9.4.3.1. Moreover, changing the type of polysaccharide from lower molecular weight (*e.g.* LMP) to higher molecular weight (*e.g.* ALG) resulted in stable macro-structural (without any oiling off) and rheological properties after drying (Section 9.4.3.2).

Practically, HIPE-templated oleogels were prepared using a two-step process by first formulating a high internal phase oil-in-water emulsion (using sunflower oil) followed by the removal of water by drying. The HIPEs are stabilized by protein–polysaccharide complexes including WPI–LMP, SC–LMP and SC–ALG. The combined effect of the protein and the polysaccharide plays a key role in the stabilization of HIPEs and oleogels against physical instability (coalescence and oiling off) owing to the stiffening of oil–water interfaces. The surface- and non-surface-active components are responsible for stabilizing the oil–water interfaces and increasing the bulk viscosity of the emulsion, respectively.

The HIPE samples prepared at different concentrations of polymers were subjected to water removal through drying at 35–65 °C for several hours (Figure 9.5). The time required for complete drying of an emulsion was followed by weighing of samples at certain time intervals, and it was found that the drying time ranged from 10 h at 65 °C to 24 h at 35 °C. These data proved that HIPE is dried faster because the water phase content is relatively low in the bulk phase. Previous research investigated the drying time of a 60% oil-in-water emulsion, which required 72 hours at 50 °C for a complete removal of the 40% water phase.[53] The effect of drying of concentrated emulsions containing 60% oil on the oxidative degradation of oleogels was also investigated.[53] At this temperature, the oleogels had a lower peroxide value (PV), *p*-anisidine value (pAV), and total oxidation value (TOTOX = 2 × PV + pAV) compared to the samples dried at higher temperatures. Furthermore, exposure

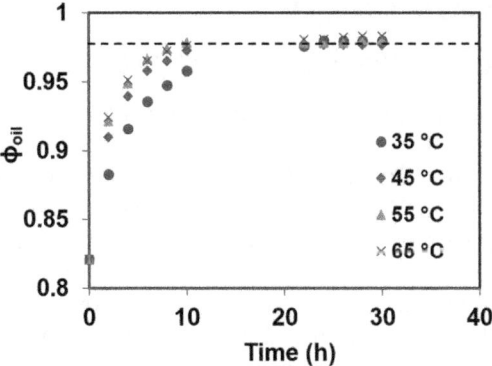

Figure 9.5 Drying of the HIPE stabilized by SC–ALG (1.0:0.2 wt%) in terms of the oil fraction (φ_{oil}) as a function of drying time at different temperatures. The dash line is equal to $\varphi_{oil} = 0.98$.

Figure 9.6   HIPE stabilized by SC–ALG (1.0 : 0.1 wt%) (A), the dried products (B), and the oleogel obtained by shearing of the dried emulsion (C).

of oils (rich in unsaturated fatty acids) to elevated temperatures for such a long time is expected to result in deterioration of the quality of the oils (*i.e.*, oxidative degradation). It was found that exposure to high temperature (80 °C) for a shorter time showed more deterioration than exposure to relatively lower temperatures (50 and 60 °C) for a longer time. The dried samples did not show any rancidity but, based on the peroxide values, they could be considered to be in a high oxidation state and thus susceptible to undergoing oxidation over time storage. Therefore, a feasible option resulting in a high yield product is using HIPEs as templates to obtain oleogels by an economical drying method, whereby due attention should be paid to the drying technique in order to minimize oil deterioration.

The dried products obtained after water removal are shown in Figure 9.6B. Moreover, the dried products could also be transformed into oleogels by simple shearing (Figure 9.6C). Owing to the presence of stiffened interfaces, the removal of water did not lead to coalescence of oil droplets resulting in a soft gel without any oiling off. Moreover, the macrostructure of the sample was retained even after shearing: the gel maintained a homogenous oil-in-gel continuous structure without any oiling off, which suggests that clusters of packed oil droplets were present scattered in the gel networks.

9.4.2   Properties of Oleogels

9.4.2.1   Oleogels of WPI–LMP and SC–LMP

Oleogels have been prepared by drying the HIPEs stabilized by WPI–LMP (1.0 : 0.02 wt%) and SC–LMP (1.0 : 0.02 wt%) at 35 °C for 24 h, followed by shearing using an Ultra Turrax (6500 rpm, 20 seconds). Oleogels stabilized by uncomplexed proteins (WPI and SC) were prepared as a control. To characterize the rheological properties of the oleogels, a range of oscillatory amplitude sweeps were conducted. As shown in Figure 9.7, all samples displayed a gel-like viscoelastic behavior (G' higher than G'' by over a decade). Generally, both HIPE samples either stabilized by uncomplexed WPI or SC showed no significant difference with HIPE samples stabilized by the complexes (WPI–LMP or SC–LMP), as indicated by a similar G' value over the applied oscillatory stress range, which is thought to be because the ratio of polysaccharide to protein was relatively small (1 : 50). However, after drying and shearing to obtain oleogels, higher moduli as well as critical stress values of the oleogels stabilized by either WPI–LMP or SC–LMP complexes were observed as compared to oleogels stabilized by uncomplexed proteins. This suggests the role of polysaccharides in strengthening the network, which also produced more stable macrostructural properties after drying and shearing, since much oiling off was present in the oleogels stabilized by uncomplexed proteins.

9.4.2.2 Oleogels of SC–ALG

Complexation between SC and ALG resulted in a stable macrostructure of the oleogels (no oiling-off). Oleogels were prepared *via* drying of the HIPE stabilized by SC–ALG (1.0 : 0.1 wt%) at 35 °C for 24 hours, followed by shearing using an Ultra Turrax (6500 rpm, 20 seconds). From the frequency sweep (Figure 9.8), the high gel strength of the oleogels was evident based on the following observations: (a) the G' of more than 30 000 Pa that was higher than G'' by more than a decade at all frequency values; (b) the G' was less dependent on the frequency as indicated by a slightly positive slope; and (c)

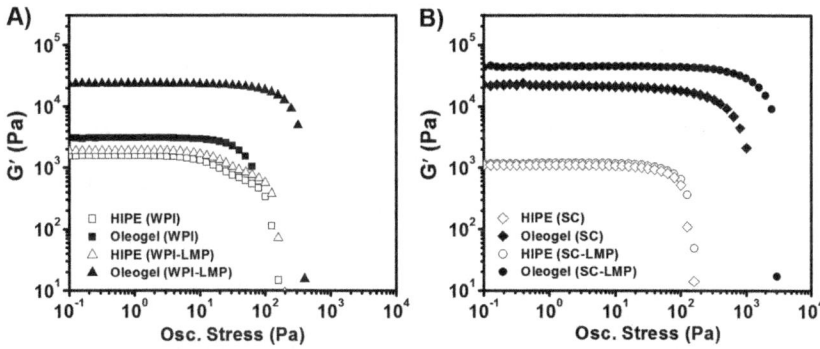

Figure 9.7 Oscillatory amplitude sweeps (at 5 °C) of HIPEs and oleogels stabilized by WPI (1.0 wt%) and WPI–LMP (1.0 : 0.02 wt%) (A); SC (1.0 wt%) and SC–LMP (1.0 : 0.02 wt%) (B).

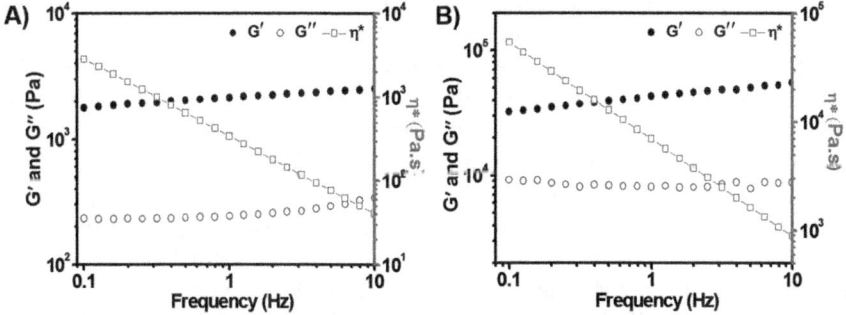

Figure 9.8 Frequency sweeps (at 5 °C) for HIPE (A) and oleogel (B) stabilized by SC–ALG complexes (1.0 : 0.1 wt%).

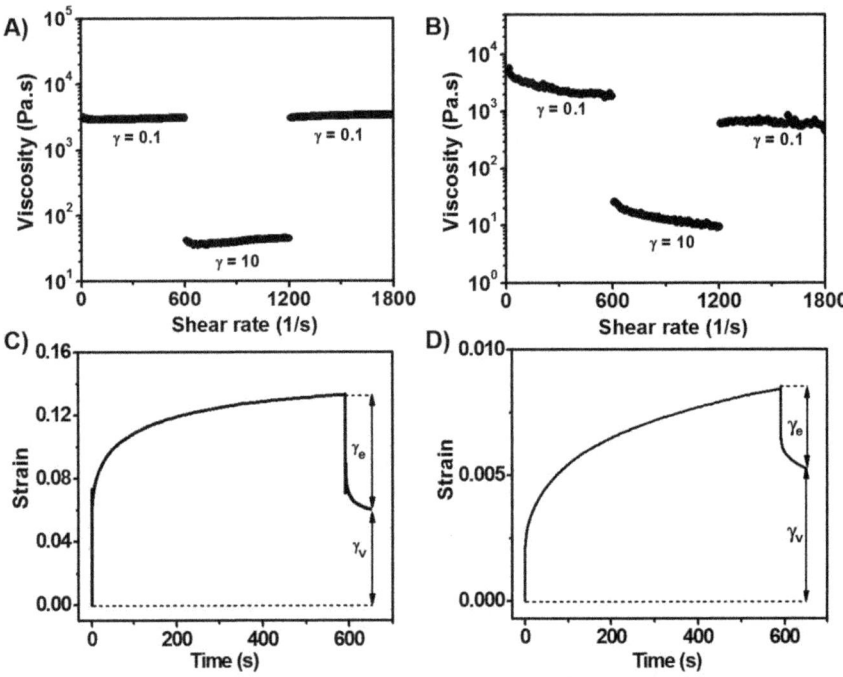

Figure 9.9   Thixotropic properties with three-interval thixotropy tests (γ = 0.1, 10, 0.1 1 s^{-1}, at 5 °C) of HIPE (A) and oleogel (B) stabilized by SC–ALG (1.0:0.1 wt%). (C & D) Creep recovery tests (at 5 °C) of HIPE and oleogel stabilized by SC–ALG (1.0:0.1 wt%), respectively.

the complex viscosity ($\eta^* = G^*/\omega$) showed a proportional decrease with the increase in the frequency.

The thixotropic behaviour of these oleogels as studied using three intervals in rotation (rot-rot-rot) is shown in Figure 9.9A. All samples showed a drop in the viscosity over time at constant shear rate, suggesting that the material response depends not only on the applied rate of shear but also on the time for which the fluid has been sheared. The structure recovery at rest (third interval) was calculated for each sample by taking the viscosity value at the end of interval 1 as 100%. The percentage of recovery for all the oleogels was in the range of 34% compared to 100% recovery of HIPEs. The thixotropic behaviour can be attributed to the microstructure of the oleogels; it could be assumed that at rest the clusters of tightly packed droplets (that were aligned in the direction of flow at higher shear rate) return back to a more random distribution, leading to an increase in the dissipation of energy or resistance to flow. Moreover, the viscoelasticity properties of the oleogels were also studied through a creep–recovery test (Figure 9.9B). Viscoelastic samples are defined by their characteristic response to a constant stress, which is called creep (and refers to the slow progressive deformation or increase in strain over time at constant stress), and when the load or

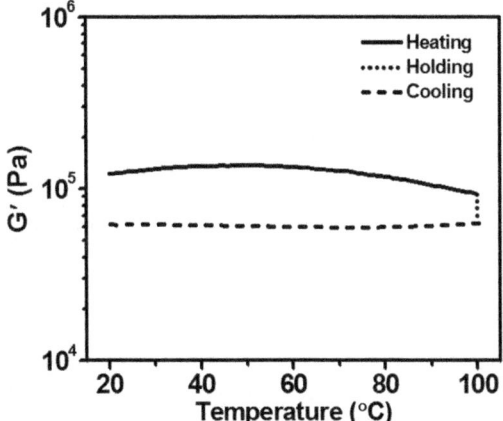

Figure 9.10 Elastic modulus (*G′*) as a function of temperature of oleogels stabilized by SC–ALG (1.0 : 0.1 wt%).

stress is removed, the material exhibits recovery or a progressive decrease of deformation (strain).[54] The extent to which the strain recovers on removal of stress gives an indication of the elastic (γ_e) and viscous (γ_v) portions of the viscoelastic behaviour corresponding to the solid- and liquid-like properties of the samples.[55] Oleogels of SC–ALG showed the best recovery of strain (γ_e = 38.5%) among all the samples while the HIPEs showed the least recovery with γ_e = 55%, indicating a clear difference in the consistency of the two bulk phases. The structure recovery and the recoverable strain of oleogels from thixotropic measurements and creep recovery tests, respectively, was lower than that of HIPEs, implying the importance of the presence of water (in addition to structured oil) in building up the consistency of the bulk phase.[56]

Figure 9.10 shows the behaviour of the oleogels during heating from 20 °C to 100 °C, holding at 100 °C, and cooling back from 100 °C to 20 °C. The oleogels did not show any structural changes during the temperature changes. The gel-like structure of the oleogel samples (indicated from *G′* > *G″* during heating, holding and cooling, *G″* data was not shown) was retained throughout the studied temperature range, suggesting that the oleogel samples are thermostable.

9.5 Potential Food Applications of Biopolymer-based Oleogels: Where Can We Use Them?

Excessive consumption of saturated fatty acids has been found to have a detrimental effect on human health, as suggested by numerous reports published in the last few years.[57–59] In response to this, dietary guidelines by governing bodies worldwide have suggested the elimination of *trans* fatty acids (*i.e.* partially hydrogenated oils) and a reduction in saturated fat consumption to no more than 10% of our daily energy, while increasing the consumption of mono- and poly-unsaturated fatty acids from vegetable (and vegetable oil) sources.[60] However, removing saturated fats from a food formulation is technically challenging for food manufacturers. Solid fats in general are responsible for the mouthfeel and textural properties in lipid-based food products. For instance, the use of solid fats in a shortening formulation contributes to the functionalities such as plasticity, solidity, and hardness. The replacement of these functionality-contributing fats may thus lead to drastic textural differences and resulting problems, such as oil leakage.

In recent years, the structuring of edible oils (*i.e.* oleogelation) has been regarded as a viable option to replace saturated fats in food products.[61] Oleogels structured by different food grade polymers have been technically applied to lower and replace saturated fats in certain food products.[53,62] Oleogels produced by food polymers (methylcellulose, MC and XG) were compared with oil, margarine and shortening in a cake batter formulation. The functionality of the oleogels as a shortening alternative was studied by comparing the batter as well as cake properties, and based on the rheological and texture measurement results the functionality of the oleogels was significantly better than that of oil and was comparable to a commercial shortening and in some cases to the margarine sample. The promising results of these oleogels warrant further studies, including the incorporation of suitable emulsifiers to enhance the functionality of polymer-based oleogels as shortening alternatives.[63] Similarly, the oleogels described in this chapter could be explored for bakery applications. In fact, as the drying of HIPE templates reported in this work is comparatively much more efficient than for the previously reported MC-XG emulsion templates, there is also a possibility of using HIPEs directly as a formulation aid in the production of layered bakery products where the removal of water at high temperature encountered during the baking process would eventually result in the formation of structured oil.

Oleogels stabilized by proteins and polysaccharides may also be potentially applied for the production of creams for various fillings, which are often made with a combination of highly saturated hard stock fat and unsaturated oils, along with other ingredients, such as sugar and flavourings.[60] These fillings, while delicious, are prone to cause quality degradation of the food products that they are used in, which is often because of oil migration. In cream-filled cookies this oil migration can lead to the softening of the cookies, which is seen as a texture defect. In cream-filled chocolates, oil migration can lead to fat bloom and cracking of the chocolate surface, which is seen as a visual defect. It can be hypothesized that by gelling the liquid oil portion of the cream, the oil would be prevented or slowed from migrating, thereby eliminating or delaying the aforementioned defects. The creams made with an oleogel of either high oleic sunflower oil or canola oil leaked very little oil compared to the traditional creams made with liquid oil.[64] These results show that organogelation can be an effective way to slow down or prevent oil migration from cream fillings, which can help prolong the shelf-life of the final products. Furthermore, the increased replacement of saturated fat with gelled oil can offer a greatly improved fatty acid profile of the creams.

The functionality of oleogels (such as high-temperature performance) can be further improved by the protein–polysaccharide interactions in polymer-based gels. In one study, the oleogels stabilized by biopolymers showed a high gel strength without any gel–sol transition during the thixotropic evaluation done at four different temperatures (20, 40, 60, and 80 °C).[65] Such high temperature stability can be very useful for applications in bake-stable cookies, where the traditional bakery fat is known to cause oil-leakage issues owing to the melting of crystalline fat network at high baking temperatures.

Besides the use in replacement of solid fats, oleogels can also be used for nutraceutical applications as delivery matrices for functional bioactives. The term nutraceutical, a food that provides medical benefits, was firstly coined by Stephen DeFelice in 1989.[66] There are many fat-soluble molecules that can be considered nutraceuticals, and could potentially be effectively encapsulated or delivered using an oleogel, such as β-carotene, lycopene, co-enzyme Q10, docosahexaenoic acid, eicosapentaenoic acid, conjugated linoleic acid, tannins, plant sterols, and isoflavones. Though they may currently be consumed in a balanced diet, the levels consumed are too low to produce a pharmacological effect; supplementation in an oleogel could enhance the expression of these effects.

Polymer oleogels that require heat to induce the gelation may degrade the incorporated nutraceuticals. However, the degradation can be minimized by adding them upon cooling, reducing their exposure to temperatures in excess of 100 °C.[67] The oleogels described in the current chapter can be processed at low temperature. Moreover, the high oil content of HIPE may ensure a high loading of bioactives in the oleogels. The abundance of available oil in the oleogels aids in the micellization of the bioactives, improving their uptake by the intestinal lumen and maximizing the effectiveness of their delivery and their eventual health benefits.[67] For instance, the micellization of β-carotene within an oleogel has been proved to control the release and enhance the bioavailability of β-carotene.[68] The results showed that the maximum release of β-carotene from oil occurred between 0 and 30 min of intestinal digestion. Conversely, the gel had a maximum release of β-carotene between 30 and 75 min. This shows that the physical matrix of the lipid component is an important factor for the kinetics of release and micellization of β-carotene during lipid digestion. β-carotene generally has a bioavailability in the range of 10–30%.[68] This experiment demonstrates the potential use of oleogels for controlled delivery and improved bioavailability of nutraceuticals.

9.6 Conclusion

Hydrophilic polymers such as proteins and polysaccharides are commonly used for structuring aqueous phases. These polymers show poor dispersibility in a hydrophobic phase. Thus, indirect approaches are necessary in order to structure the hydrophobic phase, *i.e.* liquid oils. Usually, concentrated emulsions containing 60% oil are used as templates to obtain oleogels. However, the long drying time to evaporate the water phase becomes a drawback that may cause oil deterioration. Therefore, HIPE-templated oleogels were developed as an improved strategy to obtain a high mass yield and shorten the drying time of emulsions. The process of obtaining oleogels from HIPE templates consists of the following general steps: preparation of protein–polysaccharide complexes, fabrication of a HIPE, and drying of this HIPE followed by shearing of the dried HIPE. To obtain successful stabilization of oleogels, some important parameters, such as pH, type of biopolymers, biopolymer ratios and concentrations need to be controlled. HIPEs stabilized by protein–polysaccharide complexes demonstrate that the oil–water interfaces were stabilized by the interpolymeric complexes based on the microstructure analysis. The HIPEs also showed a narrow droplet size distribution (low polydispersity index) based on the static light scattering analysis. According to the rheological analysis, oleogels stabilized by complexation of random coil proteins showed a stronger gel than those of globular proteins. Furthermore, replacement of pectin with a higher molecular weight polysaccharide, *i.e.* ALG, provided better macro-textural properties without any oiling-off.

The food applications of these oleogels have not been evaluated but it can be envisioned that such oleogels could find potential applications as fats in laminated bakery products, bake-stable pastries, cream-fillings and as a controlled release matrix for the delivery of nutraceuticals.

Acknowledgements

The first author would like to thank the BOF (Special Research Fund) of Ghent University for providing financial support.

References

1. A. K. Zetzl, A. G. Marangoni and S. Barbut, *Food Funct.*, 2012, **3**, 327–337.
2. A. J. Gravelle, S. Barbut and A. G. Marangoni, *Food Res. Int.*, 2012, **48**, 578–583.
3. A. R. Patel, P. S. Rajarethinem, N. Cludts, B. Lewille, W. H. De Vos, A. Lesaffer and K. Dewettinck, *Langmuir*, 2014, **31**, 2065–2073.
4. A. I. Romoscanu and R. Mezzenga, *Langmuir*, 2005, **21**, 9689–9697.
5. A. R. Patel, N. Cludts, M. D. B. Sintang, A. Lesaffer and K. Dewettinck, *Food Funct.*, 2014, **5**, 2833–2841.
6. I. Tavernier, A. R. Patel, P. Van der Meeren and K. Dewettinck, *Food Hydrocolloids*, 2017, **65**, 107–120.
7. A. R. Patel, Y. Rodriguez, A. Lesaffer and K. Dewettinck, *RSC Adv.*, 2014, **4**, 18136–18140.
8. E. Perrin, H. Bizot, B. Cathala and I. Capron, *Biomacromolecules*, 2014, **15**, 3766–3771.

9. I. Capron and B. Cathala, *Biomacromolecules*, 2013, **14**, 291–296.
10. L. Jourdain, M. E. Leser, C. Schmitt, M. Michel and E. Dickinson, *Food Hydrocolloids*, 2008, **22**, 647–659.
11. V. B. Tolstoguzov, *Food Sci. Technol.*, Marcel Dekker, N. Y., 1997, pp. 171–198.
12. K. W. Mattison, I. J. Brittain and P. L. Dubin, *Biotechnol. Prog.*, 1995, **11**, 632–637.
13. S. Turgeon, C. Schmitt and C. Sanchez, *Curr. Opin. Colloid Interface Sci.*, 2007, **12**, 166–178.
14. C. Schmitt, C. Sanchez, S. Desobry-Banon and J. Hardy, *Crit. Rev. Food Sci. Nutr.*, 1998, **38**, 689–753.
15. E. Dickinson, *Trends Food Sci. Technol.*, 1998, **9**, 347–354.
16. R. Adjonu, G. Doran, P. Torley and S. Agboola, *J. Food Eng.*, 2014, **122**, 15–27.
17. C. Morr and E. Ha, *Crit. Rev. Food Sci. Nutr.*, 1993, **33**, 431–476.
18. S. M. Fitzsimons, D. M. Mulvihill and E. R. Morris, *Food Hydrocolloids*, 2008, **22**, 576–586.
19. E. Dickinson and J. Chen, *J. Dispersion Sci. Technol.*, 1999, **20**, 197–213.
20. E. Dickinson and Y. Yamamoto, *J. Agric. Food Chem.*, 1996, **44**, 1371–1377.
21. K. R. Kuhn, Â. L. F. Cavallieri and R. L. Da Cunha, *Food Hydrocolloids*, 2011, **25**, 1302–1310.
22. J. Liu, E. Verespej, M. Alexander and M. Corredig, *J. Agric. Food Chem.*, 2007, **55**, 6270–6278.
23. I. Arvanitoyannis and C. G. Biliaderis, *Food Chem.*, 1998, **62**, 333–342.
24. M. Axelos and J. Thibault, *Chem. Technol. Pectin*, 1991, 109–118.
25. G. T. Grant, E. R. Morris, D. A. Rees, P. J. Smith and D. Thom, *FEBS Lett.*, 1973, **32**, 195–198.
26. X. Li, S. Al-Assaf, Y. Fang and G. O. Phillips, *Carbohydr. Polym.*, 2013, **92**, 1133–1142.
27. M. J. Fabra, P. Talens and A. Chiralt, *Carbohydr. Polym.*, 2008, **74**, 419–426.
28. A. K. Stone and M. T. Nickerson, *Food Hydrocolloids*, 2012, **27**, 271–277.
29. S. A. Fioramonti, A. A. Perez, E. E. Aríngoli, A. C. Rubiolo and L. G. Santiago, *Food Hydrocolloids*, 2014, **35**, 129–136.
30. O. G. Jones, E. A. Decker and D. J. McClements, *J. Colloid Interface Sci.*, 2010, **344**, 21–29.
31. C. Schmitt, C. Sanchez, S. Despond, D. Renard, F. Thomas and J. Hardy, *Food Hydrocolloids*, 2000, **14**, 403–413.
32. C. Bengoechea, O. G. Jones, A. Guerrero and D. J. McClements, *Food Hydrocolloids*, 2011, **25**, 1227–1232.
33. A. Ye, J. Flanagan and H. Singh, *Biopolymers*, 2006, **82**, 121–133.
34. S. Liu, N. H. Low and M. T. Nickerson, *J. Agric. Food Chem.*, 2009, **57**, 1521–1526.
35. W. Wijaya, P. Van der Meeren, C. H. Wijaya and A. R. Patel, *Food Funct.*, 2017, **8**, 584–594.
36. D. Tian, Y. Fang, K. Nishinari and G. Phillips, *Gums Stab. Food Ind.*, 2014, **17**, 52–63.

37. E. Bouyer, G. Mekhloufi, N. Huang, V. Rosilio and F. Agnely, *Colloids Surf., A*, 2013, **433**, 77–87.
38. W. Wijaya, P. Van der Meeren and A. R. Patel, *Food Hydrocolloids*, 2017, **65**, 35–45.
39. A. Teo, S. Dimartino, S. J. Lee, K. K. Goh, J. Wen, I. Oey, S. Ko and H.-S. Kwak, *Food Hydrocolloids*, 2016, **56**, 150–160.
40. J. Wu, M. Shi, W. Li, L. Zhao, Z. Wang, X. Yan, W. Norde and Y. Li, *Colloids Surf., B*, 2015, **127**, 96–104.
41. T. Dapčević Hadnađev, P. Dokić, V. Krstonošić and M. Hadnađev, *Eur. J. Lipid Sci. Technol.*, 2013, **115**, 313–321.
42. A. R. Patel, Y. Rodriguez, A. Lesaffer and K. Dewettinck, *RSC Adv.*, 2014, **4**, 18136–18140.
43. H. Tan, G. Sun, W. Lin, C. Mu and T. Ngai, *ACS Appl. Mater. Interfaces*, 2014, **6**, 13977–13984.
44. H. Yalcin, O. S. Toker and M. Dogan, *J. Oleo Sci.*, 2012, **61**, 181–187.
45. E. Dickinson, *Soft Matter*, 2008, **4**, 932–942.
46. L. Mao, L. Boiteux, Y. H. Roos and S. Miao, *Food Hydrocolloids*, 2014, **41**, 79–85.
47. B. Bhandari, A. Senoussi, E. Dumoulin and A. Lebert, *Drying Technol.*, 1993, **11**, 1081–1092.
48. Y. Cho, H. Shim and J. Park, *J. Food Sci.*, 2003, **68**, 2717–2723.
49. U. Klinkesorn, P. Sophanodora, P. Chinachoti, D. J. McClements and E. A. Decker, *J. Agric. Food Chem.*, 2005, **53**, 8365–8371.
50. Z.-M. Gao, X.-Q. Yang, N.-N. Wu, L.-J. Wang, J.-M. Wang, J. Guo and S.-W. Yin, *J. Agric. Food Chem.*, 2014, **62**, 2672–2678.
51. A. R. Patel, P. S. Rajarethinem, N. Cludts, B. Lewille, W. H. De Vos, A. Lesaffer and K. Dewettinck, *Langmuir*, 2014, **31**, 2065–2073.
52. I. Tavernier, A. R. Patel, P. Van der Meeren and K. Dewettinck, *Food Hydrocolloids*, 2017, **65**, 107–120.
53. A. R. Patel, N. Cludts, M. D. B. Sintang, A. Lesaffer and K. Dewettinck, *Food Funct.*, 2014, **5**, 2833–2841.
54. T. G. Mezger, *The Rheology Handbook: For Users of Rotational and Oscillatory Rheometers*, Vincentz Network GmbH & Co KG, 2006.
55. R. S. Lakes, *Viscoelastic Solids*, CRC press, 1998, vol. 9.
56. A. G. Marangoni, *J. Am. Oil Chem. Soc.*, 2012, **89**, 749–780.
57. S. A. Jebb, J. A. Lovegrove, B. A. Griffin, G. S. Frost, C. S. Moore, M. D. Chatfield, L. J. Bluck, C. M. Williams and T. A. Sanders, *Am. J. Clin. Nutr.*, 2010, **92**, 748–758.
58. R. P. Mensink, P. L. Zock, A. D. Kester and M. B. Katan, *Am. J. Clin. Nutr.*, 2003, **77**, 1146–1155.
59. W. C. Willett, M. J. Stampfer, J. E. Manson, G. A. Colditz, F. E. Speizer, B. A. Rosner, C. H. Hennekens and L. A. Sampson, *Lancet*, 1993, **341**, 581–585.
60. T. A. Stortz, A. K. Zetzl, S. Barbut, A. Cattaruzza and A. G. Marangoni, *Lipid Technol.*, 2012, **24**, 151–154.

61. F. C. Wang, A. J. Gravelle, A. I. Blake and A. G. Marangoni, *Curr. Opin. Food Sci.*, 2016, **7**, 27–34.

62. A. G. Marangoni, *J. Am. Oil Chem. Soc.*, 2012, **89**, 749–780.

63. A. R. Patel, N. Cludts, M. D. B. Sintang, A. Lesaffer and K. Dewettinck, *Food Funct.*, 2014, **5**, 2833–2841.

64. A. G. Marangoni, *J. Am. Oil Chem. Soc.*, 2012, **89**, 749–780.

65. A. R. Patel, P. S. Rajarethinem, N. Cludts, B. Lewille, W. H. De Vos, A. Lesaffer and K. Dewettinck, *Langmuir*, 2014, **31**, 2065–2073.

66. E. K. Kalra, *AAPS J.*, 2003, **5**, 27–28.

67. N. Garti and D. J. McClements, *Encapsulation Technologies and Delivery Systems for Food Ingredients and Nutraceuticals*, Elsevier, 2012.

68. N. E. Hughes, A. G. Marangoni, A. J. Wright, M. A. Rogers and J. W. Rush, *Trends Food Sci. Technol.*, 2009, **20**, 470–480.

CHAPTER 10

Cereal Protein-based Emulsion Gels for Edible Oil Structuring

XIAO LIU[a] AND XIAO-QUAN YANG*[a,b]

[a]Research and Development Center of Food Proteins, Department of Food Science and Engineering, South China University of Technology, Guangzhou 510640, People's Republic of China; [b]Guangdong Province Key Laboratory for Green Processing of Natural Products and Product Safety, South China University of Technology, Guangzhou 510640, People's Republic of China
*E-mail: fexqyang@scut.edu.cn, fexqyang@163.com

10.1 Introduction

Cereals as agricultural commodities can provide more than half of the total nutritional intake in the form of complex carbohydrates and proteins. The main cereals consumed are wheat, maize and rice, with regionally significant quantities of barley, sorghum, millet, oats and rye. Cereals, which provide dietary proteins with renewable and sustainable resources, are regarded for their economic aspects, widespread use and health advantages. Cereal proteins as a major part of plant proteins (if we consider that the production of milk and meat is also based on feed containing cereals) play the most important role in the protein supply.[1] The functional attributes of cereal proteins are extensively used in food processing, for example, the formation of aerated open sponge-like structure in bread that is sustained by the wheat protein network generated during fermentation and baking.

Food Chemistry, Function and Analysis No. 3
Edible Oil Structuring: Concepts, Methods and Applications
Edited by Ashok R. Patel
© The Royal Society of Chemistry 2018
Published by the Royal Society of Chemistry, www.rsc.org

Cereal proteins, which are highly complex, heterogeneous and aggregate readily, are often difficult to extract. In most cereals, storage proteins termed prolamins and glutelins are most abundant. Both protein classes are insoluble in water, salt or buffer solutions. Prolamins are soluble in aqueous alcohol blends, while glutelins are soluble only in acids, alkali, detergents, denaturants, or after disulfide bond cleavage.[2] Therefore, the insolubility and complexity of cereal proteins in aqueous solutions severely restrict their application in most liquid food systems. Fortunately, cereal proteins contain many hydrophobic amino acids, and the inherent amphiphilic property makes it a suitable material for fabricating functional colloidal particles. In recent years, our research team has found that colloidal particles prepared from cereal proteins are capable of becoming stabilizers for oil-in-water emulsions, which is a promising approach for oil structuring.[3–7] Structuring of vegetable oils to replace *trans* and saturated fats has received considerable attention in recent years owing to its potential application in foods, cosmetics, and pharmaceuticals.

It has been demonstrated that intake of excessive amounts of *trans* and saturated fats increases the risk of coronary heart disease by virtue of raising the levels of LDL (low density lipoprotein).[8] Recently, the US Food and Drug Administration (FDA) announced that partially hydrogenated oils (PHOs) no longer have GRAS (Generally Recognized as Safe) status for use in human foods and the food industry is required to phase out the use of *trans* fats by 2018.[9] Hence, structuring edible liquid oils into solid fats without PHOs has gained a lot of popularity among academic researchers, and is regarded as a hot topic in the food- and bio-scientific communities for its sustainable manufacturing and human health benefits, especially as substitutes for margarine, spreads, and shortening.[10]

Oleogelation can be simply defined as a strategy for transforming liquid oil into soft solid-like structured gels, which can be done by the addition of low-molecular-weight organogelators, such as 12-hydroxystearic acid, waxes, fatty acids, lecithins, and phytosterol–oryzanol mixtures, which self-assemble into crystalline or tubular networks in the bulk oil phase, forming self-supported oil gels.[11–14] Unfortunately, some of the above materials for oil structuring are either expensive or not granted regulatory approval in food formulation, and most food-approved polymers are inherently hydrophilic in nature.[15] The emulsion-templated approach is an indirect and valid method of structuring edible liquid oils. A variety of proteins, polysaccharides, and surface-active molecules, which can lower interfacial tension and form strong films at the interface, have been introduced to facilitate the formation, improve the stability and produce desirable physicochemical properties of emulsions.[16] In the case of low oil volume fraction, an oil-in-water emulsion with a low-viscosity continuous phase is expected to show liquid-like flow behavior. In another case where the continuous phase is a viscoelastic biopolymer solution or a higher oil volume fraction, the emulsion can be converted into soft-solid-like materials by common food processing, such as heating, acidification, and enzyme action.[17] Furthermore, the water in the emulsion can be removed by freeze or oven drying until the system contains more than 95 wt% oil and a physical network of oil gels is formed.[18,19]

An ideal plastic fat may exhibit simultaneously good elastic properties and low-viscosity behavior. However, the conventional emulsion-templated approach is liable to arouse physical and chemical instability, oil leakage and weak rheological properties. Compared with conventional emulsions, Pickering emulsions are known to display long-term physical stability. Particles, which are located at the interface and wetted by both dispersed and continuous phases, usually form rigid structures that can arrest droplets rupture and minimize their coalescence.[20] Generally, the initial rigidity of the emulsion interface and the stability of the primary emulsions are conducive to developing firmly structured oil gels using the emulsion-templated approach, which increases the potential for texture modification in structuring oil.[3] In terms of food-grade emulsifiers, cereal proteins have considerable functional properties, including emulsification, gelation, foaming and their applications as ingredients in the food industry.[21]

Therefore, in this chapter, we will provide an overview of the recent literature available on cereal protein-based emulsion gels and edible oil structuring, especially about cereal protein (mainly zein, kafirin, gliadin, and wheat gluten)-stabilized Pickering emulsion gels or as emulsion-templated induced oil gels. Some novel methods have been reported for dissolving cereal proteins and fabricating nanoparticles, which are able to overcome the water insolubility limitation that is caused by their high amount of proline. This makes it possible for the cereal proteins to become the solid oil stabilizer. However, there are still considerable challenges with respect to food processing and large-scale production. In this chapter, we intend to introduce this strategy for preparing novel functional cereal protein-based emulsion gels and edible oil structuring in the food industry. Furthermore, the mechanisms are also discussed.

10.2 Zein

Zein, a natural cereal protein from corn gluten meal or maize seeds, comprises 45–50% of the protein in corn. More than 50% of the amino acids of zein are nonpolar (*e.g.*, leucine, proline, alanine, phenylalanine, isoleucine, and valine), which is responsible for its low solubility in water.[22] Moreover, the deficiency in basic and acidic amino acids, particularly the absence of tryptophan and lysine, results in the negative nitrogen balance of zein. A combination of imbalanced amino acid profile and water insolubility results in poor applications in food products for human consumption. Instead, its solubility is known to increase with sufficient concentrations of alcohols, glycols, extreme alkali conditions (pH > 11), and anionic detergents.[23] Fortunately, the capability of forming self-assembled nanoparticles through antisolvent or direct dissolution in the above solvents owing to its amphiphilic character provides zein with the potential for serving as a food-grade Pickering emulsion stabilizer. Generally, the most common strategy for preparing zein nanoparticles is a liquid–liquid dispersion or antisolvent method. In this method, zein is firstly dissolved in aqueous alcohol (55–90% v/v) solution, and then the solution is poured into water with stirring to abruptly decrease the alcohol concentration, causing phase separation and formation of zein nanoparticles.

Pickering emulsions stabilized by zein nanoparticles at pH 4.0 (lower than the pI of zein) were firstly reported by Folter and colleagues.[24] However, creaming or coalescence of the above emulsions occurred within a few days, which was due to the aggregation and low diffusive mobility of zein colloidal particles. The sparse interfacial adsorption and insufficient surface coverage led to instability of the Pickering emulsions. Subsequently, we employed a novel strategy to prepare zein–sodium stearate (SS) complex colloidal particles by simple ultrasonication, and attempted to affect their interfacial wetting properties.[3] Finally, these particles were used as stabilizers, and the resulting Pickering type emulsions exhibited superior stability against both coalescence and creaming. After removing the water by freeze-drying, this liquid emulsion turned into a solid-like oil gel (Figure 10.1A). Moreover, the stability of these emulsion and oil gel depended on the critical complexation concentration (CCC) of SS (about 5 mM) (Figure 10.1B). When the concentration of SS was below the CCC (2.5 mM), the natural conformation of zein might not be disturbed completely. In this case, the intact and inherently hydrophilic zein colloidal particles migrated to the oil–water interface, resulting in an unsaturated particle monolayer at the interface owing to the inefficient accumulation of particles (low surface protein loading) and weak adsorption at the interface. Therefore, the particles failed to provide a packed steric barrier for the oil droplets. As a result, the oil gel obtained after freeze-drying had a small amount of oil leakage. However, when the SS concentration was above the CCC (10 mM), the C-16 alkyl chain of SS might be incorporated into the hydrophobic structure of zein, forming the zein–SS complexes with exposed hydrophobic microdomains. This not only improved the diffusive mobility of water-insoluble zein particles but also endowed the particles with equilibrium interfacial wetting properties, leading to a significant increase of surface particle coverage and efficient packing of zein colloidal particles at the oil–water interface. Finally, a homogeneous and translucent gel without oil leakage was obtained after freeze-drying, meaning that the oil content entrapped in this matrix was more than 92 wt%. In contrast, the emulsion only stabilized by 10 mM SS yielded a turbid flowing liquid, indicating that freeze drying failed to produce an encapsulated oil gel. Additionally, interface manipulation in multiple-layer structures *via* combination of proteins and polysaccharides is another widely used approach for improving emulsion stability. In this manner, a protein-stabilized interface is deposited electrostatically with an oppositely charged polysaccharide, which acts as a physical barrier, resulting in an increased steric hindrance.[25] Since it is a substrate for enzymatic oxidation-induced cross-linking reactions, sugar beet pectin was chosen to interact with zein particles. An emulsion with good stability for a long period had been obtained by using these pectin-enriched zein particles

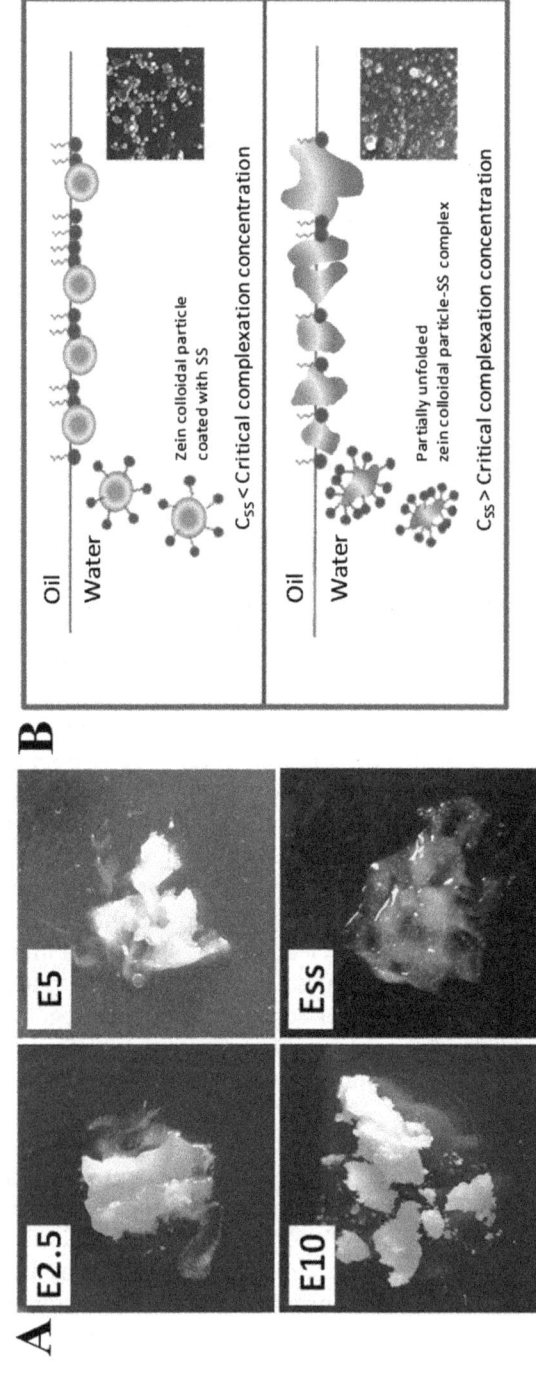

Figure 10.1 (A) Visual aspect of the freeze-dried emulsions stabilized by zein–SS and sole SS: E2.5, emulsion stabilized with 0.5 wt% zein and 2.5 mM SS; E5, emulsion stabilized with 0.5 wt% zein and 5 mM SS; E10, emulsion stabilized with 0.5 wt% zein and 10 mM SS; Ess, emulsion stabilized with 10 mM SS. (B) Schematic diagrams showing enhanced adsorption and targeted accumulation of zein particles at the interface with the synergism of SS.[3] Reprinted with permission from Z. M. Gao, X. Q. Yang, N. N. Wu, L. J. Wang, J. M. Wang, J. Guo and S. W. Yin, Protein-based Pickering emulsion and oil gel prepared by complexes of zein colloidal particles and stearate, *J. Agric. Food Chem.*, 2014, **62**, 2672–2678. Copyright 2014 American Chemical Society.

as stabilizers. Sequential enzymatic and calcium cross-linking of pectin transformed the Pickering emulsions to self-standing emulsion gels with feasible applications as fish oil- and calcium-carrying delivery systems.[26]

It is concluded that adsorption and conformation aspects of protein at the oil–water interfaces play key roles in the formation and stability of protein-based Pickering emulsions. However, the modifiers of SS and enzymes are either synthetic or have a poor organoleptic acceptance, which limits their application in food formulations. Tannic acid (TA), a natural polyphenol classed as GRAS by the FDA for the use as a direct additive in food products, contains sufficient hydroxyls and other groups such as carboxyls that could be used as a substitute.[27] The perception of astringency is derived from the affinity between tannins and human salivary proline-rich proteins (PRP),[28] and cereal proteins (*e.g.* zein, gliadin, kafirin) are also rich in proline. Therefore, TA was used to control the self-assembly behavior of zein by controlling strong hydrogen bonding interactions between the hydroxyl groups on TA and the carbonyl groups of the proline-rich domain in zein, consequently forming stable and edible zein–TA colloidal particles and Pickering emulsions thereafter.[4] Owing to the protonation and ionization of TA, hydrogen-bonding interactions between zein and TA can be influenced by changing the pH and TA concentration. At pH 3.0 and 7.0, partial ionization of the hydroxyl groups in TA led to the formation of electrostatic complexes with low surface charge and unstable colloidal dispersions. However, at the intermediate TA concentration (zein/TA (w/w) = 1:0.2), partially protonated TA facilitated strong hydrogen-bonding interactions at pH 5.0, which generated a conformational change in zein without altering its supramolecular structure. Zein colloidal particles coated by TA further suppressed the ionization of TA. This interaction caused an increase in the hydroxyl groups on the surface of zein particles at acidic pH, thereby lessening their hydrophobic characteristics without decreasing surface charge significantly, and facilitating the multilayer interfacial architecture on the oil–water surface of the oil droplets. Finally, TA could tune near neutral wettability and enhance the interfacial reactivity of zein particles *via* modulation of the hydrogen-bonding interactions, which allowed the stabilization of oil droplets and the formation of stable Pickering emulsion gels (Figure 10.2).

Osborne reported a method for dissolving zein by using glycerol as the solvent at the temperature of 150 °C, and noted that such solution was heat stable and could be heated to 200 °C without apparent denaturation of the protein.[29] The distinct conformation of zein leads to the formation of "rectangular tri-blocks" with a hydrophobic core, accompanied by two parallel hydrophilic patches at the surface. This unique geometry of zein resembles a tri-block, which provides an amphiphilic character.[30] Recently, de Vries and co-workers reported a one-step process of homogenizing to obtain a gel based on the orientation-controlled assembly behavior of "Janus particle" zein.[31] Compared with the above preparation of conventional zein-based emulsions and oil gels,[3,4] this strategy did not involve the use of organic solvents and other surfactants, which made it suitable for possible application

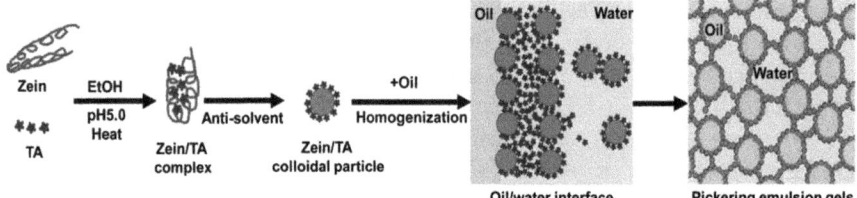

Figure 10.2   Schematic mechanism for the formation of Pickering emulsion gels prepared with zein/TA complex colloidal particles.[4] Reprinted from Y. Zou, J. Guo, S. W. Yin, J. M. Wang and X. Q. Yang, Pickering emulsion gels prepared by hydrogen-bonded zein/tannic acid complex colloidal particles, *J. Agric. Food Chem.*, 2015, **63**, 7405–7414. Copyright 2015 American Chemical Society.

in the food industry. More importantly, the resulting gel samples could be obtained even after several cycles of heating and cooling, confirming the self-standing and thermo-reversibility of the emulsion gel. Although viable strategies for structuring liquid oils into semi-solid fats that take advantage of the zein gelation behavior have been reported, the edible-related applications for delivery of bioactive compounds and as margarine or shortening alternatives have received very little attention. Based on the research of Vries *et al.*, β-carotene was used as an efficient bioactive to fabricate fortified edible zein-based oil-in-glycerol emulsion gels in our study.[5] The results showed that β-carotene incorporated in zein-based emulsion gels could be successfully utilized as an easily spreading fat materials owing to the hydrophobic interaction between zein and β-carotene (Figure 10.3). In addition, compared with liquid oil, zein-based emulsion gels significantly enhanced the UV photo-stability of β-carotene and retarded oil oxidation. Furthermore, cakes prepared using zein-based emulsion gels as a margarine alternative showed comparable functionalities (texture and sensory attributes) to the standard cake. This work provides in-depth characterization of zein-based oil-in-glycerol emulsion gels incorporated with β-carotene, and it has the potential to contribute to the use of natural cereal protein zein in food systems and functional food products fortified with lipophilic bioactive components.

Recently, we also demonstrated firstly that the use of acetic acid as a solvent for zein followed by the addition of water led to the formation of zein particles by phase separation. These particles could be utilized for the fabrication of stable emulsion gels by one-step homogenization. Their large particle size and low zeta potential might decrease the interparticle forces and assembly rate to provide enough time for the formation of ordered interfacial networks. The presence of acetic acid in the continuous liquid phase could decrease the surface tension of the oil–water interface and disperse oil droplets with smaller size. This study not only provides a promising way to form solid-like viscoelastic emulsion gels, which can serve as suitable alternative to solid fats and avoid the use of *trans* fat and saturated fat, but also facilitates the widespread use of zein in the food, cosmetic, and pharmaceutical

Figure 10.3 Physical appearance of zein-based emulsion gels incorporated varying concentrations of β-carotene. Panel B presents the photograph of
zein-based emulsion gels enriched with β-carotene (3.0% zein, 0.030%
β-carotene).[5] Reprinted from *Food Chemistry*, **211**, X. W. Chen, S. Y. Fu,
J. J. Hou, J. Guo, J. M. Wang and X. Q. Yang, Zein based oil-in-glycerol
emulgels enriched with β-carotene as margarine alternatives, 836–844,
Copyright 2016, with permission from Elsevier.

industries without any modification and pretreatment, especially as an oral
delivery vehicle for functional ingredients owing to its super-stability in the
low pH conditions of the stomach environment. Additionally, we also found
that amphiphilic zein hydrolysate (ZH) could induce protein precipitation
during heat treatment owing to the hydrophobic interactions between the
peptide and the hydrophobic region of protein. In this case, ZH contributed
to the formation of a cross-linking bridge between the surface-active proteins
and those in the bulk phase, which could be used for modification and regulation of structured protein-based emulsions.

10.3 Kafirin

Kafirin is the major storage protein in cereal sorghum grain. The protein
content of sorghum grain is in the range of 7.3–15.6% (dry matter basis),
and the main prolamin protein kafirin represents approximately 50–60%
of total protein. Compared with other cereals, kafirin is rich in proline
and amide nitrogen derived from glutamine, since these two amino acids
together account for over 30% of the total residues. Consequently, kafirin is
insoluble in water, but soluble in alcohol–water mixtures. It is defined as a
member of the prolamin family with the highest hydrophobicity, as shown

by its solubility and calculated hydration free energies.[32] The protein is also important in malting and beer brewing with functional characters.[33,34]

Kafirin and zein are homologous proteins and have very similar properties. Kafirin nanoparticles are also prepared by phase separation from a solution of protein in acetic acid with water addition. Pickering emulsions were then prepared through high-speed homogenization of vegetable oil with the water suspension of kafirin nanoparticles.[35] The resultant emulsions exhibited resistance against coalescence with internal oil phase fraction ranging from 58.8% to 78.6%. Their droplet size distribution and rheological properties were influenced by the applied particle concentration, oil fraction and ionic strength in aqueous phase. Increasing particle concentration decreased the oil droplet sizes, and increased the emulsified phase volume and the storage modulus for the viscoelastic responses. As the oil phase fraction increased, the emulsified phase volume fraction and droplet size increased, and the rheological properties shifted from fluid dominant to elastic dominant. The addition of salt in the aqueous phase promoted the migration process and enhanced the stiffness of the gel-like emulsions. The Pickering emulsions stabilized by kafirin nanoparticles were also found to be more stable in acidic than alkaline environments. However, elevated temperature induced their structural instability. Kafirin nanoparticles offer promise for fulfilling the demands of "surfactant-free" emulsions with tunable characteristics.

10.4   Gliadin

Gliadins are mainly monomeric proteins with molecular weights (MWs) around 28 000–55 000 Da and can be subdivided into different types (α-, γ-, and ω-gliadins) according to their primary structure. Although cysteine residues are absent in ω-gliadins, α- and γ-gliadins can form three and four intramolecular disulfide bonds, respectively.[36] Gliadin is similar to zein or kafirin, which is not soluble in water or oil, but is soluble in an aqueous alcohol solution. For a short N-terminal domain of gliadin, a repetitive domain contains repeated motifs rich in proline, glutamine and phenylalanine, whereas a nonrepetitive domain includes all intramolecular disulfide bonds. The terminals of gliadin molecules are generally more hydrophobic than the repetitive domain, making gliadins amphiphatic.[37] Amphiphilicity is one of the main driving forces for self-assembly. However, gliadin carries only a slight charge at neutral pH since it has an isoelectric point of around pH 7.8 owing to high levels of glutamine and proline and low content of basic amino acids.[38] The size evolution of the aggregates was found to be related to the aqueous pH value as well as the amino acid composition of gliadins. It is difficult to scale up the process since rapid extensive sedimentation of particles occurs under neutral conditions. Therefore, stable self-assemble gliadin nanoparticles can be made into stable colloidal particles by adding stock solution into 1% acetic acid solution with stirring. The obtained gliadin colloid particles (GCPs) could be used as a particulate stabilizer in oil-in-water emulsions.[6]

However, the stability of emulsions showed pH-dependent behavior. At pH 3.0, the emulsions were unstable. This situation might be associated with the relatively hydrophilic properties of the GCPs at highly acidic pH where the protonation of amino groups in gliadin happened; thus, they could be preferentially wetted by water. Moreover, for Pickering emulsions, the surface charges of particles play an important role in the formation and stability of emulsions. Previous studies have reported that highly charged particles could not stabilize emulsions because a charged particle saw an "image charge" at the oil–water interface and experienced a repulsive energy barrier when it came close to the interface.[39] Therefore, it is difficult for highly charged particles (pH 3.0) to adsorb at the oil–water interface, resulting in sparse coverage at the oil–water interface. Therefore, hydrophilic GCPs may form weak steric barriers because they are preferentially wetted by water and most of the particles' surfaces protrude into the continuous phase of the emulsions. Consequently, the interfacial absorption of GCPs was restrained to some degree. However, the relatively stable emulsions had micrometer-scale droplet sizes under experimental mixing conditions at pH 4 and above. Concomitantly, the above-mentioned "image charge" effect might disappear when less charged GCPs came close to the interface, facilitating the interfacial absorption of the particles. In principle, the formation of an emulsion gel from particle-covered droplets and free particles in the aqueous continuous phase could provide additional stabilization against coalescence. Furthermore, the high detachment energy of adsorbed particles contributed to the stability of the emulsion made at pH 5.0 owing to the few surface charges and partial wettability of the particles. It is suggested that the GCPs could be an effective particle emulsifier at pH 5.0, and it is a facile approach to produce stable food-grade Pickering emulsions, especially for Pickering HIPEs. When the pH of the GCPs dispersions was further increased (pH 8.0 or 9.0), the resulting emulsions were in an aggregated state. In addition, the adjacent oil droplets were closely packed with each other to form a network structure. The combination of the Pickering mechanism and the droplet-based network is responsible for the stabilization of the emulsions. When the pH of the GCPs dispersions was adjusted to 6.0 and above, the absolute value of the zeta potential decreased to a few mV, leading to hydrophobic GCPs aggregates. This situation enabled the absorption of GCPs at the oil–water interface and increased their interfacial coverage, producing stable Pickering emulsions with a thick interfacial membrane. However, the particles quickly aggregated to form visual agglomerates during the pH adjustment, and the random aggregation behavior of the GCPs at this pH range affected the homogeneity of the products since free protein-based aggregates were detected in the emulsions produced at pH 9.0. Hence, we fabricated GCPs through a facile anti-solvent precipitation procedure and demonstrated their use in the formation of Pickering emulsions as well as HIPEs at pH 5.0. The Pickering HIPEs show potential for replacing solid fats in food formulations, which outlines new directions for future fundamental research. This study opens a prospective route based on Pickering HIPEs to transform liquid oils into viscoelastic emulsion gels with no *trans* fat and less saturated fat.

10.5   Wheat Gluten

Wheat gluten (WG) can be extracted from wheat dough by washing with distilled water. The dry solid contains 75–85% proteins, 5–10% lipids, and most of the remaining ingredients are starch and non-starch carbohydrates. In general, gluten as wheat protein is closely related to flour quality and baking performance. Gluten contains hundreds of protein components that are present either as monomers or linked by interchain disulfide bonds as oligomers and polymers. They are unique in terms of their amino acid compositions, which are characterized by high glutamine and proline contents, and low contents of amino acids with charged side groups.[40] The molecular weights (MWs) of native proteins range from around 30 000 Da to more than 10 000 000 Da. Both gluten and zein contain a high amount of proline, and they are insoluble in water. However, compared with zein, gluten proteins exhibit remarkable elastomeric properties, which is ascribed to the ability of gluten proteins to form very long polymers. This can be regarded as a real asset in the optics of producing protein-based food-products and biomaterials. The repetitive domains of gluten proteins are rich in glutamine, which is prone to hydrogen bonding.[41] Traditionally, gluten proteins are divided into roughly equal fractions according to their solubility in alcohol–water solutions: the soluble gliadins and the insoluble glutenins. Both fractions are important contributors to the rheological properties of dough, but their functions are divergent. Gliadins, which account for half of the gluten proteins, are monomeric species, and glutenins as the other half consist of a concatenation of disulfide bond-stabilized polypeptides whose size can reach several millions of Da. So far, the complexity of gluten proteins and the lack of knowledge on their structure are clearly scientific bottlenecks that restrict the use of gluten. Moreover, the water-insolubility of gluten severely restricts its more extensive use in aqueous emulsion systems. Chemical or enzymatic modifications can improve its solubility and broaden its applications. Takeda and co-workers successfully improved the solubility of wheat gluten at pH 4 or lower, where it showed good emulsifying activity.[42] In addition, the emulsifying and foaming properties of Alcalase-assisted hydrolysates of wheat gluten with relatively low degrees of hydrolysis (5.0%) were remarkably higher.[43] However, the above methods may disrupt the characteristics of aggregation and interfacial assembly. Actually, gluten proteins participate in the formation of the film or the matrix, surrounding and separating individual gas cells in bread dough. Moreover, glutenin was only found in the bulk matrix of bread dough, whereas gliadin was seen both in the dough bulk and at the gas cell surface.[44] The highly active surface indicates that gluten has great potential as an emulsifier to stabilize emulsions.

Therefore, to the best of our knowledge, we firstly discovered hot glycerol (130 °C) as a suitable solvent to dissolve gluten.[45] Hot glycerol might destroy the important protein–protein interactions of WG protein chains, and de-aggregate the WG networks into monomers and oligomers of gliadin and glutenin subunits, forming stable WG-glycerol suspensions. Gliadin, like zein as a monomeric protein, could easily dissolve in aqueous ethanol or hot glycerol. For polymeric glutenins, hot glycerol (above 130 °C) might partially break down the key intra- or inter-molecular disulfide bond cross-linking of glutenin chains, resulting in the dissolution of glutenin subunits, which is similar to gliadin in hot glycerol. Hydrothermal treatment above 100 °C in alkaline pH can initiate the β-elimination reaction of WG, glutenin and even gliadin. This reaction is initiated by abstracting the proton from the chiral C atom of cysteine, followed by elimination of a persulfide in the β-position of the chiral C atom, thus opening the SS bond of the protein. Generally, this reaction triggered further cross-linking of WG protein chains, which was attributed to the oxidation of SH groups or SH–SS interchange reactions.[46] Probably, glycerol as a strong hydrogen donor might prevent the occurrence of the SH oxidation or SH–SS interchange reaction, and finally could partially de-aggregate or "dissolve" the glutenin fraction of WG.[47] WG was firstly dissolved in edible glycerol at 130 °C, and then the WG-based emulsion gels with high corn oil loading (60%) were successfully fabricated *via* one-step homogenization.[7] The self-supported emulsion gels with only WG concentration (0.5 wt%) were formed (Figure 10.4). Meanwhile, gel trapping technology combined with interfacial absorption behavior was applied to characterize the adsorbed particles monolayer at the oil–glycerol interface. These results revealed that gliadin was the dominant component of gluten in stabilizing oil-in-glycerol emulsions, but also synergistic interfacial effects of glutenin and gliadin. Morphologically, confocal laser scanning microscopy (CLSM) observation showed that the structure of the emulsion gel resembled a protein percolating 3D network, where the WG protein constituted the cell walls and the oil phase was trapped inside.

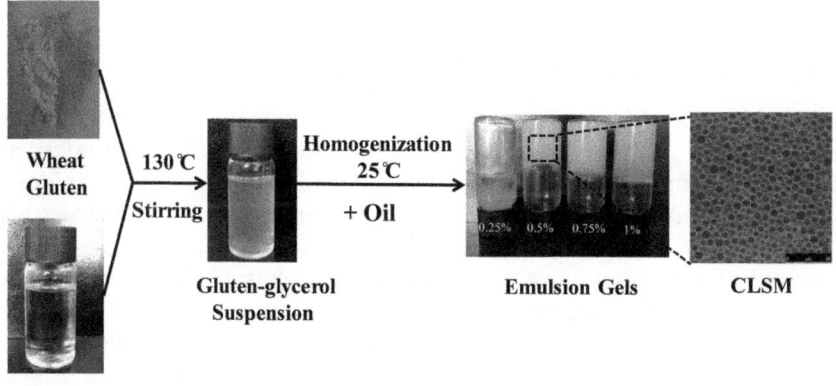

Figure 10.4 Schematic representation of the fabrication process of wheat gluten-based emulsion gels.[7] Reprinted from *Food Hydrocolloids*, Volume 61, X. Liu, X. W. Chen, J. Guo, S. W. Yin and X. Q. Yang, Wheat gluten based percolating emulsion gels as simple strategy for structuring liquid oil, 747–755, Copyright 2016, with permission from Elsevier.

It seemed that the WG protein films, which were originally adsorbed at the interface of an oil-in-glycerol emulsion template, finally converted into a percolating 3D network that confined the oil within closely packed droplets. The resultant gluten-based emulsion gels showed higher elasticity and shear thinning. Meanwhile, WG emulsion gels, especially with high WG concentrations (1 wt%), showed good thixotropic recovery. The applicability of these thixotropic emulsion gels is broader and better suited for a typical product where reversible structure breakdown and recovery are desired. Moreover, the property of heat stability could be beneficial for baking, especially considering edible applications where oil leakage needs to be avoided at high temperature. As gluten and edible glycerol are available and low-cost food ingredients, this study may help to develop a simple and economic approach to transform liquid oil into solid fat without the occurrence of PHOs.

It is worth noting that wheat gluten is a cysteine-rich cereal protein; gliadin and glutenin as two major components of gluten polymerize through a heat-induced sulfhydryl–disulfide exchange mechanism during baking, which is apparently responsible for the wheat flour functionality in bread making. Moreover, for wheat protein, the mechanism of β-elimination reactions during heating under mild alkaline conditions and the formation of covalent cross-linking have been systematically studied. Therefore, imperative cross-linking disulfide bonds at the interface is an effective method to improve functional stability of wheat protein systems. Importantly, cross-linking of the interfacial protein can form strong films, which prevents the oil droplets from coalescing and improves the oxidation stability. Construction of interfacial crosslinking, especially disulfide cross-linking of wheat protein, which plays a prominent part in wheat protein-based emulsion gels and edible oil structuring, deserves our further attention and discussion.

10.6   Conclusions and Outlook

As described in this chapter, cereal proteins could be used for fabricating nanoparticles owing to their amphiphilic structural characteristics, and the prepared multitudinous nanoparticles could be tailored for the design of Pickering emulsions. With the rapidly growing interest in oil structuring without PHOs and considering widely available low-cost in food formulation, cereal proteins show great potential for preparing Pickering emulsion gels and edible oil structuring. Additionally, cross-linking adsorbed proteins makes emulsions that are good templates for structuring of liquid oil into capsules, gels, powders, and HIPEs, and utilizing disulfide bonds or cysteine-rich proteins to prepare emulsions *via* intermolecular disulfide bonds is a promising strategy for avoiding the use of chemical cross-linkers. The facile construction of cross-linking in cereal proteins, especially for wheat proteins, without any thermal treatment or chemical cross-linkers at the interface will offer novel and intriguing options for structured emulsions in the future.

Nevertheless, to date, there are few studies about cereal protein (except for zein, kafirin, gliadin, and wheat gluten)-based emulsion gels, and further studies are necessary to extend these findings. Although the field of edible oil structuring has made a lot of progress in the last few years and it is following the current trend, it can be foreseen that improvements in the compatibility of structured emulsions with real food matrices and the processing pressure they should withstand during food manufacture will play a key role in increasing its utilization in functional foods.

Acknowledgements

This work was supported by the grants from the Chinese National Natural Science Foundation (Serial numbers: 31371744), and the National High Technology Research and Development Program of China (863 Program: 2013AA102208-3).

References

1. R. Lásztity, *The Chemistry of Cereal Proteins*, CRC, 1996.
2. S. R. Bean, J. A. Bietz and G. L. Lookhart, High-performance capillary electrophoresis of cereal proteins, *J. Chromatogr. A.*, 1998, **814**(1), 25–41.
3. Z. M. Gao, X. Q. Yang, N. N. Wu, L. J. Wang, J. M. Wang, J. Guo and S. W. Yin, Protein-based pickering emulsion and oil gel prepared by complexes of zein colloidal particles and stearate, *J. Agric. Food Chem.*, 2014, **62**, 2672–2678.
4. Y. Zou, J. Guo, S. W. Yin, J. M. Wang and X. Q. Yang, Pickering emulsion gels prepared by hydrogen-bonded zein/tannic acid complex colloidal particles, *J. Agric. Food Chem.*, 2015, **63**, 7405–7414.
5. X. W. Chen, S. Y. Fu, J. J. Hou, J. Guo, J. M. Wang and X. Q. Yang, Zein based oil-in-glycerol emulgels enriched with β-carotene as margarine alternatives, *Food Chem.*, 2016, **211**, 836–844.
6. Y. Q. Hu, S. W. Yin, J. H. Zhu, J. R. Qi, J. Guo, L. Y. Wu, C. H. Tang and X. Q. Yang, Fabrication and characterization of novel Pickering emulsions and Pickering high internal emulsions stabilized by gliadin colloidal particles, *Food Hydrocolloids*, 2016, **61**, 300–310.
7. X. Liu, X. W. Chen, J. Guo, S. W. Yin and X. Q. Yang, Wheat gluten based percolating emulsion gels as simple strategy for structuring liquid oil, *Food Hydrocolloids*, 2016, **61**, 747–755.
8. D. Mozaffarian and R. Clarke, Quantitative effects on cardiovascular risk factors and coronary heart disease risk of replacing partially hydrogenated vegetable oils with other fats and oils, *Eur. J. Clin. Nutr.*, 2009, **63**, S22–S33.
9. A. R. Patel and K. Dewettinck, Edible oil structuring: an overview and recent updates, *Food Funct.*, 2016, **7**(1), 20–29.

10. M. Davidovich-Pinhas, S. Barbut and A. G. Marangoni, Development, Characterization, and Utilization of Food-Grade Polymer Oleogels, *Annu. Rev. Food Sci. Technol.*, 2016, 7(1), 65–91.

11. N. E. Hughes, A. G. Marangoni, A. J. Wright, M. A. Rogers and J. W. E. Rush, Potential food applications of edible oil organogels, *Trends Food Sci. Technol.*, 2009, 20(10), 470–480.

12. A. J. Wright and A. G. Marangoni, Formation, structure, and rheological properties of ricinelaidic acid-vegetable oil organogels, *J. Am. Oil Chem. Soc.*, 2006, 83, 497–503.

13. F. R. Lupi, D. Gabriele, N. Baldino, L. Seta and B. de Cindio, A rheological characterisation of an olive oil/fatty alcohols organogel, *Food Res. Int.*, 2013, 51, 510.

14. A. Bot and W. G. M. Agterof, Structuring of edible oils by mixtures of γ-oryzanol with β-sitosterol or related phytosterols, *J. Am. Oil Chem. Soc.*, 2006, 83, 513–521.

15. I. Tavernier, A. R. Patel, P. Van der Meeren and K. Dewettinck, Emulsion-templated liquid oil structuring with soy protein and soy protein: κ-carrageenan complexes, *Food Hydrocolloids*, 2007, 65, 107–120.

16. I. Tavernier, W. Wijaya, P. Van der Meeren, K. Dewettinck and A. R. Patel, Food-grade particles for emulsion stabilization, *Trends Food Sci. Technol.*, 2016, 50, 159–174.

17. E. Dickinson, Emulsion gels: The structuring of soft solids with protein-stabilized oil droplets, *Food Hydrocolloids*, 2012, 28(1), 224–241.

18. A. R. Patel, N. Cludts, M. D. Bin Sintang, B. Lewille, A. Lesaffer and K. Dewettinck, Polysaccharide-Based Oleogels Prepared with an Emulsion-Templated Approach, *ChemPhysChem*, 2014, 15, 3435–3439.

19. A. R. Patel, P. S. Rajarethinem, N. Cludts, B. Lewille, W. H. De Vos, A. Lesaffer and K. Dewettinck, Biopolymer-based structuring of liquid oil into soft solids and oleogels using water-continuous emulsions as templates, *Langmuir*, 2015, 31, 2065–2073.

20. S. Melle, M. Lask and G. G. Fuller, Pickering emulsions with controllable stability, *Langmuir*, 2005, 21, 2158–2162.

21. Z. L. Wan, J. Guo and X. Q. Yang, Plant protein-based delivery systems for bioactive ingredients in foods, *Food Funct.*, 2015, 6, 2876–2889.

22. (a) R. Shukla and M. Cheryan, Zein: the industrial protein from corn, *Ind. Crops Prod.*, 2001, 13, 171; (b) F. A. Momany, D. J. Sessa, J. W. Lawton, G. W. Selling, S. A. H. Hamaker and J. Willett, Structural Characterization of α-Zein, *J. Agric. Food Chem.*, 2005, 54, 543–547.

23. Y. Luo and Q. Wang, Zein-based micro-and nano-particles for drug and nutrient delivery: A review, *J. Appl. Polym. Sci.*, 2014, 131, 40696.

24. J. W. J. de Folter, M. W. M. van Ruijven and K. P. Velikov, Oil-in-water Pickering emulsions stabilized by colloidal particles from the water-insoluble protein zein, *Soft Matter*, 2012, 8, 6807–6815.

25. E. Dickinson, Interfacial structure and stability of food emulsions as affected by protein–polysaccharide interactions, *Soft Matter*, 2008, 4(5), 932–942.

26. S. Soltani and A. Madadlou, Two-step sequential cross-linking of sugar beet pectin for transforming zein nanoparticle-based Pickering emulsions to emulgels, *Carbohydr. Polym.*, 2016, **136**, 738–743.

27. T. J. Kim, J. L. Silva, M. K. Kim and Y. S. Jung, Enhanced antioxidant capacity and antimicrobial activity of tannic acid by thermal processing, *Food Chem.*, 2010, **118**, 740–746.

28. D. Zanchi, C. Poulain, P. Konarev, C. Tribet and D. I. Svergun, Colloidal stability of tannins: astringency, wine tasting and beyond, *J. Phys.: Condens. Matter*, 2008, **20**, 494224.

29. T. B. Osborne, The amount and properties of the proteids of the maize kernel. 2, *J. Am. Chem. Soc.*, 1987, **19**(7), 525–532.

30. Q. Wang, P. Geil and G. Padua, Role of hydrophilic and hydrophobic interactions in structure development of zein films, *J. Polym. Environ.*, 2004, **12**(3), 197–202.

31. A. de Vries, C. V. Nikiforidis and E. Scholten, Natural amphiphilic proteins as tri-block Janus particles: Self-sorting into thermo-responsive gels, *Europhys. Lett.*, 2014, **107**, 58003.

32. P. S. Belton, I. Delgadillo, N. Halford and P. Shewry, Kafirin structure and functionality, *J. Cereal Sci.*, 2006, **44**, 272–286.

33. J. Taylor, Effect of malting on the protein and free amino nitrogen composition of sorghum, *J. Sci. Food Agric.*, 1983, **34**(8), 885–892.

34. J. Taylor and H. K. Boyd, Free α-amino nitrogen production in sorghum beer mashing, *J. Sci. Food Agric.*, 1986, **37**(11), 1109–1117.

35. J. Xiao, X. A. Wang, A. J. P. Gonzalez and Q. Huang, Kafirin nanoparticles-stabilized Pickering emulsions: Microstructure and rheological behavior, *Food Hydrocolloids*, 2016, **54**, 30–39.

36. J. H. Woychik, J. A. Boundy and R. J. Dimler, Starch gel electrophoresis of wheat gluten proteins with concentrated urea, *Arch. Biochem. Biophys.*, 1961, **94**, 477–482.

37. A. Banc, B. Desbat, D. Renard, Y. Popineau, C. Mangavel and L. Navailles, Structure and orientation changes of ω-and γ-gliadins at the air–water interface: A PM–IRRAS spectroscopy and Brewster angle microscopy study, *Langmuir*, 2007, **23**(26), 13066–13075.

38. Y. V. Wu and R. J. Dimler, Hydrogen ion equilibria of wheat glutenin and gliadin, *Arch. Biochem. Biophys.*, 1963, **103**(3), 310–318.

39. H. Wang, V. Singh and S. H. Behrens, Image charge effects on the formation of Pickering emulsions, *J. Phys. Chem. Lett.*, 2012, **3**, 2986–2990.

40. A. Van Der Borght, H. Goesaert, W. S. Veraverbeke and J. A. Delcour, Fractionation of wheat and wheat flour into starch and gluten: overview of the main processes and the factors involved, *J. Cereal Sci.*, 2005, **41**(3), 221–237.

41. A. S. Tatham, L. Hayes, P. R. Shewry and D. W. Urry, Wheat seed proteins exhibit a complex mechanism of protein elasticity, *Biochim. Biophys. Acta, Protein Struct. Mol. Enzymol.*, 2001, **1548**, 187–193.

42. K. Takeda, Y. Matsumura and M. Shimizu, Emulsifying and surface properties of wheat gluten under acidic conditions, *J. Food Sci.*, 2001, **66**(3), 393–399.

43. X. Kong, H. Zhou and H. Qian, Enzymatic preparation and functional properties of wheat gluten hydrolysates, *Food Chem.*, 2007, **101**, 615–620.
44. W. Li, B. J. Dobraszczyk and P. J. Wilde, Surface properties and locations of gluten proteins and lipids revealed using confocal scanning laser microscopy in bread dough, *J. Cereal Sci.*, 2004, **39**, 403–411.
45. B. Lagrain, B. G. Thewissen, K. Brijs and J. A. Delcour, Mechanism of gliadin–glutenin cross-linking during hydrothermal treatment, *Food Chem.*, 2008, **107**(2), 753–760.
46. I. Rombouts, B. Lagrain, K. Brijs and J. A. Delcour, β-Elimination reactions and formation of covalent cross-links in gliadin during heating at alkaline pH, *J. Cereal Sci.*, 2010, **52**(3), 362–367.
47. Y. L. Gu and F. Jérôme, Glycerol as a sustainable solvent for green chemistry, *Green Chem.*, 2010, **12**, 1127–1138.

Section V

Edible Applications

CHAPTER 11

Edible Applications of Wax-based Oleogels

E. YILMAZ* AND S. OK

Çanakkale Onsekiz Mart University, Department of Food Engineering, Faculty of Engineering, Çanakkale, 17020, Turkey
*E-mail: eyilmaz@comu.edu.tr

11.1 Introduction

Natural solid fats with plastic consistency display an essential role in many lipid-based food products. Traditionally, butter, margarines, shortenings, palm stearin, coconut oil, hydrogenated fats, and fat/oil blends have been used to fulfil the hard fat stock needs. These hard fat stocks provide structure, hardness, spreadability, air incorporation, starch and gluten network protection, melting and cooling effects, flavor release, protection and delivery of bioactives, controlling oil migration and other functions of the products. These sources also contain large amounts of saturated fats and some *trans* fatty acids, which are considered to have a negative effect on human health. Since the introduction of oleogels as an alternative to structured fats, there has been growing interest in utilizing oleogels in various food systems. Among the different kinds of oleogels, wax oleogels have received attention owing to their availability, low cost, easiness of oleogel production, stability of the produced gels, similarities of the melting behavior to saturated fats, and applicability in emulsion-type products. While natural food grade waxes are permitted as coating or glazing agents in candy and confectionery sectors in

Food Chemistry, Function and Analysis No. 3
Edible Oil Structuring: Concepts, Methods and Applications
Edited by Ashok R. Patel
© The Royal Society of Chemistry 2018
Published by the Royal Society of Chemistry, www.rsc.org

most countries, their direct use in oil structuring requires additional regulatory approval. Furthermore, waxy taste in the products must also be resolved with product formulation studies. Most importantly, some clinical studies might be needed to observe long term consumption effects of various concentrations of the waxes in the oleogels applied to food products, although some potential health benefits like the lowering of cholesterol levels have already been published.[1-9] Various food applications of wax oleogels are discussed below.

11.2 Wax Oleogels in Margarine and Spread Production

Margarines and spreads are some major fat products of the edible fats/oils industry in addition to the regular liquid cooking and salad oils.[6] Generally, margarines are emulsion products containing not less than 80% fat, whereas spreads may contain fat at 30% or lower. There are currently many types of margarines and spreads on the market, including regular, stick, whipped, soft tub, diet, liquid, *trans*-free margarines, low calorie, whipped spreads, and butter blends.[1,6] Margarine was first developed by the French chemist Hippolyte Megè-Mouriès upon the prize offer of Emperor Louis Napoleon III to meet butter shortages. He called his invention 'oleomargarine' and received French Patent Number 86480 for his development. After the 1930s, margarines spread all over the world with the available technologies of hydrogenation and cooling equipment.[1]

Essentially, margarines and spreads are emulsion products composed of oil and aqueous phases. The aqueous phase may include water, milk or milk products, salt, potassium chloride, antimicrobials (sorbic and benzoic acids and their salts), antioxidants (tocopherols, propyl, octyl, and dodecyl gallates, butylated hydroxytoluene, butylated hydroxyanisole, ascorbyl palmitate, and ascorbyl stearate) and metal scavengers (lecithin, isopropyl citrate, and calcium disodium ethylenediaminetetraacetic acid), flavoring substances (diacetyl, lactones, butyric acid and its esters, ketones, aldehydes), vitamins (A and D), and colorants (β-carotene, natural extracts). The types and amounts of each ingredient depend upon the product specifications and regulatory status of the producing country.[1,6]

The oil phase is composed of fat and additives dissolved in it. The quality and functionality of margarines and spreads are mostly governed by the fats used to produce them. The type, level, composition, and crystal habit of the oil/fat used determine the functional properties of the product, like consistency and texture, spreadability, oil separation, melting behavior, appearance and color, mouthfeel and taste, and product shelf-life and stability. Generally, hard fat stocks are defined by the solid fat index (SFI) to estimate the feasibility. Usually, a SFI of around 10–20 at serving temperature is required. Furthermore, polymorphism is important, and oils with β′ crystal habit (cottonseed, palm, tallow, rapeseed, *etc.*) are preferred.

Usually, blends of common vegetable oils, hydrogenated or partially hydro-genated fats, palm oil, and, when available and permitted, tallow and lard are used in formulations. In spreads, additionally hydrocolloids such as gelatin, pectin, carrageenan, agar, xanthan gum, starch, alginates, and methylcellu-lose derivatives are added as gelling or thickening agents to improve the con-sistency and body of the product.[1,6,7]

The final product characteristics are not only dependent on the formulae but are also affected by the processing parameters. Basically, margarine and spread processing consists of formulating of the two phases, emulsification, cooling and crystallization, working and tempering, packaging, storage and distribution steps. The most important stage is the cooling step, and it is car-ried out with a tubular swept-surface heat exchanger referred to as a votator, chemetator, perfector or combinator. The cooling speed, duration and rest-ing times determine the crystal type and stability, and hence product texture and consistency, to a large extent beyond the product formulae used. This brief introduction is essential during the discussion of oleogels in marga-rine/spread formulations.[6,7]

It has been stated many times that oleogels could be used in margarine/spread production and/or in similar products.[2–4,7,10,11] In an early attempt, crude rice bran oil was solvent fractionated and the separated fractions were characterized for spread production.[12] Although researchers have not used oleogelation terminology, it can be considered as an oleogel formation of crude rice bran oil with the rice bran wax present in the oil. Compositional analysis indicated that molar proportions of wax from 0.33 to 0.37 could yield solid fat contents of 27.64% to 29.10% in the final products. Furthermore, thermal and rheological analyses of the fractions indicated that these rice bran oil fractions (or oleogels) could be used as spread products. Sunflower (SW), candelilla (CW) and rice bran wax (RBW) oleogels were used to produce margarine at 80% level with other regular margarine ingredients (skim milk, mono-diacylglycerols, lecithin, salt, citric acid, calcium disodium EDTA, and potassium sorbate) and were compared with two commercial margarines and four commercial spreads.[10] It was shown that while all oleogels had certain levels of firmness, the CW margarines showed phase separation and no firm-ness was measured. Since there is no relation between the minimum gelling concentration and the gel firmness of oleogels, there was no direct relation between the gelation efficiency of an oleogel and effectiveness in margarine formulation. The firmness of 18–30% hydrogenated oil-containing marga-rines and oleogel margarines containing 2–6% SW oleogel was fairly similar, while the oleogel margarine showed higher melting points. Hence, this study suggested that SW oleogels could successfully replace solid fat stocks in com-mercial margarine and spread products, and more studies about sensory and consumer perceptions are needed. Hwang *et al.* expanded their studies to 12 different vegetable oils to prepare SW oleogel margarines with 3%, 5%, and 7% wax-containing oleogels.[13] The oil:water ratio was 80:20 and the rest of the formulation was the same for all produced margarine samples. It was shown that the fatty acid composition of the base oil did not correlate with margarine firmness, while the concentration of polar compounds in the oil inversely affected margarine firmness. Generally, 3% SW oleogel margarine yielded higher firmness than commercial spreads, and 7% SW margarine was softer than commercial stick margarine. The dropping points of the SW oleogel margarines were usually higher than those of commercial products. The results suggested that healthy vegetable oils could be used to produce margarines and spreads with appropriate SW oleogels as a hard fat stock.[13]

In another study, shellac wax oleogels were used in spread preparations.[14] It was shown that shellac wax oleogel could completely replace fat phase in spread formulations (water content up to 60%) without any emulsifier addition. The surface active fatty alcohols with C28–C32 chain length were responsible for the emulsifying property of the shellac wax at high temperature and on cooling the remaining wax molecules crystallized on the droplet surface as well as in the bulk phase to form the emulsifier-free spreads. In a similar study, virgin olive oil–beeswax oleogel emulsion products were prepared with a simultaneous emulsification and oleogelation approach, and their properties were compared with those of the breakfast margarine.[15] The production process is illustrated in Figure 11.1.

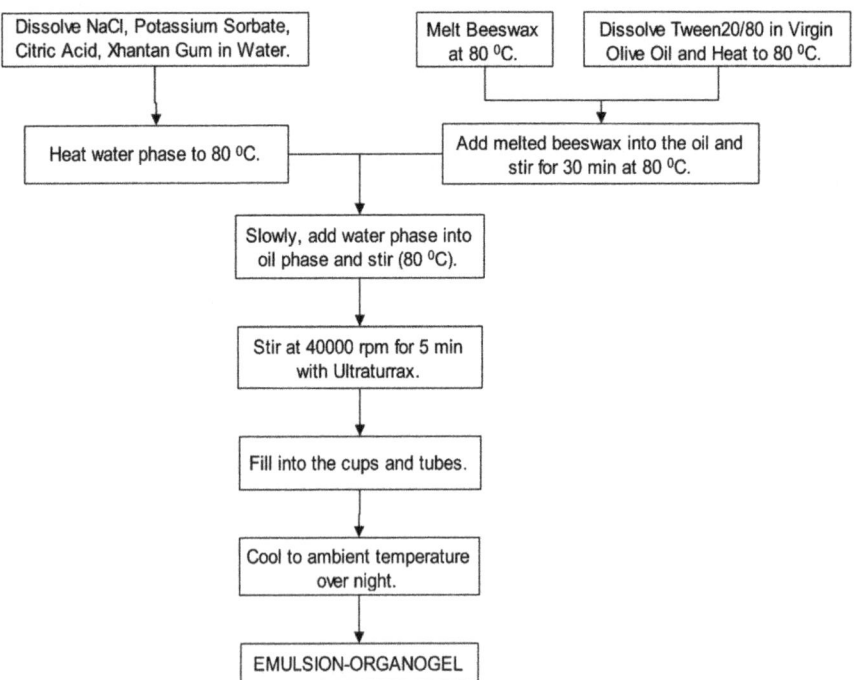

Figure 11.1   Flowchart of the oleogel emulsion preparation process. *Journal of the American Oil Chemists' Society*, Preparation and Characterization of Virgin Olive Oil Beeswax Oleogel Emulsion Products, **92**, Issue 4, M. Öğütcü, N. Arifoğlu and E. Yılmaz, Copyright 2015. With permission of Springer.

In this study, four different stable emulsion oleogels were prepared with Tween 20 and 80 as emulsifiers and xanthan gum as the thickening agent, while 20 other trial formulations with different emulsifiers and thickening agents were found to be unstable over a week of storage. In all experiments, the level of beeswax was 5%, and the oil:water ratio ranged from 90:10 to 75:25. This study suggested that beeswax as an organogelator could be used to prepare stable emulsions only with appropriate emulsifiers. Since beeswax could accumulate at the oil–water interface and the emulsification behavior is unclear in this matrix, the proper emulsifiers were only determined by trial. The melting temperatures of the four emulsion oleogels ranged from 52.29 to 57.52 °C, while it was 40.36 °C for the control breakfast margarine. The oleogel emulsions contained β′ crystal polymorphs, and the hardness and stickiness of the samples were stable for at least 90 days of storage. The polarized light microphotographs of the samples revealed water droplets coated with emulsifiers and needle-shaped crystals of beeswax in the continuous phase of the emulsion. These emulsion oleogel products were found to be quite similar to the breakfast margarine, which is a commercial spreadable product.

There are other studies carried out with different waxes to create spreadable products to replace breakfast margarines, butter or other spreadable products. In one of them, carnauba wax and monoglyceride were used to prepare oleogels with virgin olive oil.[11] Generally, monoglyceride oleogels were better than carnauba wax oleogels in terms of thermal and textural similarities to commercial margarines. Virgin olive oil and hazelnut oil were gelled with beeswax and sunflower wax, and aromatized with strawberry, butter and banana aromas.[15-17] Furthermore, these aromatized oleogels were enriched with vitamin E and D_3, and β-carotene. All physicochemical, textural and thermal properties as well as sensory descriptions were completed. These studies were the first reports of complete sensory descriptive analysis of the oleogel samples. In Table 11.1, the panel defined sensory descriptors, their definitions, and reference materials, which were used to analyze the aromatized virgin olive oil-beeswax and sunflower wax oleogels are presented.

These aromatized wax oleogel products were stored for 3 months, and their structural, oxidative stabilities as well as volatile aromatic compositions were analyzed. The oleogels were structurally stable, uniform, homogeneous, and retained most of the added aromas during storage. The main volatile components of the strawberry flavored organogels were ethyl acetate, ethyl butanoate, ethyl-2 methyl butanoate, D-limonene, ethyl caproate; banana flavored organogels contained isoamyl acetate, isoamyl valerianate, ethyl acetate; and butter flavored organogels showed the presence of 2,3-butanedione and 3-hydroxy 2-butanone. These volatile components were not only detected in the fresh samples but also at the end of the 3 months storage period. Sensory definition terms were matched with the sensory descriptors of the detected volatiles. Furthermore, the oxidative stability of the samples during the storage period was very good, like the morphological and crystalline stability detected by the X-ray diffraction and polarized light microscopy analyses.

Table 11.1 Panel-defined sensory descriptive terms and their references used. *Journal of the American Oil Chemists' Society*, Influence of Storage on Physicochemical and Volatile Features of Enriched and Aromatized Wax Organogels, **92**, Issue 10, M. Öğütcü, E. Yılmaz and O. Güneser, Copyright 2015. With permission of Springer.

Descriptor	Definition	Reference
Appearance properties		
Roughness	Smoothness of a cut surface	Anhydrous milkfat, ice cream
Gloss	Surface brightness, overall gloss	Liquid oil, margarine
Homogeneity	Observe the spread homogeneity	Cream cheese, butter
Texture properties		
Firmness	Force required to push a knife	Cream cheese, butter
Knife stickiness	Amount of residuals on the knife	Cream cheese, butter
Spreadability	Easiness to spread it over bread	Cream cheese, ice cream
Liquefaction	Amount of oiliness on bread surface	Liquid oil, margarine
Aromatic properties		
Fruity	Hazelnut aroma	Hazelnut
Added aroma	Strawberry, diacetyl (butter), bananas	Liquid aromas used
Sweet aromatic	Aromatics associated with honey, maple syrup	Honey
Rancid	Aroma of oxidized oil or popcorn	Stored oil (at least a year)
Flavor properties		
Sweet	Taste of sugar	5% sucrose in water
Bitter	Bitter taste of caffeine or quinine	0.05% and 0.1% caffeine solution
Fatty	Melted vegetable shortening	Shortening and anhydrous milkfat
Waxy	Candlewax or paraffin flavor	Paraffin, sunflower wax
Mouthfeel properties		
Melting rate	Speed of melting in mouth	Breakfast margarine, shortening
Sandiness	Dry or floury texture in mouth	Anhydrous milkfat, cake cream
Mouthcoating	Thin layer of fat deposit on tongue	Liquid oil, shortening, cream cheese

These aromatized wax oleogels could be used to prepare margarines, spreads and similar products. It was also suggested that the aromatized oleogels by themselves could be used as spreadable breakfast products.

In another oleogel study, fish oil–beeswax oleogels were aromatized with strawberry and lemon aroma and their structural, textural, thermal characterization as well as consumer preferences were tested against control samples (no added aroma).[18] Since liquid fish oils or regular fish oil oleogels have a distinct fishy smell that is not appreciated by most consumers, the approach in this study was to aromatize the oleogels and assess the consumer behaviors. It was found that all fish oil oleogel samples were stable and fairly spreadable, and the strawberry flavored sample was best liked by the consumers in terms of appearance, odor, flavor and spreadability characteristics.

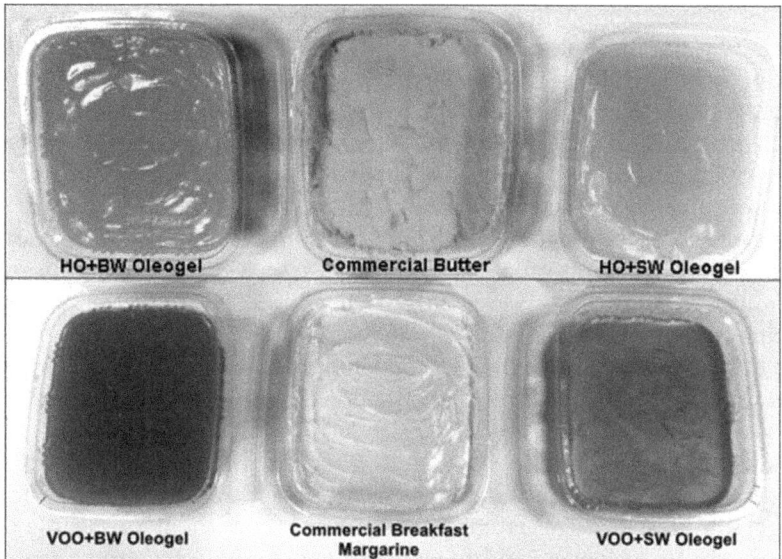

Figure 11.2 Hazelnut oil and virgin olive oil oleogels prepared with BW and SW, and commercial breakfast margarine and commercial butter samples. Reproduced from ref. 19 with permission from The Royal Society of Chemistry.

As a result, fish oil consumption through aromatized oleogels could be enhanced in different product formulations.

Yılmaz and Öğütcü determined the suitability and consumer perceptions of virgin olive oil (VOO) and hazelnut oil oleogels prepared with beeswax and sunflower wax as direct alternative products to spreadable margarine and butterfat.[19] Pictures of the samples are shown in Figure 11.2.

Since VOO is an aromatic and cold pressed oil, no flavor/aroma was added to VOO, and 5% wax oleogels were prepared. Full physicochemical and thermal characterization as well as sensory descriptive analysis and consumer preference tests were completed against the breakfast margarine control. The aim was to estimate consumer acceptance of spreadable VOO prepared as an oleogel. Sensory descriptive properties of the samples were determined by a trained panel with 13 sensory descriptive terms. The results indicated that except grassy, milky and waxy terms, most descriptors were quite similar to those for the commercial margarine. Furthermore, there were fair similarities of texture and thermal properties with the margarine. Most importantly, the consumer hedonic attribute scores showed that appearance, odor, flavor and spreadability scores were close to 3.0 values out of the 5.0 maximum score for acceptance. Hence, the consumers indicated a lower acceptance except for spreadability, which was highly accepted. Further research is needed, especially to improve the waxy odor characteristics of the product. Spreadable VOO would be a good new product invention, especially in the

Mediterranean region where the consumption of VOO is very high.[20] In the same study, hazelnut oil oleogels were prepared with 0.5% w/w diacetyl as a butter flavor. The aim was to observe the consumer behavior for these new products as butter alternatives. Sensory descriptive properties such as hardness, spreadability, grassy and waxy were significantly different to those for commercial butter, while thermal properties were rather similar. The consumers indicated high acceptance for spreadability, but appearance, odor and flavor scores were under 3.0 (the neutrality point). Hence, some improvements are needed to optimize the quality of these systems. Furthermore, the hazelnut oil–beeswax (HBW) and VOO–beeswax oleogel (OBW) samples were selected based on their higher hedonic scores and tested with consumers for their buying decisions. Among 120 consumers, around 57% and 43% indicated that they would definitely buy the product; 24% to 29% indicated that they would try once and then decide; and 12% and 25% indicated that they would definitely not buy the HBW and OBW samples, respectively. This study was the first in the literature to report consumer tests results for the wax oleogel samples. More research is needed to improve the flavor/aroma properties of the wax oleogels and to estimate consumer expectations from the products prepared from the wax oleogels.

In conclusion, studies published to date prove that oleogels can successfully replace hard fat stocks in margarine/spread type products, especially in terms of textural and thermal properties. Furthermore, spreadable oleogels may find some usage as breakfast spreads if aromatized or enriched. However, there are still some problems that need to be solved before commercial exploitation of these systems. Firstly, wax oleogels may exhibit somewhat higher melting points than the expected melting points for commercial margarines/spreads. Secondly, although the rheological properties of the wax oleogel-based margarines were found to be appropriate, some spreadability and/or stickiness problems may be encountered in some of these products. Therefore, hybrid systems based on oleogel and hard fat stock mixtures might offer a better solution. In emulsion oleogel products, the stability of the emulsion might be another big issue, especially during prolonged storage periods and within distribution chain. More research is also needed to find compatible wax-emulsifier-thickening agent combinations in emulsion oleogel type products. Waxy flavor and aroma notes could be a serious problem for consumer acceptance, especially if the amount of wax in oleogels exceeds the 5% level. Product formulations with minimum possible wax concentrations and proper flavor/aroma additives may resolve the problem. Constant supply, availability and price considerations for food grade, high quality plant waxes must also be considered. One of the important issues is the possible physiological and health effects on humans of wax consumption within the oleogel-based products in the long-term. There is not much data in the literature, and this issue must be thoroughly evaluated. Lastly, margarine/spread products with wax oleogels will need new regulatory approval before commercialization of developed products in most countries.

11.3 Wax Oleogels in Bakery Products

Solid fats with plastic consistency are one of the most important components in many bakery products. Specifically, fats used in the bakery sector are called 'shortenings', a term derived from the tenderizing or shortening effect of the lard and butter in the final products, which is achieved by preventing the cohesion of flour gluten during preparation. Today, bakery shortening applications are expanded to all baked products with many shortening types. Shortenings and margarines are distinguished from each other by the purpose of usage and their compositions. Although most shortenings are full fat products, some specific types, such as puff pastry and roll-in shortenings, include some moisture. Shortenings are processed fat products that provide tenderness, lubrication, aeration, long-term softness, stability and storage quality, calories and eating characteristic functions to the dough batter and bakery foods through emulsification, structure formation, protection of dough structure, retention and release of flavors, a moisture barrier and heat transfer.[6-8]

Many different types of shortening products are present commercially, but generally three different categories are distinguished for the food sector; namely plasticized semisolid shortenings, liquid or pumpable shortenings, and flakes, beads, or powder shortenings.[6] Since this section is for baking shortenings, the other types will not be discussed. Baking is applied to all food products in which flour is the main ingredient. Specifically, baking is involved in items such as breads, cakes, pastries, biscuits, crackers, cookies, pies, and toppings, frostings, fillings. In all baked goods, quality production with the lowest cost and consistent product performance is required. Hence, shortening functionality is an essential issue in baking and determines how well a shortening performs in a special baked food through the expectations of the quality requirements. Usually, in baked foods, shortening use results in the formation of a creamy texture and rich flavor, tenderness, enough aeration, moisture retention, size expansion, flakiness in pie crust and puff pastries, lubrication of gluten particles and storage stability.[6,21] Baking shortenings are plastic solids that are defined as products appearing solid until a certain force and after some point permanent deformations occur to cause flow like a viscous liquid. For a shortening to be a plastic solid, it must contain both solid and liquid phases, the solid phase must be homogeneously dispersed and bound together with cohesive forces, and the proportion between the phases must be proper. In these plastic shortenings, the liquid phase must be within the three-dimensional matrix that is formed by the fat crystals. Furthermore, crystal size and polymorphy affect the rheological properties and storage stability of baked products. Baking shortenings should have a balanced ratio of liquid to solid fat (or solid fat content), the right plasticity and enough oxidative stability to impart their functional roles. Usually, a moderate level of solid fat content with β' type crystal polymorphs would be adequate to entrap air bubbles during mixing to prevent hard texture and oiling out problems. Likewise, the right solid fat and crystal polymorphy yield optimum creaming, tenderness and eating quality.[6,8,21-23]

The potential of organogel application in baked foods and many other areas has been discussed, especially in the context of oil migration or leakage prevention.[2,3,24] Jacob and Leelavathi investigated the effects of fat type (bakery fat, margarine, hydrogenated fat and sunflower oil) on doughs and cookies.[25] Sunflower oil had the least initial farinograph consistency or more resistance to mixing, while bakery fat reduced its consistency. Oil-containing cookies had harder texture and cookies containing non-emulsified hydrogenated fat had the least spread. This study showed that fat type is essentially important in bakery products quality.[25]

Shellac wax (SW) oleogels have been used in various edible applications, including cakes.[14] The SW oleogel emulsion containing 20% water and 10% shellac wax in total was used as the shortening in a 4/4 sponge cake formulation against cake margarine as the control. Comparison of the cake batter indicated that the SW oleogel emulsion yielded a runny consistency batter (10-fold lower G' value), and a much lower cross-over point. Furthermore, the decrease in the batter firmness resulted in higher batter density owing to air incorporation. It was indicated that these findings were as a result of the absence of true fat crystals in the SW oleogel emulsion. Although the cake batters were fairly different, the cakes exhibited comparable texture and sensory results. Except for springiness index and chewiness, the textural properties like firmness, cohesiveness and adhesiveness were statistically not different. Springer index is defined as the ratio of sample height that springs back after application of deformation maximum force, and the higher value in SW oleogel emulsion was related to the increased elasticity owing to higher protein crosslinking in the batter. It was also explained that since the SW oleogel emulsion contained only 9.9% crystallized fat compared to 28% in the cake margarine, that amount of fat crystals might not be enough to create a barrier to the gluten network to prevent protein crosslinking. Six sensory definition terms (volume, cell size, moistness, stickiness, sponginess and crumbliness) were used by 20 panelists to compare the two cakes. Except for cell size and crumbliness, all other sensory properties were quite similar. The difference in cell size was attributed to the uneven air distribution in the batter. Furthermore, the non-uniform air distribution in the batter might have caused the higher crumbliness. Since the other sensory properties were comparable, the authors suggested the incorporation of some foam stabilizers in the SW oleogel emulsion cake formulation to enhance the overall quality.

Canola oil oleogels with candelilla wax at 3% and 6% by weight addition levels were prepared and their impact on cookie quality was compared with a commercial shortening.[26] The rheological and textural properties of the prepared oleogels, cookie doughs and cookies were measured. The firmness of the shortening was highest (around 10 N) followed by the 6% wax oleogel (around 3 N) and the 3% wax oleogel (around 0.5 N). The viscosity of all samples decreased non-linearly with increasing temperature. The storage (G') and loss (G'') moduli increased with frequency, and the storage moduli were higher than the loss moduli in the samples, showing a more elastic structure. The dough with commercial shortening had higher values for both moduli, indicating the relation with the higher firmness measured with the shortening sample. Fatty acid composition analysis of the cookies showed that oleogel-incorporated cookies had 90–92% more unsaturated fatty acids, and this would be very important for health considerations. When the physical parameters such as diameter, height, spread ratio and snapping force were analyzed, it was observed that cookie diameter increased and height decreased in the samples prepared with the oleogels as a result of enhanced spread factor. Hence, cookie physical properties were positively affected by oleogel use. Similarly, snapping, which is defined as sudden break under an applied force, decreased with oleogel applications. The reduced snapping observed in the oleogel cookies could be attributed to softer eating characteristics. This study pointed out comparable textural properties and improved fatty acid composition in terms of health for cookies prepared with candelilla wax oleogels, though sensory and consumer preference studies were not performed.

In another study, sunflower wax oleogel (SWO) and beeswax oleogel (BWO) (5% wax by weight) were tested against commercial bakery shortening in cookie preparation, and the textural, sensory and stability properties of the cookies were compared.[23] The standard cookie dough recipe included 24.07% of each oleogel or the shortening and all other ingredients were the same. The BWO and SWO had 3.30% and 3.64% of solid fat content at 20 °C, compared with 29.73% of the commercial shortening. Similarly, the melting temperatures of the samples were 47.67 °C, 61.02 °C, and 45.92 °C, respectively. Hardness values of 2.73 N, 4.18 N and 14.21 N and stickiness values of 2.12 N, 1.68 N and 8.75 N were reported for BWO, SWO and commercial shortening, respectively. These findings clearly indicated that although there was some similarity in terms of melting points of the oleogels and commercial shortening, the textural characteristics were fairly different. The measured crystal sizes of the oleogels were significantly lower than those of the commercial product, and their polymorphs were β' type as measured, indicating better fit for creamy texture expectancies in baked foods. Unfortunately, the cookie dough properties were not evaluated in this study, but the cookie products were comprehensively analyzed.

A picture of the three types of the cookies can be observed in Figure 11.3 together with the consumer hedonic data.

Cookie color measurements showed that cookies prepared with commercial shortening had a lighter surface (highest *L* value), and cookies prepared with BWO had darker surface. While color parameter *a** (redness-greenness) was not different, the *b** values (yellowness-blueness) were different, indicating that wax type in the oleogel could influence the product color. Cookie weight and thickness were different between the shortening and oleogel containing samples. In addition, the spread ratio, described as the diameter:thickness value, indicated that cookies prepared with commercial shortening had the lowest (5.62) value compared to the BWO- and SWO-containing cookies (7.18 and 6.85). It was also stated that the higher the spread ratio, the lower the aeration in cookie dough. Hence, dough aeration was better

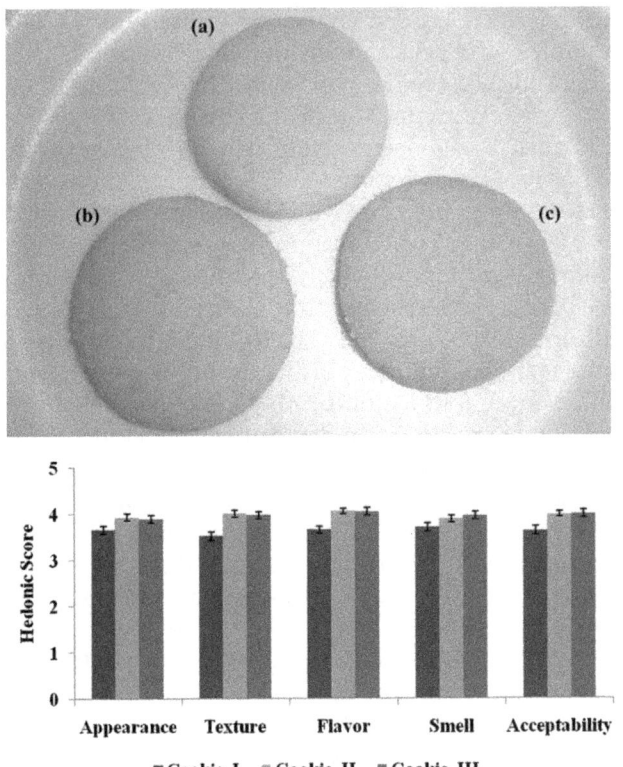

Figure 11.3   Cookies produced with (a) commercial baking shortening, (b) BW oleogel, (c) SW oleogel, and their consumer hedonic scores (1: dislike extremely to 5: like extremely; mean ± SE; n = 200) (reproduced from ref. 23 with permission from The Royal Society of Chemistry).

when commercial shortening was used owing to the presence of actual fat crystals, as indicated by the higher solid fat content value. Cookie hardness and fracturability values exhibited different results. Cookie hardness was highest and fracturability was lowest in the cookie made with commercial shortening. It was discussed that it was possible to produce less hard more fracturable cookies with oleogels owing to the presence of actual higher amounts of liquid oils in the oleogels. The cookie moisture, water activity and ash content were a little higher in the cookies prepared with commercial shortening, while fat content and combustion caloric value were higher in the cookies prepared with the wax oleogels. Since all cookie recipes included 24.07% fat, the difference in the energy values was owing to the composition of the commercial shortening, which was an emulsion containing 79% fat. Contrarily, the oleogels were composed of 95% hazelnut oil and 5% wax. Since water activity lower than 0.3 is accepted to be enough for microbial and enzymatic stability of cookies and similar dry products, the cookies prepared with the oleogels were fairly stable in this respect. There were no significant differences among cookie pH and peroxide values, and the peroxide values ranged from 0.40 to 0.45 meqO$_2$ kg^{-1} sample. This range is quite low and is acceptable in terms of stability and quality.

A detailed texture/flavor profile analysis was performed on the aforementioned three types of cookies by 12 panelists with 15 sensory descriptive terms and a five-point measurement scale from 1 for minimum intensity to 5 for maximum intensity was used. The sensory attributes were grouped into four categories, namely surface properties (glossiness, color, smoothness), cross-sectional properties (compactness, pore distribution, crust thickness, internal color, color difference), mouthfeel properties (bite hardness, fracturability, sandiness, dispensability, meltdown), and flavor properties (flavor, sweetness). Except for compactness, bite hardness, fracturability and meltdown, there was no significant difference among the samples. Most sensory attributes took scores over 3.5 out of a maximum 5.0 score. Compactness was defined as the amount of pores present in the interior of the cookies, and higher compactness indicated smaller pores. Cookies prepared with commercial shortening showed better air incorporation, as evidenced from the lower compactness scores. Bite hardness was measured as the resistance of the cookie structure against the first bite by teeth, and the cookies made with the oleogel samples had higher scores. Contrarily, instrumental hardness values were in the opposite direction, but the sensory fracturability defined as the brittleness level of the cookies during chewing was well correlated with instrumentally measured fracturability values. The fracturability of the cookies prepared with commercial margarine was lower than that of the cookies prepared with the oleogels. The sensory meltdown is a measure of the actual diffusion or solubility of the cookies in the mouth, and was found to be highest in the cookies prepared with SWO. This might be as a result of the higher melting point of the SWO. This sensory descriptive analysis revealed that most sensory attributes of the cookies resemble each other, and hence wax oleogels could be used as baking shortenings in cookies.

Furthermore, the same cookies were tested by 200 volunteer consumers using a five-point hedonic scale (1: dislike extremely to 5: like extremely) for appearance, texture, flavor, smell and acceptability. There were no significant differences among the samples, and all scores were above 3.80, indicating a good level of consumer preference. This result proved that cookies prepared with BWO and SWO were well accepted and preferred by the consumers compared to cookies prepared with commercial shortening. In the last part of this study, the 30 day storage stability of the cookies was monitored at room temperature. The cookie samples were fairly stable, and quality changes were minimal.[23]

In conclusion, this study, proved that wax oleogels can be successfully used in cookies as shortening alternatives, and they provide healthy fat without any *trans* fatty acids and high amounts of saturated fats.[23]

In another recent study, carnauba wax (CRW) and candelilla wax (CDW) oleogels of sunflower oil (SFO) were prepared at 2.5 and 5.0% addition levels and were utilized as shortening alternatives in cookie formulations against shortening- and only oil-containing samples.[27] The fatty acid and solid fat contents of the fat samples, as well as their temperature and frequency sweep rheologies, were measured. After preparation of the standard cookie dough, the biaxial extensional viscosity, hardness and extensibility of the dough samples, and dimensions, moisture content and texture of the cookies were measured. Since the oleogels had significantly lower solid fat contents, the effects of this deficiency were observed during dough formation. Although the oleogels by themselves stand as a solid material, during dough mixing, as a result of the applied shear force, they quickly lost their solid-like structure, leading to decreases both in elastic and viscous moduli. The doughs with oleogels had higher extensibility values than the dough with shortening and when the amount of the waxes used in the oleogel increased the extensibility values decreased owing to the reduced gluten network formation by fat crystals. However, CDW or CRW oleogels still had a significant amount of liquid oil that was not as effective as solid fat in forming short dough. CDW or CRW oleogels significantly increased the hardness values of the dough samples compared to the SFO-containing sample, but they were still softer than the dough containing shortening. Hence, clearly wax-based samples had complex rheology during dough formation owing to their shear- and time-dependent nature. A similar trend was also observed for the biaxial extensional viscosity values of the dough samples. Lowest extensibilities with dough prepared with shortening indicated the minimum stretching distance. Although the dough samples prepared with the oleogels had higher extensibilities than the shortening-containing sample, they had structure comparable to the SFO-containing sample. As the amount of wax increased in the oleogels, the extensibility values increased owing to the presence of wax crystals. CDW and CRW oleogels still contained significant amounts of liquid oil, which was not as effective as the solid fat present in the shortening in dough formation. Likewise, dough with SFO had the lowest hardness, followed by the samples prepared with the oleogels. Fats with more solid fat yield dough with a firm texture. The biaxial extensional viscosity values also showed a similar trend with the hardness values. The biaxial extensional viscosity of dough is an important property since it is determined at strain rates at the same order of magnitude as those encountered during baking. The highest biaxial viscosity values were observed in the doughs prepared with shortening, while the lowest biaxial viscosity values were recorded for the SFO-containing dough. The oleogels led to increases in biaxial viscosity values. Also the biaxial extensional viscosity values were closely correlated with the spread ratio of the cookies. The spread ratio of the cookies prepared with SFO was higher than the other samples. Likewise, the breaking strength of the cookies with SFO was the highest, and therefore the cookies with the oleogels were softer than the SFO-containing sample. Hence, as the amount of wax increased in the oleogel, the resulting cookies became softer, but the type of wax did not have an effect. Although a sensory study was not conducted, no significant differences were observed in flavor and aroma between SFO- and oleogel-containing samples. It was concluded that although wax oleogels were better than liquid oil in cookies, the functional properties were not sufficient compared to shortenings.

The same research group conducted another study with candelilla wax (CDW) oleogels and commercial shortening blends for cookie preparation as a shortening replacement.[28] The researchers evaluated the quality characteristics of commercial shortenings in cookies with those of the oleogel–shortening mixtures. Canola oil and candelilla wax oleogels were prepared at 3% and 6% (w/w) levels, and oleogel/shortening blends at 30:70 and 60:40 were evaluated. All fat products were analyzed for rheological properties, fatty acid composition and solid fat contents (SFC). The cookie dough was also analyzed for hardness and extensibility. Finally, the prepared cookies were analyzed for physical, chemical, and textural properties. Fatty acid composition analysis showed that compared with commercial shortening (63.4% saturated), the total saturated fatty acids in the blends ranged from 32% to 49.2%, indicating a significant reduction in the saturated and *trans* fatty acids. Similarly, the SFC was highest in commercial shortening and lowest in the oleogels, and the blends were between them. In the dough in the unperturbed state, the oleogels and blends formed self-standing structures comparable with shortening, but under shear, the oleogels, and to a lesser extent oleogel–shortening blends, transformed from solid-like structures to viscous liquid-like structures. As the shortening level (and hence SFC) increased in the blends, better shear sensitivity was observed. Likewise, much higher extensibility and lower hardness values were observed with the oleogel- and blend-prepared dough compared with the dough prepared with shortenings. Compared with the shortening, oleogel and blend samples, the cookies prepared with liquid oil showed higher spread ratios and hardness values. Although oleogels were not enough to produce cookies with the same quality as shortening, the blends improved cookie quality significantly. This study suggested that partial replacement of shortening with wax oleogels may help to improve cookie quality to an acceptable level.

Hwang *et al.* examined the quality of cookies made with conventional margarine *versus* those made with organogels prepared from sunflower wax, rice bran wax, beeswax and candelilla wax.[29] They used olive, soybean and flaxseed oils as different sources of oil. Organogel properties were affected heavily by the wax and oil type. The dough hardness was found to be in the order of soybean oil > flaxseed oil > olive oil > and rice bran wax oleogel > candelilla wax oleogel > beeswax oleogel. However, neither wax type nor oil type had significant effects on the hardness, spread ratio, and fracturability of the cookies. They concluded that cookies made with several combinations of wax and vegetable oil oleogels had comparable quality attributes to cookies made with commercial margarine, and hence oleogels could replace solid fat in cookies to reduce the amounts of saturated and *trans* fatty acids.

As indicated earlier, the bakery sector requires multifunctional, healthy and differently composed fats or shortenings. Puff pastry products or laminate fats (roll-in shortenings) for croissant- and Danish pastry-like products are examples requiring special shortenings. Puff pastry is a product characterized with laminated structures of baked dough layers separated with thin layers of fat at a ratio of approximately 2 : 1. The lamination process includes separating each layer of dough by one layer of fat. Puff pastry margarines must be very plastic to allow the fat to be processed, folded and extruded without breaking and becoming greasy. Since breaking leads to insufficient lift and flaky texture, fat must be very plastic to stay during the extrusion process.[30] Currently, one study has been published about wax use in puff pastry-like shortenings.[31] They modified the commercial shortening alternative Coasun™ with rice bran wax (RBX), carnauba wax (CRX), candelilla wax (CLX) and sunflower wax (SFX) addition at 1–15% (w/w) and prepared emulsions with other ingredients. After measuring the rheological properties, solid fat content and water activity of the modified emulsions, Danish pastry and croissant with Coasun laminate containing 7.5% RBX, 35% water, 6% C-18 monoglyceride, 0.3% SSL, 0.1% potassium sorbate, and canola oil was prepared. In the first part of the study, the rheology of the developed different Coasun emulsions was evaluated. It was found that addition of wax at greater than 10% (w/w) level, the use of palm oil or the gelation of the liquid oil phase with 5–7.5% RBX, the use of C-18 saturated monoglyceride at 6% (w/w), or C-22 saturated monoglyceride at 4% (w/w) were effective modifications capable of improving the rheology of the emulsion so that it can be used as a zero *trans* and reduced saturated fat laminating shortening substitute for puff pastry products. In the second part of the study, the Coasun laminate was used as the laminate fat for the Danish pastry and croissants. The surface of the products was characteristic flaky golden-brown, and there was no waxy taste in the mouth, while both products were a bit denser than the traditional products prepared with regular Coasun. These formulations approached the rheological requirements of the standard lamination fat, demonstrating that the combination of oil phase, water content, and monoglyceride length and concentration are critical parameters to be modified to tailor the behavior of Coasun to a laminating fat.

In conclusion, current literature points out that oleogels could replace commercial shortenings to some extent in baked foods. Especially, some level of solid fat crystals in the baking fats is a prerequisite for dough structure development and total product quality; hence, oleogel–hard fat stock blends seem suitable alternatives to balance both saturated and *trans* fat reductions and the lubricating effects of dough networks. For other types of baking fats, which require different levels of plasticity, moisture content, and emulsifier presence, more research is needed to incorporate or replace the commercial products with oleogels. Furthermore, the long-term clinical effects of wax oleogels should be studied with different food products to determine any possibly biochemical effects on the human body. Lastly, prior to commercial use, regulatory issues should be resolved for baked foods.

11.4 Wax Oleogels in Chocolate and Confectionery Products

Chocolate and confectionery is one of the largest food industry sectors. Most of the time, the name confectionery is used to imply both, and in fact chocolate is a sub-branch of confectionery by definition. The term "confectionery" comprises sweet foods, divided broadly into bakers' and sugar confectionery. In this section, our emphasis will be especially on chocolate and other sugar confectioneries, since bakery shortenings were covered previously. Beyond chocolate, sugar confectionery includes sweets, candied nuts, candied fruits, chewing gums, sweetmeats, pastils and others.[6,32] Chocolate is made mainly from the beans of the *Theobroma cacao* tree. The Aztecs living in Mexico prepared a drink from the beans and it was called '*Chocolatl*', the source of the current name for chocolate. After Columbus, cocoa beans were introduced to Europe and have been used since the 1600s. The first chocolate fat, known as chocolate butter, was removed from the pressed beans in 1828 by Van Houten of Holland, which gave the opportunity to start the chocolate industry. Basically, cocoa beans yield three fundamental products used in various manufactures: chocolate liquor, cocoa butter and cocoa powder. It is unnecessary to discuss cocoa bean processing here, but chocolate production basically consists of mixing the ingredients, particle size reduction or refining, conching and crystallization steps. Within each step, a certain texture should be achieved for fully flavored, well accepted, truly melting and stable chocolate and chocolate products. Although there are many different types of chocolates composed of different proportions, the main ingredients of chocolate are chocolate liquor (15–42%), cocoa butter (8–20%), sugar (40–50%), skim milk or other solids (1–15%), and lecithin (0.4%). There are many defined quality criteria for chocolate, but two major characteristics are clearly distinguished: chocolate flavor and texture. Although many differently flavored chocolates exist, the flavor provided by the processed cocoa beans must be free from objectionable tastes and consumers should associate well with the products. The most important aspect of chocolate texture is that it must be solid at normal room temperature (20–25 °C) and yet rapidly melt in the mouth at around 37 °C.[9,32]

Fats and oils are the fundamental components of confectionery products including chocolate. They provide flavor, texture, aroma, eye appeal, mouthfeel, stability and the expected eating quality. Fats and oils can be incorporated into the confectionery products as direct ingredients, as well as in coatings and centers, solid and hollow molded pieces, as release agents, as vehicles for flavorants, color and aromas, as a frying medium, and as polishing and glazing agents. Generally, confectionery fats and oils are expected to provide a specific flavor like typical cocoa butter, provide structure, shortness or lubricity, provide gloss, surface finish and eye appeal, provide a moisture barrier, serve as a heat transfer medium, and delay oil migration within the product.[9,32]

Cocoa butter is the key component in successful confectionery. It provides the characteristic brittle nongreasy texture, rapid melting in the mouth and excellent keeping quality at room temperature. Usually, it consists of 98% triglycerides, 1% free fatty acids, 0.3–0.5% diglycerides, 0.1% monoglycerides, 0.2% sterols, 150–250 ppm tocopherols, and 0.05–0.13% phospholipids.[9] Stearic acid (36%), oleic acid (34%) and palmitic acid (25%) are the dominant fatty acids, and most oleic acid is esterified at the sn-2 position. Generally, palmitic–oleic–stearic (36–42%), stearic–oleic–stearic (23–29%), and palmitic–oleic–palmitic (13–19%) configurations occur, and this structure is said to be responsible for the special melting attitudes of chocolate fat. Most melting occurs over 27–33 °C and essentially completes at 35 °C, but it does not melt at room temperature (25 °C). It was stated that any fat remaining solid above 35 °C is recognized as waxy and chewy, which is not appropriate for flavor release and eating quality. Hence, any fat used as an alternative to cocoa butter must resemble at least these melting attributes. For some other applications, chocolate analogs may offer some technical advantages as well. Usually, the term 'cocoa butter alternatives' is preferred for any fat used instead of cocoa butter in confectionery. Cocoa butter alternatives are classified into three groups, namely 'cocoa butter equivalents' (CBE), 'cocoa butter replacers' (CBR), and 'cocoa butter substitutes' (CBS). All cocoa butter alternatives are produced from specific fat sources by fractionation, dry fractionation, hydrogenation and interesterification technologies or by blending suitable source materials. In the European Union, the permitted six fat sources to be utilized as cocoa fat alternatives are palm (*Elaeis guineensis, Elaeis olifera*), shea (*Butyrospermum parkii*), illipé (Borneo tallow, *Shorea stenoptera*), kokum (*Garcinia indica*), sal (*Shorea robusta*), and mango kernels (*Mangifera indica*). The CBE or cocoa butter extenders are compatible with cocoa fat in any proportion and do not alter the melting, processing and rheological properties of the cocoa butter. Most commonly, palm, illipe, shea, sal and kokum fats are used for this purpose. Since their fatty acid compositions and triglyceride distributions are very similar, they tend to show quite similar melting behaviors, solid fat contents and X-ray diffraction patterns. CBRs are usually produced by partial hydrogenation or partial hydrogenation and fractionation of common soft or liquid oils like palm olein, soybean oil, cottonseed oil, canola and rapeseed oils. They have high contents of *trans* fatty acids up to 55%, and constitute some health problems. Hence, lower *trans* fat levels were developed with hydrogenation–fractionation technologies to be used in toffees and fillings. They provide poor eating quality, poor snap, and low coefficient of contraction in chocolate. Thus, they can be compatible and may be tolerated up to only 20% levels. Lauric CBS are derived mainly from lauric fat sources like palm kernel oil and coconut oil. Both include lauric acid as high as 40–50%. The fat could be used as unmodified or modified by hydrogenation, fractionation and interesterification, or a combination of them. Coconut oil has significant amounts of medium-chain fatty acids (C8:0 and C10:0), and is a soft fat, even when fully hydrogenated. Hydrogenated palm kernel oil yields harder fats with rapid and clear melting profiles.

Hence, it provides suitable coating fats for almost all climates and/or seasons. Generally, CBS exhibits good mouthfeel and flavor release, good gloss, good mold and belt release, and excellent oxidative stability. For bakery and confectionery coatings, they are highly suitable. However, mixtures with cocoa butter are incompatible and results in severe eutectic problems. Since CBS includes shorter acyl chains and a stable beta-2 polymorph, only up to 5% cocoa butter can be tolerated in CBS; beyond this limit the fat will separate out into two phases. This situation produces fat bloom in compound chocolate. Hence, CBS are usually used with cocoa powder and some emulsifiers. Unfortunately, this formulation yields less intensity in cocoa flavor. The palm kernel stearin fraction offers firmness, snap and steep melting, and produces excellent eating properties, texture, mold and belt release, and gloss with the negative images of more snap and brittleness. Hence, it is preferred in butterscotch, peanut butter and chocolate-flavored baking chips.[9,32,33]

Application of wax oleogels in confectionery and chocolate products is starting to be investigated, but the literature is quite poor in this area. In an early review of edible applications of oleogels, restriction of oil mobility and migration *via* organogelation was postulated.[3] They summarized the work describing prevention of oil migration in different bakery and chocolate confectioneries by oleogels other than wax oleogels in particular, which were available at that date. It was indicated that especially in filled chocolate confections, liquid oil diffuses into the cocoa butter and migrates through the coating layer to the outer surface. During this diffusion, some cocoa butter dissolves and diffuses as well. The diffused fat could easily be recrystallized on the outer surface to yield the fat bloom, a well-known defect leading to consumer dissatisfaction and rejection. Hence, a potential strategy to limit oil migration would be organogelation of any liquid fat in the confectionery products, as suggested.

Shellac oleogels were used to replace solid fat in chocolate paste.[14] Chocolate pastes are cocoa powder and sugar dispersions in oil continuous medium composed of solid fat crystals. Rheologically, chocolate pastes resemble both solid and liquid foods since they prevent sedimentation of dispersed particles and separation of liquid oil (oiling out) at the same time. This situation is a result of crystallization of solid fat within the oil continuous medium to yield what is known as the viscoelasticity. Hence, solid fats like palm oil or hydrogenated vegetable oils up to 20% are used in chocolate paste formulations as oil binders. Shellac wax oleogel (1.5%, w/w) was replaced as the oil binder in the chocolate paste formulations. In addition, palm oil was replaced by around 27% rapeseed oil, regarded as a healthier oil owing to its low saturated fatty acid content. The solid fat content (SFC) of the paste was significantly reduced by the oleogel replacement and the paste was not likely to have any greasy taste since it had quite a low SFC at 35–40 °C. In addition, the samples were stored at 30 °C for 4 weeks and showed no oiling out, suggesting that shellac wax oleogel was a successful oil binder replacement in chocolate paste. The rheological properties of the chocolate paste samples were determined with flow measurements and dynamic shear tests.

The oleogel and reference pastes showed the desired low plastic viscosity. Yield stress values, on the other hand, were higher in the oleogel paste, probably owing to the enhanced inter-particle interactions between the sugar particles at lower SFC. Shellac wax oleogel pastes showed some differences in the shear rate at 5 s^{-1} and showed thixotropy. Thixotropy is defined as the decrease in apparent viscosity with time at a constant shear rate. When the shear is discontinued, the apparent viscosity recovers, but this results in a hysteresis loop. Since oleogels usually present shear sensitivity and partial structure recovery, thixotropy occurred. The critical stress value of the reference paste was almost 10-fold higher in the LVR region, as indicated by the amplitude sweeps. Likewise, the $G*$ values were always higher for the reference paste. Furthermore, viscoelasticity compared by the G' and G'' values were higher for the reference paste, indicating a more solid-like behavior. This study proved the possibility of creating a chocolate paste with a much lower level of saturated fatty acids and no visible oiling out problem.

In another recent study, beeswax oleogels (BW) were used in the partial replacement of confectionery fillings.[34] Fillings fall into the general cocoa butter alternative classes described above (CBE, CBR, and CBS), with higher levels of unsaturated fatty acids or softer structures. Often in confectionery fillings, the compatibility between the coating and the filling fat is considered owing to oil migration and fat bloom problems. Hence, in proper formulations, the fat in the center should match the fat in the coating.[33] Doan *et al.* prepared beeswax (BW) oleogels in rice bran oil (RBO) at 1.5, 2.0, 2.5, 3.0, and 3.5% wt and used these oleogels to replace palm oil (PO) at 17, 33 and 50% wt levels to produce low-saturated fat hazelnut fillings.[34] Both PO–BW oleogel blends and the hazelnut fillings prepared from them were evaluated. SFC measurements indicated that at body temperature, the solid mass fractions of the blends were below 2.0%, suggesting a non-waxy mouthfeel. Microstructure studies indicated spherulitic morphology for PO and platelet crystals with an average length of 15–30 mm for BW oleogels, and both were in β' polymorph, respectively. Following the applied temperature programming, all the wax-based palm blends had a melting range of 31–44 °C. At the same replacement ratios, the samples containing higher amounts of BW had a higher T_m, offset as compared to the others. Owing to the dilution effects of the BW oleogel replacement, the crystallization and melting temperatures, and melting enthalpies of the wax-based blends and hazelnut fillings shifted to lower temperatures. Fillings containing 17% BW oleogel yielded higher crystalline density and gel strength similar to full palm oil. Also, wax-based fillings started to melt at higher temperatures than wax-based blends. The complex modulus $|G*|$ of the wax-based fat mixtures was approximately 300 Pa, and G' gradually increased during isothermal storage at 5 °C for 1 h. In the wax-based fillings, $|G*|$ was 300–2000 Pa even at 85 °C, the temperature at which all the fat crystals were melted. The $|G*|$ value of the wax-based filling substantially soared to nearly 200 000 Pa at the end of the cooling stage, and $|G*|$ was maintained after 1 h at 5 °C. There was no difference in $|G*|$ between the reference and the wax-based fillings at 5 °C. When the temperature was increased to 20 °C, there was a small drop in $|G^*|$ of all the samples. In conclusion, this study proved that low-saturated fat products with defined structure and rheology could be used in hazelnut fillings and similar confectionery.

Since different types of wax oleogels were produced by different wax sources, and many possibilities of wax oleogels–plastic fat blends can be prepared, more studies are expected in the area of fat replacement by oleogels for chocolate butter alternatives and other confectionery products. A fairly limited number of studies exist, as discussed above, and there is a huge research need for incorporation of wax oleogels into imitation chocolate products, chocolate liquor, fillings, cream toppings, candy fats, and similar.

11.5 Wax Oleogels in Dairy Products

Ultimately, wax oleogels or oleogel emulsions are fat products. Hence, their use in dairy products would be for the replacement or co-addition of milk fat (butter). At this point, it is worth briefly describing the main properties of butter. It is assumed that butter-making is as old as the domestication of farm animals, but industrial production started around 1864 after the invention of a machine similar to a laboratory centrifuge in Germany. Today, industrial and large-scale butter production by continuous cream separators, heat exchangers and packaging machines are common. Fresh cream, butter, anhydrous butter, and butter–vegetable oil mixtures are commercially available. Around 98% of milk fat is composed of triglycerides, and the rest includes free fatty acids, phospholipids, cerebrosides, sterol esters, tocopherols, pigments and flavor compounds. About 70.6% of butterfat is composed of saturated fatty acids, mainly palmitic, myristic, stearic, lauric and butyric acids. The presence of butyric acid and other medium-chain fatty acids is common for butterfat. Oleic and linoleic acids constitute 29.4% of the total unsaturated fatty acids. Usually, butterfat has a saponification number around 210–250, iodine number of 30–33, melting point of 38 °C, titer of 34 °C, and 0.4% unsaponifiable matters. It starts melting at – 30 °C and completely melts at 37 °C. At any temperature between these two, it is a mixture of solid and liquid triacylglycerols. The solid:liquid ratio at a given temperature determines its rheological properties. Milkfat crystallizes in an aggregate fashion to form a network in which both solid and liquid phases may be regarded as continuous. This network largely influences the perceived firmness of the fat. Hence, at refrigerator temperatures butter would be spreadable. Quality butter has a fine and close texture, with a firm and waxy body and a sufficiently plastic texture to be easily spreadable. Different butter products may have solid fat contents of around 20–50% at around 20 °C. One of the most important properties of butter is its special flavor, which is mostly due to the presence of diacetyl, dimethylsulfide, and several lactones, esters, ketones, aldehydes and free fatty acids.[35]

Kerr *et al.* investigated the effects of sunflower oil wax (SFO) addition on the functional properties and crystallization behavior of anhydrous milk fat (AMF).[36] The experimental samples were AMF with 0.1% to 0.25% added SFO wax by weight. Laser-polarized light turbidity, polarized light microscopy, texture analysis, thermal analysis and SFC were measured. The results showed that there is a decrease in the induction times of nucleation and smaller crystal formation by the addition of waxes. As a result, the hardness of the fat crystal network was improved. Crystal growth after tempering was also promoted by wax addition. The changes were dependent on both the temperatures used during processing and the level of wax addition. In conclusion, this study proved the possibility of modifying milk fat for better rheological properties without any increase in *trans* fatty acid levels.

Another study aimed to create a butter fat alternative from solely diacetyl added to hazelnut oil–sunflower wax and hazelnut oil–beeswax oleogels (Figure 11.2).[19] The 5% (w/w) oleogels were formed and enriched with 0.5% diacetyl (w/overall w). After physicochemical, textural and thermal analysis, the sensory descriptive analysis and consumer tests were performed. The color values were significantly different from those for commercial butter (CB) since no color additives were used. While CB had an iodine value of 29.7, the oleogel samples showed 88.43 and 92.02 g 100 g^{-1} iodine values, indicating that the unsaturation level was significantly higher in the oleogels. The peroxide values of the oleogels were a little higher than that of the CB, but all were very low (below 1.0). The peak melting temperature of CB was 49.61 °C, and those of the beeswax and sunflower wax oleogels were 49.39 °C and 62.02 °C, respectively. Clearly, in terms of melting temperature, the beeswax oleogel most closely resembles CB. Likewise, the SFC of the samples was 3.50%, 3.55%, and 17.40% for the beeswax oleogel, sunflower wax oleogel and CB, respectively. Instrumental hardness and stickiness values of CB were higher than those for the oleogel samples. 13 different sensory definition terms were used to define the samples, and especially hardness, spreadability, grassy and waxy notes were significantly different among the samples. Consumer hedonic scores for appearance, odor, flavor and spreadability indicated that, except for spreadability, for all attributes CB had higher scores, but the spreadability of the oleogel samples was much better. The consumer preference test showed that around 57% of the consumers would 'definitely buy' the hazelnut oil–beeswax oleogel as a butter alternative. This study showed that diacetyl aromatized hazelnut oil–wax (beeswax and sunflower wax) oleogels are structurally and thermally suitable samples as butter alternatives for direct consumption. Consumer tests indicated that the samples are usually accepted, but some improvements, especially for color and aroma, are needed. Further studies of butter–oleogel mixtures, oleogels with different gelators at much lower addition levels, and oleogels with suitable color and aroma additives should be conducted as butter alternatives for direct consumption as well as in food applications. There seems to be good potential to improve the use of oleogels as butter alternatives by carrying out further studies.

Another interesting application of wax oleogels was for cream cheese production.[37] Soybean oil–rice bran wax (RBW) oleogels at 8% and 10% (w/w) and ethylcellulose (EC) oleogels at 10% (w/w) were prepared and used against control samples including full-fat and fat-free commercial cream cheese samples. Skim milk and the oleogels were mixed and then processed for cream cheese production. Cream cheese samples with RBW and EC oleogels, with liquid soybean oil, and a fat-free sample were produced with the same processing techniques. All cheese samples were then analyzed for composition, microscopy, texture, rheology, fatty acid composition, and sensory analysis. The compositional analysis indicated that RBW oleogel cream cheeses had around 15.65% fat and 23–25% non-fat solids, while full-fat cream cheese had 20.36% fat and 13.81% non-fat solids. The pH of the oleogel-added cream cheese samples were within the acceptable range of 4.4–4.9. Confocal laser scanning microscopy studies revealed that the structural networks of the commercial products and oleogel cream cheeses (OCC) were very similar. All samples had numerous small lipid globules dispersed in a protein continuous matrix. The lipid globule area for the full-fat control was 9.52 μm, and RBW oleogel sample had 8.74 μm. The RBW oleogel samples appeared to have a more dense protein network than the others. Texture analyses indicated that all oleogel cream cheese samples had comparable hardness values with the control sample, while the non-gelled control sample (including liquid oil) had significantly higher hardness values than all the samples. Likewise, the spreadability of the oleogel cheeses was comparable and similar to that of the control sample. While stickiness values were similar, the adhesiveness values were fairly different; all OCC samples showed lower values. Usually, samples prepared with RBW mimicked the control sample more successfully in terms of spreadability, hardness, stickiness and adhesiveness results. Rheology studies showed that RBW samples demonstrated comparative G' values to the control sample, while the non-gelled sample showed a significantly higher G' value. Furthermore, all OCC samples had higher storage modulus than the fat-free sample. While commercial cream cheese samples displayed a steep decrease in storage modulus when it was heated from 5 °C to 30 °C, a more gradual decrease was observed in OCC samples. These results were expected since the OCC samples did not experience the same melting profile of the milk fat in the control sample in the measurement temperature range. However, a decline in G' values was said be likely to be due to some hydrophobic interactions of casein particles to reduce the modulus of elasticity. In fact, the increase in temperature from 60 °C to 80 °C increased the storage modulus of all samples. Since the RBW sample had comparable results for texture and rheology, it was selected for sensory analysis in the 'Just About Right' test. There were no significant differences between the RBW sample and the control cream cheese for hardness, spreadability, and mouthfeel, while the strength of flavor and bitterness for the RBW sample were rated too strong and need improvements. Overall, this study indicated the potential for the use of food-grade organogelators in a variety of low-fat dairy products and to enhance their fatty acid profile. More studies for the utilization of different wax oleogels in dairy products such as various cheeses, quark, yoghurt and others are needed.

Ice cream, a very special dairy product, might be an open area for oleogel application for milk fat replacement. Ice cream is a universal term to generally name dozens of different frozen dairy desserts that are characterized by containing milk solids, being consumed in a frozen state, and frequently aerated. Ice cream and other related products are classified and named differently according to their origin. Generally, these types of products contain seven main groups of ingredients: non-fat milk solids, fat, sweeteners, stabilizers, emulsifiers, water, and flavor. Once produced, air becomes the other important component. The manufacturing is very briefly based on blending the ingredients, pasteurization, homogenization, cooling, aeration and aging at or around 4 °C. There is no need to detail the product types, production processes and quality parameters, but the importance of fat components will be described.[38,39]

The main properties of milk fat have already been discussed above. It is a complex component that includes almost 400 fatty acids. In ice cream, the emulsion droplets are usually about 0.8 mm, and after freezing the fat globule clusters form as results of partial coalescence of individual globules. These clusters may vary in size from 5 μm to 80 μm, and their ratio determines the shape retention and meltdown rates of the products. The fat triglycerides are usually in a crystalline state in the ice cream mix, while some low melting point triglycerides could be in a liquid state. The ratio of solid and liquid portions is also very important in determining ice cream stability and texture properties. During freezing, the fat emulsion partially coalesces as a result of emulsifier action, air incorporation, water crystallization and shear forces applied by dasher and scraper blades in a dynamic freezer barrel. In fact, this partial coalescence is good for structure and texture formation in ice cream. More than half of the volume of the final ice cream is air, with overrun values that can range from 25% to 150%. Air is dispersed in very fine cells between 20–25 μm and provides a light texture to the products. Stability of the air cells is also influenced by the fat crystals and other components. Hence, the fat content and composition in these types of products is essential for structure, stability, overall mouthfeel and flavor, and nutrition. Thus, fat intensifies flavor, carries added flavor and aromas, produces a smooth texture, helps to give the body, and leads to desirable melting properties. Furthermore, fat helps in lubricating the freezer barrel during processing. However, excessive use of fat may lead to increased manufacturing costs, hindered whipping ability, decreased consumption and high caloric value.[39]

The best source of fat in ice cream is fresh cream. It provides the best flavor and a unique texture. However, good quality unsalted butter could replace the fresh cream. In addition, anhydrous milk fat (butter oil) can be used for ice cream or frozen dairy dessert formulations. Plant fats and oils are also used in ice cream production. In the selection of a proper fat source, the rate of fat crystallization, the structure of the crystal, the melting profile, and the solid fat content are considered. Usually fats/oils with a high content of solid fats are preferred for the combination of structure formation, melting behavior and flavor release properties. Hence, fat replacers were used in ice cream studies and up to 5–6% of total fat was replaced successfully. The current trend towards low calorie foods is also driving the fat replacement in ice cream.[38,39] In this section, fat replacement with oleogels will be discussed.

Rice bran wax (RBW) oleogels (10%, w/w) were used as the fat source in ice cream production against two controls, regular milk fat and unsaturated oil (high oleic sunflower oil, HOSO), and routine analyses were done.[40] Centrifugal separation of the lipid phases of the samples revealed that while HOSO lipid is liquid, the other samples' lipid phases were solid. Oil droplet size analysis indicated that the RBW oleogel sample had a bimodal distribution, one lower than 10 μm, and the other composed of larger sized particles, while control samples had monomodal particle distribution with 2 μm size. Protein per surface oil analysis showed that as a concentration of the emulsifier increased, the protein decreased, and this value was lower than both the milk fat and HOSO samples. Hence, a higher instability of the emulsion during whipping/freezing stage may occur. The final products, ice cream samples, were analyzed for amount of air incorporation, destabilization of fat and meltdown stability. The overrun of the oleogel ice cream was higher than the HOSO and lower than the milk fat sample. Furthermore, the oleogel sample was better than the HOSO sample in terms of oil spreading at the air interface, which results in better ice cream foam structure. However, the overrun of the oleogel ice cream was not as high as that of the milk fat control, suggesting that air incorporation and air stability were not fully promoted by the gelled oil compared to the crystalline milk fat. While no improvements in meltdown resistance were observed for the RBW oleogel sample compared to HOSO, all samples melted much faster than the milk fat control. The destabilization phenomenon with the oleogel sample was not sufficient compared to the milk fat sample. The cryo-SEM images suggested that fat and air cell interfaces aggregate in the oleogel sample. Large numbers of spherical droplets protrude out of the air bubble interface in the milk fat sample. Both oleogel and milk fat samples had air cells with small diameters in large populations, though the mean diameters were statistically different. The difference was very large for the HOSO sample, which had very large air bubbles. Furthermore, more surface distortion was observed in the oleogel ice cream sample. Overall, this study proved the possibility of replacing milk fat in ice cream, but the fat structure was not sufficient to delay the structural collapse. More studies are envisioned.

The same research group published another similar study.[41] In this study, they utilized RBW, candelilla wax (CDW) and carnauba wax (CBW) oleogels at 10% wax level with HOSO as the oil. Further, they used glycerol monooleate (GMO) as the emulsifier and applied different rates of freezing. The GMO emulsifier significantly improved the structural integrity, but still was not as good as milk fat-created structures. The overrun of the RBW oleogels with GMO was significantly different from that of the RBW oleogel with other emulsifiers, but it was still lower than that of milk fat, and much higher than HOSO itself. These parameters suggest that GMO possess some unique features to structure fat properties. Moreover, addition of GMO to the RBW oleogel created uniform spherical shape and interior crystals. The meltdown rate was significantly larger in the CBW and CDW oleogel ice creams compared to the RBW sample. In addition, the overrun of these two samples was lower than that of the RBW sample, and similarly their fat destabilization was significantly less. Hence, the CBW and CDW oleogels were evaluated as less suitable for ice cream. Ice cream with 15% oleogel and GMO showed a reduction in melt rate compared to HOSO, but it was significantly higher than that for milk fat ice cream. Additionally, increase in oleogel concentration in the formulation from 10% to 15% stabilized the system to reduce structural collapse. Furthermore, these samples had better shape retention. The air bubble sizes of the 15% oleogel and milk fat samples showed no significant difference. Usually, there was an inverse relationship between the meltdown rate and fat destabilization. This study proved that RBW with GMO emulsifier added at higher concentrations (15%) creates better ice cream structures. More studies with different wax oleogels and oleogel–milk fat mixtures with different emulsifiers are needed for better fat replacement in ice cream or similar frozen dairy dessert products.

Unfortunately, to date, there are no other published studies in the literature for oleogel applications in other dairy products. It is well recognized that studies involving wax oleogels in different dairy products, like various cheeses, cream cheese products, quark, ice cream, yoghurt and others, are needed. Whipped dairy creams and butter products are other potential research areas for application of oleogels with healthier fatty acid and minor component compositions.

11.6   Wax Oleogels in Comminuted Meat Products

Reformulated (comminuted, ground, restructured, *etc.*) meat products contain different proportions of minced meat, fat, ice (water) and flavoring agents. There have been efforts to alter the components of meat products to achieve a more healthy and/or tasty product by removing, increasing, adding or replacing different components. The fat content in meat products is our subject, and three main goals related to fats in meats are identified: reduction of total fat (and energy), reduction of cholesterol, and modification of fatty acid profile.[24,42] Oil bulking systems, structured emulsions, liquid–solid fat mixtures and oleogels have been used to modify the fat content of meat suspensions, meat sauces, meat batters, sausages, frankfurters and meat patties.[7]

It has been stated that replacing animal fat with edible liquid oils (olive, corn, soybean, peanut, fish oils, *etc.*) in comminuted meat products improves the fatty acid profile, but substantially negatively affects the textural properties. Since solid animal fats in meat products provide mouthfeel, texture, juiciness, bite, heat transfer, shape and stability to the products, it would be reasonable to replace then only with another fat having with a fatty acid profile and solid-like rheology. In this respect, interesterified fats and oleogels seem to be suitable new fat sources. Especially, oleogels provide solid-like fats with the same fatty acid composition as the starting liquid oil.

Hence, incorporation of oleogels into comminuted meat products formulations may yield both benefits at the same time.[24,42]

Applications of oil bulking systems in meat products are excellently reviewed by Jimenez-Colmenero *et al.* and will not be considered any more here, since this chapter is especially devoted to wax oleogels.[42] It is enough to state that konjac- and alginate-based oil bulking systems were quite successful in reducing the fat content in the meat products. Hydrogelled emulsions, Pickering emulsions and interesterified fats have also been used in meat products for various purposes.[42]

The use of true oleogels in meat products is a very new and unexplored area of research. There are very limited numbers of studies done with oleogels formed with organogelators other than waxes. The research needs for wax oleogels replacements in meat products are immediately apparent. Although not directly related, the applications of other oleogels will be discussed shortly. Dry-ripened venison sausages (venison salchichon) were formulated with 25% pork meat (control) and replacement of total fat with olive oil oleogel at 15, 25, 35, 45, and 55% levels was done. The olive oil organogel was prepared by emulsification of olive oil with soy protein. All samples were adequate in terms of pH, water activity, moisture loss, color changes, levels of lipolysis and oxidative stability. As the proportion of olive oil was increased, the amounts of unsaturated fatty acids increased, as expected for health benefits. Furthermore, the consumers evaluated all samples as acceptable.[43] Canola oil–ethylcellulose oleogel was used to replace 25% of animal fat in frankfurters.[44] Both the hardness and chewiness of the oleogel-added sample were similar to those of the control sample and much better than those of the canola oil (liquid oil)-added sample. It was stated that addition of oleogel enhanced the size of the fat/oil globule to improve the frankfurter texture compared to the liquid canola oil-added sample. In another similar study, canola oil–ethylcellulose oleogel was used to replace 26% fat in all-beef frankfurters and 20.8% fat in pork sausages.[45] Except for some deficiencies in the textural properties, it was stated that oleogel replacement was promising to yield sausages with healthier lipid profiles with similar processing parameters, economic feasibility and acceptable sensory properties. Panagiotopoulou *et al.* utilized γ-oryzanol–phytosterol structured sunflower oil oleogels in the production of frankfurters to partially replace pork backfat.[46] Three organogels and three organogel emulsions with various γ-oryzanol:phytosterol ratios (all at 10% addition level) were used in frankfurter formulations. There was 10% fat replacement and 10% pork backfat in all products. The pH and oxidation levels were equal among the samples, but some color differences occurred. While frankfurters with the oleogels were similar to the control samples, in terms of texture, the oleogel emulsion products had lower chewiness, hardness and gumminess values. Moreover, except for some emulsion treatments, all samples had similar sensory acceptability values. In one study, olive and sunflower oil oleogels prepared with Myverol and lecithin were used to prepare meat suspensions.[47] Myverol is a commercial distilled food emulsifier containing glycerol monostearate and monopalmitate. The meat phase and oil phase were ground at various proportions and evaluated. It was observed that both additives exerted stabilizing action, even if one of them (monoglycerides of fatty acids) is more effective than the other at very low concentrations. The adopted rheological approach has proved to be very useful in determining the type and amount of the proper organogelator to be used as an oil suspension stabilizer. In conclusion, these studies showed that various oleogels could very successfully replace animal fat in various meat products.

Although there are no studies directly using wax oleogels in meat products, there is one study in which polycosanol was used as the organogelator.[48] It is a gelator that resembles beeswax. Hence, this study indirectly indicated how wax oleogels may act in meat products. Polycosanol and Myverol were incorporated at 0.5% and 2.5% addition levels into olive and sunflower oils, and these prepared oleogels were used to produce meat sauces. It was shown that polycosanol yielded a product that was very similar to the control sample in terms of rheology and stability. This promising result indicated that various wax oleogels may be utilized in meat products to replace the animal fat at various degrees. More research is needed in this area.

11.7 Wax Oleogels in Other Food Applications

One possible food products application of oleogels is the delivery and controlled release of nutraceuticals. By definition, a nutraceutical is a molecule that has some health benefits or medical expectations to prevent and/or cure a disease. Nutraceuticals could be water soluble or lipid soluble. It has been shown by several studies that lipid-soluble nutraceuticals can be easily entrapped within an oleogel matrix. Likewise, it is possible to encapsulate a water-soluble nutraceutical in water-in-oil oleogel emulsions. The most commonly encapsulated lipid-soluble food grade nutraceuticals are omega fatty acids, carotenoids, conjugated linoleic acid, isoflavones, co-enzyme Q, phytosterols, and tocols.[3,24]

Encapsulation of the nutraceuticals within an oleogel matrix aims to protect the molecules during food processing as well as during the digestion processes. Furthermore, bioavailability (uptake from intestine into bloodstream) could be enhanced significantly, since most limited solubility nutraceuticals may be easily dissolved in the oleogel. In addition, control of the nutraceutical release rate can be possible to give it a longer half-life in the bloodstream. Hence, by releasing the nutraceutical into the bloodstream at a steady or controlled rate, the possible negative and/or side effects of sharp concentration changes might be prevented. Generally, it is said that nutraceuticals within a gel can be prevented from precipitation and slow lipolysis, and release can be provided owing to the crystalline or fibrillary networks of the oleogels.[3,49,50] Some studies have been published with oleogels prepared by oleogelators other than waxes, but since this chapter is dedicated only to wax oleogel applications, we will consider only studies with wax oleogels.

As we described throughout this chapter, wax oleogels have been investigated in many food applications, but the use for nutraceutical encapsulation and delivery is mostly neglected for some unknown reasons. We can locate only two studies using wax oleogels for delivery purposes. In one study, fish oil–beeswax (10%) oleogels were shown to delay the release of betamethasone dipropionate, a drug, applied for transdermal drug delivery by *in vivo* experiments.[51] In the other study, lanolin wax-based oleogel was located into alginate microparticles by emulsification and internal gelation method.[52] The drugs salicylic acid and metronidazole were entrapped within the oleogel. It was shown that the release of the drugs from the microparticles was dependent on the pH of the medium. The results suggested that encapsulation of the drugs in oleogel could prolong the release of the drug in a controlled manner. Truly, these two studies used wax oleogels, but they were not examples of food applications of nutraceuticals; hence, the research need for nutraceutical delivery and encapsulation with wax-based oleogels is appreciated.

Recently, we investigated the potential of *in situ* oleogelation to limit oil migration in tahini halva products.[53] Tahini halva is prepared by mixing sugar, water, citric or tartaric acid, emulsifier and soapwort (*Saponaria officinalis*) extract with tahini (sesame paste). It is a common and preferred dessert in the Mediterranean, Middle East and Balkan countries. It has a characteristic solid, homogeneous, fibrous and crunchy texture. Since its main ingredient is sesame paste, it contains around 57–65% sesame oil. Hence, with storage time, some oil migration occurs as a serious problem of manufacturing. Oil migration or separation could be caused by mechanical effects, temperature fluctuations, and loss of natural product texture. In tahini halva, the oil migration problem leads to changes in the color, texture and sensory characteristics of the products, and contamination of the packaging material with released oil. To prevent this problem, traditionally some emulsifiers, stabilizers, hydrocolloids, hydrogenated fats and other additives have added. However, addition of these materials not only increases cost of manufacturing, but also raises questions regarding health effects. Hence, sunflower wax, shellac wax and beeswax were added at levels of 1, 3, and 5% of total product weight during tahini halva processing. The control groups were addition of hydrogenated palm stearin (HPS) at the same levels, and no addition. The physicochemical, textural and sensory evaluation of the halva samples revealed that sunflower and shellac wax at 3 and 5% addition levels led to a reasonable reduction of the oil migration problem and these products were quite similar to the control (HPS) sample.[54] The oil leakage level was below 7% in all samples. The structure was fairly stable after 6 months of storage at room temperature. The prepared tahini halva products with 3% addition of the organogelators and HPS stored for 6 months at room temperature are shown in Figure 11.4.

Clearly, the control (no addition) sample had considerable amounts of oil leakage into the packaging material, while the HPS-added sample and the oleogelled samples were quite stable and had no oil separation at all.

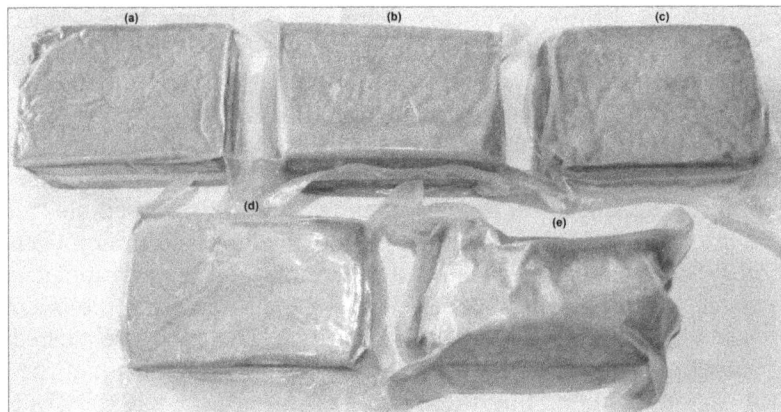

Figure 11.4 Tahini halva products stored for six months at room temperature. (a) Sunflower wax, (b) beeswax, (c) shellac wax, (d) hydrogenated palm stearin, and (e) no addition (unpublished data from ref. 53).

This situation proves that *in situ* oleogelation of sesame oil by wax organogelators at even the 3% addition level is quite possible. Hence, without saturated and *trans* fatty acid-containing solid fat addition, the liquid sesame oil could be oleogelled to limit oil migration. Furthermore, the waxes did not create any offensive sensory perceptions, as proved by the consumer tests. Sunflower wax and shellac wax could successfully replace hydrogenated palm stearin in tahini halva production to prevent the oil separation or migration problem.[53,54] In conclusion, the oleogelation technique can be used for similar purposes in similar products like peanut butter and nut spreads.

11.8 Conclusions and Recommendations

Various plant and animal waxes have been shown to be successful organogelators for creating fairly stable, thermoreversible, viscoelastic oleogels. This chapter briefly described the food applications of the wax oleogels. The most important issues regarding the use of wax oleogels in food products must be resolved beforehand. Firstly, in most countries waxes are permitted as coating or glazing additives only, and there is a definite need for regulatory clearance to apply them as direct food ingredients. Secondly, most waxes contain a distinct waxy flavor, which may create some sensory problems for consumers. Although there are some studies indicating that this is not a significant problem, more studies are clearly needed. Thirdly, the long-term effects of wax consumption in oleogel form in food on the body functions and health must be thoroughly studied. In fact, we know that natural sources of plant waxes such as sunflower seeds are totally healthy. The fate of wax in the human intestine and tissues should be investigated to declare any possible good health effects or risks. Lastly, the expected functionality of wax oleogels in the specific foods must be at least near to that of commercial products.

Studies have indicated some loss of functional properties, such as mechanical strength, aeration properties, coating properties, water sensitivity and mouthfeel, when solid fats are totally replaced by oleogels. Hence, hybrid systems composed of wax oleogel plus solid fat could enhance the expected food functionality. In this context, oleogels produced from different waxes (mixtures of waxes) or from wax and other organogelator mixtures might be evaluated. Likewise, different proportions of wax oleogels and different natural solid fat sources might be studied in different food applications for better functionality with lowered levels of saturated and *trans* fatty acids. Throughout this section, it was clearly noted that applications of wax oleogels in some foods are fairly rare compared to other oleogels made from ethyl cellulose or polymer gelators. More studies of wax oleogels in meat products, cheeses, pastry fats, and confectionery products are needed. Furthermore, delivery and encapsulation systems based on wax oleogels for various food applications are a virgin research area. More research and actual industrial applications of wax oleogels are expected.

References

1. M. M. Chrysam, in *Bailey's Industrial Oil & Fat Products*, ed. F. Shahidi, Wiley-Intersience Pub, New Jersey, US, 6th edn, 2005, vol. 4, ch. 2, Margarines and Spreads, pp. 33–82.
2. E. D. Co and A. G. Marangoni, *J. Am. Oil Chem. Soc.*, 2012, **89**, 749.
3. E. N. Hughes, A. G. Marangoni, A. J. Wright, M. A. Rogers and J. W. E. Rush, *Trends Food Sci. Technol.*, 2009, **20**, 470.
4. H.-S. Hwang, *INFORM*, 2016, **27**, 708.
5. A. G. Marangoni and N. Garti, in *Edible Oleogels: Structure and Health Implications*, ed. A. G. Marangoni and N. Garti, AOCS Press, Urbana, Illinois, 2011, An Overview of the Past, Present, and Future of Organogels, pp. 1–18.
6. R. D. O'Brien, in *Fats and Oils: Formulating and Processing for Applications*, ed. R. D. O'Brien, CRC Press, New York, 3rd edn, 2009.
7. A. R. Patel and K. Dewettinck, *Food Funct.*, 2016, **7**, 20.
8. C. E. Stauffer, in *Bailey's Industrial Oil & Fat Products*, ed. F. Shahidi, Wiley-Intersience Pub, New Jersey, US, 6th edn, 2005, vol. 4, ch. 7, Fats and Oils in Bakery Products, pp. 207–227.
9. R. E. Wainwright, in *Bailey's Industrial Oil & Fat Products*, ed. Y. H. Hui, Wiley-Intersience Pub, New York, US, 5th edn, 1996, vol. 3, Chapter: Oils and Fats in Confections, pp. 353–408.
10. H. Hwang, M. Singh, J. K. Winkler-Moser, E. L. Bakota, S. Kim and S. X. Liu, *J. Am. Oil Chem. Soc.*, 2013, **90**, 1705.
11. M. Öğütcü and E. Yılmaz, *Grasas Aceites*, 2014, **65**, 1.
12. E. L. Bakota, J. K. Winkler-Moser, H.-S. Hwang, M. J. Bowman, D. E. Palmquist and S. X. Liu, *Eur. J. Lipid Sci. Technol.*, 2013, **115**, 847.
13. H. Hwang, M. Singh, J. K. Winkler-Moser, E. L. Bakota and S. X. Liu, *J. Food Sci.*, 2014, **79**, C1926.

14. A. R. Patel, P. S. Rajarethinem, A. Gredowska, O. Turhan, A. Lesaffer, W. H. De Vos, D. Van de Walle and K. Dewettinck, *Food Funct.*, 2014, **5**, 645.

15. M. Öğütcü, N. Arifoğlu and E. Yılmaz, *J. Am. Oil Chem. Soc.*, 2015, **92**, 459.

16. M. Öğütcü, E. Yılmaz and O. Güneser, *J. Am. Oil Chem. Soc.*, 2015, **92**, 1429.

17. E. Yılmaz, M. Öğütcü and Y. Karagül Yüceer, *J. Food Sci.*, 2015, **80**, S2035.

18. E. Yılmaz, M. Öğütcü and N. Arifoglu, *J. Oleo Sci.*, 2015, **64**, 1049.

19. E. Yılmaz and M. Öğütcü, *RSC Adv.*, 2015, **5**, 50259.

20. D. Boskou, in *Olive Oil: Chemistry and Technology*, ed. D. Boskou, AOCS Press, Champaign, Illinois, 2nd edn, 1996, ch. 1, History and Characteristics of the Olive Tree, pp. 1–11.

21. D. J. Metzroth, in *Bailey's Industrial Oil & Fat Products*, ed. F. Shahidi, Wiley-Intersience Pub, New Jersey, US, 6th edn, 2005, vol. 4, ch. 3, Shortening: science and technology, pp. 83–123.

22. B. S. Ghotra, S. D. Dyal and S. S. Narine, *Food Res. Int.*, 2002, **35**, 1015.

23. E. Yılmaz and M. Öğütcü, *Food Funct.*, 2015, **6**, 1194.

24. T. A. Stortz, A. K. Zetzi, S. Barbut, A. Cattaruzza and A. G. Marangoni, *Lipid Technol.*, 2012, **24**, 151.

25. J. Jacob and K. Leelavathi, *J. Food Eng.*, 2007, **79**, 299.

26. A. Jang, W. Bae, H. Hwang, H. G. Lee and S. Lee, *Food Chem.*, 2015, **187**, 525.

27. B. Mert and I. Demirkesen, *LWT–Food Sci. Technol.*, 2016, **68**, 477.

28. B. Mert and I. Demirkesen, *Food Chem.*, 2016, **199**, 809.

29. H. Hwang, M. Singh and S. Lee, *J. Food Sci.*, 2016, **81**, C1045.

30. C. Christensen, Palsgaard, *Puff Pastry Margarine-Focusing on Functionality and Fat Reductions*, 2011, Technical Paper, pp. 1–7.

31. A. I. Blake and A. G. Marangoni, *Food Struct.*, 2015, **3**, 30.

32. G. Talbot, in *Industrial Chocolate Manufacture and Use*, ed.S. T. Beckett, Blackwell Publishing Ltd., West Sussex, UK, 4th edn, 2009, ch. 19, Vegetable Fat, pp. 415–433.

33. K. Smith, in *Science and Technology of Enrobed and Filled Chocolate, Confectionery and Bakery Products*, ed. G. Talbot, CRC Press, New York, 2009, ch. 15, Ingredient preparation: the science of tempering, pp. 313–343.

34. C. D. Doan, A. R. Patel, I. Tavernier, N. De Clercq, K. Van Raemdonck, D. Van de Walle, C. Delbaere and K. Dewettinck, *Eur. J. Lipid Sci. Technol.*, 2016, **118**, 1903.

35. D. Hettinga, in *Bailey's Industrial Oil & Fat Products*, ed. F. Shahidi, Wiley-Intersience Pub, New Jersey, US, 6th edn, 2005, vol. 2, ch. 1, Butter, pp. 1–59.

36. R. M. Kerr, X. Tombokan, S. Ghosh and S. Martini, *J. Agric. Food Chem.*, 2011, **59**, 2689.

37. H. L. Bemer, M. Limbaugh, E. D. Cramer, W. J. Harper and F. Maleky, *Food Res. Int*, 2016, **85**, 67.

38. W. S. Arbuckle, in *Ice Cream*, ed. W. S. Arbuckle, Springer Science+Business Media, LLC, New York, 4th edn, 1986.

39. H. D. Goff and R. W. Hartel, in *Ice Cream*, ed. H. D. Goff and R. W. Hartel, Springer Science+Business Media, New York, 7th edn, 2013.
40. D. C. Zulim Botega, A. G. Marangoni, A. K. Smith and H. D. Goff, *J. Food Sci.*, 2013, **78**, C1334.
41. D. C. Zulim Botega, A. G. Marangoni, A. K. Smith and H. D. Goff, *J. Food Sci.*, 2013, **78**, C1845.
42. F. Jimenez-Colmenero, L. Sandoval-Salcedo, R. Bou, S. Cofrades, A. M. Herrero and C. Ruiz-Capillas, *Trends Food Sci. Technol.*, 2015, **44**, 177.
43. M. C. Utrilla, A. G. Ruiz and A. Soriano, *Meat Sci.*, 2014, **97**, 575.
44. A. K. Zetzl, A. G. Marangoni and S. Barbut, *Food Funct.*, 2012, **3**, 327.
45. J. M. Wood, MSc Thesis, The University of Guelph, Guelph, Ontario, Canada, 2013.
46. E. Panagiotopoulou, T. Moschakis and E. Katsanidis, *LWT–Food Sci. Technol.*, 2016, **73**, 351.
47. F. R. Lupi, D. Gabriele, N. Baldino, L. Seta, L. B. de Cindio and C. De Rose, *Eur. J. Lipid Sci. Technol.*, 2012, **114**, 1381.
48. F. R. Lupi, D. Gabriele, L. Seta, N. Boldino and B. de Cindio, *Eur. J. Lipid Sci. Technol.*, 2014, **116**, 1734.
49. A. K. Zetzl and A. G. Marangoni, in *Trans Fat Replacement Solutions*, ed. D. Kodali, AOCS Pub, Urbana, Illinois, 2014, ch. 10, Structured Emulsions and Edible Oleogels as Solutions to Trans Fat, pp. 215–244.
50. C. M. O'Sullivan, S. Barbut and A. G. Marangoni, *Trends Food Sci. Technol.*, 2016, **57**, 59.
51. M. F. D. Huri, S. Ng and M. H. Zulfakar, *Int. J. Pharmacol. Pharm. Sci.*, 2013, **5**, 458.
52. S. S. Sagiri, J. Sethy, K. Pal, I. Banerjee, K. Pramanik and T. K. Maiti, *Des. Monomers Polym.*, 2013, **16**, 366.
53. M. Öğütcü, N. Arifoglu and E. Yılmaz, *Solution of Oil Migration Problem in Tahin Halva Through Organogelation Technique*, TÜBİTAK Project Number: 115O027, 2016.
54. M. Öğütcü, N. Arifoglu and E. Yılmaz, *Eur. J. Lipid Sci. Technol.*, 2017, **119**, 160.

CHAPTER 12

Edible Applications of Ethylcellulose Oleogels

K. D. MATTICE AND A. G. MARANGONI*

Department of Food Science, University of Guelph, 50 Stone Road E,
Guelph, Ontario, Canada N1G 2W1
*E-mail: amarango@uoguelph.ca

12.1 Introduction

Solid fats play a distinct and essential role in a large variety of food products. This includes functional roles, as well as providing characteristic mouthfeel and texture. Their semi-solid nature is due to the presence of a large proportion of high melting triacylglycerols (TAG), which form crystalline structures at room temperature, resulting in a network that confines the lower melting TAGs within it. The ratio of crystalline to liquid TAGs at a given temperature is expressed by the percent of solid fat content (SFC). In general, the high melting TAGs are made up of a combination of saturated and/or *trans* fatty acids. On the other hand, lipid materials high in low melting TAGs, commonly composed of unsaturated fatty acids, exist in liquid state as oils at room temperature owing to the absence of this fat crystal network.[1]

While lipids are an important component of the diet, it is recommended to limit consumption of saturated and *trans* fatty acids owing to their well-documented association with adverse effects on cardiovascular health.[2,3] In contrast, there are recognized beneficial health effects from increasing unsaturated fatty acid consumption.[4] For this reason, the concept

Food Chemistry, Function and Analysis No. 3
Edible Oil Structuring: Concepts, Methods and Applications
Edited by Ashok R. Patel

of replacing saturated and *trans* fatty acids with unsaturated fatty acids in foods is gaining popularity. However, direct replacement of solid fat with oils is usually not an option owing to differences between their physical and sensory properties. For some time, liquid edible oils have been hydrogenated to create solid fats; however, this led to the production of *trans* fatty acids, the consumption of which is more concerning than the saturated fatty acids they were created to replace.[3] Trends then returned to using naturally occurring highly saturated fats, which are solid at room temperature, including palm, milkfat and animal fats.

Recently, oleogels have emerged as a novel means of employing liquid, edible oils in solid fat applications in food systems. This concept has the potential for creating foods with desirable physical and sensory properties, which also meet evolving regulations and health concerns. Canola oil is one of the many oils that can be gelled and contains only 6% saturated fatty acids, with mono- and poly-unsaturated fatty acids comprising the remainder. This is one of the lowest proportions of saturated fatty acids naturally occurring in edible oils. Using gelled canola oil as a replacement of common solid fat sources, a significant reduction in *trans* and saturated fatty acid consumption is possible, without sacrificing the characteristic properties of foods.

Ethylcellulose (EC) is a polymer capable of structuring oils into solid gel networks, which are potential alternatives to solid fat sources with improved fatty acid profiles. EC is a semi-crystalline cellulose polymer derivative, consisting of a cellulose backbone with ethoxyl substitutions at hydroxyl groups such that the substitution ratio is 2.5/2.6.[5,6] It is this degree of substitution that causes the necessary lipophilicity at high temperatures. It is also very soluble in organic solvents, giving EC many industrial applications.[7] Although multiple other oil gelators have been documented and researched, including certain mono-acylglycerides, hydroxylated fatty acids, sterols and plant waxes, EC possesses important advantages. EC is currently the only known food-grade polymer capable of structuring oil. This is remarkable, given that most food biopolymers are hydrophilic in nature and therefore are used solely in aqueous gelation applications. Additional beneficial properties of EC oleogels include being tasteless, transparent (which is maintained during gelation), and the strength of the gels is easily manipulated by adjusting known parameters, including gel composition or manufacturing conditions.[5,8] This chapter will discuss the principles, considerations and practicality of using EC oleogels as a solid fat replacement in high-fat food applications normally utilizing shortenings, animal fats or other solid fats.

12.2 Ethylcellulose Oleogels

EC is obtained through the ethylation of cellulose, a main structural component in the cell walls of many plants, comprising approximately one third of total plant mass. These organisms produce such carbohydrate polymers through their photosynthetic ability to fix carbon dioxide. The ability of cellulose to produce EC is owing to the presence of three hydroxyl groups,

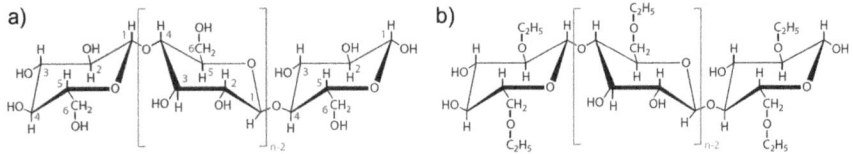

Figure 12.1 Molecular structure of (a) cellulose and (b) ethylcellulose building blocks. Adapted with permission from M. Davidovich-Pinhas, S. Barbut and A. G. Marangoni, *Annu. Rev. Food Sci. Technol.*, 2016, 7, 65–91.[65]

which can be derivatized (Figure 12.1).[9] The ethylation process occurs after cellulose is dissolved in an alkali solution and ethyl chloride gas is added, causing replacement of the hydroxyl end groups with ethoxy end groups in the cellulose structure (Figure 12.1).[10] Alkalizing cellulose is critical to form alkoxide ions, which are then derivatized with an ethyl end group. The degree of substitution (DS), or ethoxy content, is used to characterize the final EC, where lipophilicity is achieved with a DS of 2.4–2.5.[7] On its own, EC demonstrates high flexibility, thermoplasticity, film forming ability, mechanical strength and transparency, allowing for many applications.[7]

EC can be obtained commercially for use as a food additive. Multiple EC products exist and are differentiated according to the viscosity in centipoise (cP) of the solution that they will create when dispersed in 80% toluene and 20% ethanol.[11] Commonly used EC include 10, 20 or 45 cP, which is based on the average molecular weight (MW) of polymers present, where a higher MW corresponds to a greater viscosity.[11] While the exact molecular weights of commercially available EC varieties are not disclosed, results of recent studies expect that 10, 20 and 45 cP EC correspond to MWs of 28.6 ± 6.2, 51.9 ± 10, and 72.8 ± 1.5, respectively.[12,13]

EC forms a network within liquid oils through intermolecular forces. EC gels are prepared by heating the EC–oil mixture to above the glass transition temperature of the EC, approximately 130–140 °C, at which point the EC dissolves fully in edible oils.[5,6] The exact glass transition temperature is dependent on the MW of the EC used.[5,13] Solubility occurs above this temperature because some crystalline regions of the polymer become amorphous, resulting in exposure of the ethoxy groups. Upon cooling, the gel will then form as a network of hydrogen bonds form between EC strands and trap oil within.[6,8] Secondary structures, such as helices or sheets, are not formed upon gelation.[5] The minimum concentration of EC required to induce the gelation of oils is 4–6%.[14] The exact amount depends on the MW of EC used and the processing parameters, described further in the following section.

12.3 Physical Properties of Ethylcellulose Oleogels

Different foods require solid fats possessing different physical properties for proper functionality, including thermal behaviour, mechanical strength and rheology. For example, a laminated dough shortening must have a high

plasticity in order to be rolled into a dough without leaking at room temperature.[15] However, a cream filling in desserts must be very soft at room temperature, and have a lower SFC for a pleasant mouthfeel upon consumption.
Oleogels must therefore be produced to match the functionality of the fat
source they are intended to replace in order to achieve all desired organoleptic properties. Of these, mechanical strength is the parameter of the most
significance for oleogels, and the most easily controlled. This can be manipulated by selectively controlling variables known to directly affect mechanical
strength. Variables of interest in this case are the type of vegetable oil used,
the polarity of the oil, the polymer concentration and its MW, as well as the
addition and concentration of small molecules or surfactants.[14,16]

12.3.1 Techniques for the Analysis of Oleogel Physical Properties

A common technique to examine the mechanical strength of oleogels is back
extrusion, which involves penetration of a set oleogel sample with a stain-
less steel probe, having a cylindrical shaft and a truncated spherical tip. The
rate and depth of penetration are controlled. The gel strength is commonly
reported as the average force over the final 25% of the penetration depth.
This parameter is chosen owing to the large amount of signal noise inherent
in back extrusion. Another technique is a compression test known as texture
profile analysis (TPA), where an oleogel sample of specific height and width
is compressed between two parallel plates to 50% of its original height. Once
again, the rate and depth of compression are controlled. Gel strength is then
evaluated as the force exerted, or resistance to deformation, at the point of
maximum deformation during the first compression cycle. Deciding which
method to use depends on the firmness of the oleogel in question. In the
case of TPA, there is a lower limit for gel firmness, which restricts the ability
to test softer gels that do not retain their shape. While this is not a factor
for back extrusion, there is a contrasting upper limit for gel firmness as the
probe must actually be able to penetrate the gel.

Small and large deformation rheometry are used to characterize the viscoelastic properties of oleogels by applying a specific stress or strain to the
material and measuring the corresponding response. Small deformation rheology probes the linear viscoelastic region, while large deformation rheology
probes the non-linear region. These techniques can be used to determine the
storage modulus (G'), which describes the elastic or solid properties, the loss
modulus (G''), which describes the viscous or liquid properties, and Young's
modulus (E), which describes the relationship between stress and strain for
the given sample. Given their solid gelled state, it is to be expected that G''
$\ll G'$ for all oleogels. Temperature sweep and time sweep rheology experiments can be used to understand the gelation process, where measurements
are taken as the gel solidifies with decreasing temperature, and increasing
time.[5] TPA can also be used to identify the elastic constant (K), a component of Young's modulus (E), by taking the gradient of the plotted force *vs.*

distance from the first 5% of the compression as this region demonstrates linear deformation. While back extrusion and TPA are effective and are often used for a general comparison, a complete assessment of the mechanical properties of oleogels using multiple techniques will aid further in the development of successful oleogel solid fat replacement.

12.3.2 Parameters Affecting the Physical Properties of Oleogels

The inherent differences in the fatty acid compositions of oils are responsible for differences in gel strength.[14,17] It was hypothesized that a higher density oil, such as highly polyunsaturated oils, would produce a denser gelled oil network and a firmer consistency.[12,14,17] However, Gravelle *et al.* showed that it is the polarity of an oil rather than the density.[18] This was demonstrated by increasing the strength of EC in soybean oil oleogels with the addition of castor oil, a polar component, and decreasing strength with the addition of the same amount of mineral oil, a non-polar component. Given that the densities of castor oil and mineral oil were similar, the only difference between the oil mixtures was their polarity, and therefore, the hypothesis that an oil's density affects its strength was invalidated. An oil of greater polarity is more effective at solubilizing EC, causing the EC polymer chain to experience greater unfolding, increasing contact and therefore hydrogen bonding between the polymers upon gelation. Essentially, increasing its polarity will increase the oil's affinity for EC, leading to an increase in strength.

Aside from differences in polarity caused be compositional differences, the polarity of oils will increase depending on the extent of oxidation that has occurred owing to the polarity of the oxidation products. This means processing conditions, including temperature, will affect the strength of the gels.[5,8,19] Researchers have noticed that an increase in gel mechanical strength will occur as the holding time at the glass transition temperature of EC is increased.[19] It was proposed that this phenomenon results from an increase in the extent of oxidation. It appears that hydrogen bond formation, critical for EC gelation of oils and their strength, is dependent on temperature conditions through the heating and cooling processes. An increase in gel strength and gel brittleness has also been documented upon increasing gelation setting temperatures.[8] Brittleness is evident from the appearance of a fracture point during mechanical testing, that being the point during mechanical testing at which the force applied causes the material to fracture, rather than just deform. Lower setting temperatures cause the gel to set at a rapid rate, resulting in less ordered hydrogen bonds and therefore weaker gels. Another factor affecting the final strength of a gel is the temperature at which EC is dispersed into the oil. Although commonly dispersed at the glass transition temperature, dispersion above the melting temperature of EC, approximately 180 °C, will also contribute to an increase in gel hardness.[8] This is hypothesized to be caused by the melting of crystalline regions within EC, resulting in an increased capability of hydrogen bonding.

Several studies have shown that increasing the EC concentration results in an exponential increase in gel strength.[8,12,14,20] Addition of surfactants, including sorbitan monostearate (SMS), sorbitan monooleate (SMO), glycerol monostearate (GMS) and glycerol monooleate (GMO), can also increase a gel's strength.[8,12,18] These compounds are suspected to interact with the polysaccharide backbone of EC and plasticize it, leading to an increase in strength. However, they do not interfere with network formation.[18] Davidovich-Pinhas *et al.* showed a significant increase in EC oleogel mechanical strength with the addition of 3.77% GMO, GMS, SMO or SMS.[21] Further, Gravelle *et al.* were able to demonstrate an increase in mechanical strength for soybean oil EC oleogels with just 0.25% of oleic acid or oleic alcohol by weight.[16]

Increasing the viscosity, and MW, of the EC used results in increasing gel strength. Therefore, selection of EC by viscosity (in cP) or MW can be a simple method for controlling the hardness of an EC oleogel. The gel strength of oleogels prepared with 10, 20 and 45 cP EC was investigated by small deformation rheology.[5] The storage modulus, G', an indication of gel stiffness, was found to increase with increasing polymer viscosity (higher cP values).[22,23] However, there is an upper limit to this principle, as the glass transition temperature, and therefore the temperature to which the oil must be heated, increases with increasing MW.[12] Beyond 45 cP (approximate MW of 72.8 ± 15 kDa), solubilisation in oil is limited unless heated to a higher temperature for a longer amount of time, which leads to oil oxidation.

12.4 Health Implications of Ethylcellulose Oleogel Consumption

12.4.1 *In Vitro* and *In Vivo* Digestion of Ethylcellulose Oleogels

By structuring oil *via* polymer gelation, a solid, structured lipid material is created, composed of primarily unsaturated fatty acids. This is a healthy alternative to traditional solid fats, which commonly contain saturated fatty acids at levels up to approximately 60%. However, replacement of solid fat with novel ingredients in food products must take into account the digestibility of the new component. Because the concept of edible oleogels is still relatively new, no oleogel has been studied extensively in this regard. A study by Duffy *et al.* on oleogel digestibility through *in vitro* models found that lipolysis occurring during the digestion of phytosterol gelled oleogels was decreased as compared to un-gelled edible oils.[24] They hypothesized that lipase activity was hindered by either gelation limiting migration of the oil to the surface, the oleogel structure entrapping and protecting oil, or by enzyme amount restriction caused by adsorption of the lipase to the hydrophobic surfaces within the oleogel structure. Essentially, this study concluded that the oleogel structure in some way limited interactions between lipases and oil.

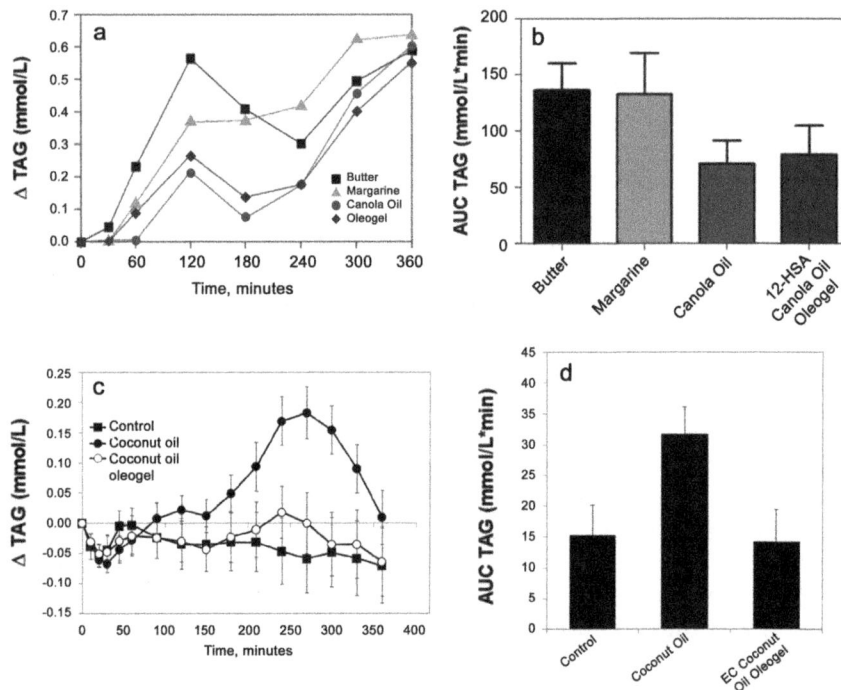

Figure 12.2 Post prandial serum TAG response curves (a = baseline corrected) and area under the curve values (±SEM) for (a–b) butter, margarine, canola oil and 12-HSA in canola oil oleogels; and (c–d) control meals (containing no fat), meals containing coconut oil and meals containing coconut oil oleogel. Reprinted from *Trends in Food Science and Technology*, **20**, N. Hughes, A. G. Marangoni, A. J. Wright, M. A. Rogers and J. W. E. Rush, Potential food applications of edible oil organogels, 470–480, Copyright 2009, with permission from Elsevier and from ref. 33 with permission from the Royal Society of Chemistry.

However, contradictory results were reported by Hughes *et al.* after an investigation of the effects resulting from consumption of oleogels formed with 12-hydroxystearic acid (12-HSA). 12 healthy, young individuals were fed meals containing 48 g of either 12-HSA in canola oil organogel, butter, margarine or un-gelled canola oil.[25,26] Based on blood serum levels of TAGs and free fatty acids over 6 hours after consumption, this study found that the post-prandial increase in serum lipids resultant from digestion of the 12-HSA oleogel was not significantly different from that of liquid oil, but it was significantly different from consuming solid butter or margarine (Figures 12.2a and b). In another study, Yu *et al.* also determined that there was no significant difference in the extent of lipolysis between gelled and liquid oil systems after a 30 minute *in vitro* digestion experiment with medium TAG systems gelled by monoglycerides and containing sorbitan monolaurate.[27] The conclusions from these studies indicate that the digestion and lipolysis of an oleogel could depend on the gelator used, and not simply the gelation phenomenon. Therefore, to describe the impact of consuming EC oleogels on digestion, results from specified experimentation are necessary.

A comprehensive study by O'Sullivan *et al.* investigated lipid digestibility in EC oleogels by *in vitro* methods.[28,29] This investigation looked at gels made with 10%, 12% and 14% 10 cP EC in canola oil, in comparison with un-gelled canola oil. The samples were subjected to an *in vitro* digestion simulation and lipolysis was measured at specified time points. Analysis revealed that the only difference in lipolysis between the samples occurred during the initial phase, where the gel structure is assumed to inhibit the release of oil, delaying contact with lipase. After this fact, there was no significant difference in the extent of lipolysis, suggesting that these EC gel structures are broken by the shear forces involved in digestion, releasing the oil. However, gels made with 10 cP EC were notably weaker than gels made with higher MW EC. As described in the previous section, the hardness of an EC oleogel can be manipulated in order to best mimic the solid fat that they will replace. The hardness is also expected to be a significant factor with regards to digestion of the oleogel as this influences mastication breakdown, the particle size swallowed, and subsequently the eventual breakdown in the body by lipases.[30,31] This could have an impact on the residence time of the material in the digestive tract.[32] When the authors prepared much harder gels using 10% 45 cP EC, *in vitro* digestion experiments showed a significantly lower extent of lipolysis when compared to un-gelled oil control samples.[28] In this case, the oleogel structure appeared to be more resistant to breakdown, partially retaining its structure through the simulation, and consequently lipase was unable act to the same extent. This leads to the conclusion that, while weaker oleogels retain the digestibility of edible oils, digestion of harder oleogels is affected by gelation. Though the implication of this effect is largely unknown, there are potential benefits where harder oleogels could act, for example, as carriers for highly lipophilic molecules protected by the structures that remain. In their study, the authors also took into account ingredient interactions with the gelator molecules as another possible factor affecting digestibility, given that the samples in this experiment contained 10% to 14% EC.[28] Given the results, it can be said that interactions such as EC surrounding lipid droplets or direct interference with molecules involved in lipid digestion did not occur.

Recently, Tan *et al.* investigated the *in vivo* digestion of coconut oil and an EC coconut oil oleogel when ingested with a carbohydrate rich meal.[33] This was the first study to evaluate the *in vivo* digestibility of EC oleogels. The authors fed 16 adult males meals with added coconut oil, or 11% 45 cP EC in coconut oil oleogels. The results indicated that fats that were entrapped in an EC polymer network experienced delays in digestion, and showed that postprandial levels of glucose and insulin of oleogel containing meals were more similar to meals containing no fat than meals containing ungelled coconut oil. Postprandial TAG levels were also found to decrease in the consumption of EC oleogels when compared to ungelled oil. These results contradict those of Hughes *et al.*, where no significant difference was determined between ingestion of 12-HSA oleogels and canola oil.[26] These conclusions are evident by the respective response curves and areas under the curves in Figure 12.2. Tan *et al.* explained these differences simply by the greater mechanical strength of the gels prepared with 11% 45 cP EC. These results match those by O'Sullivan *et al.*, where EC oleogels that have a greater mechanical strength just do not break down as easily.[28] However, this study was not conducted on a representative group of individuals (only a small group of Chinese men was included) and they did not measure lipid digestibility directly, but rather only postprandial TAG levels, indicating that our conclusions here are still speculative in nature. However, similar to the concept of "glycemic index" for carbohydrates, the potential to control the "lipidemic index" of an oil or fat by encapsulation within an EC oleogel is exciting, and could open up the field of controlled release of triglycerides, a macronutrient.

12.4.2 Ethylcellulose Oleogels for the Controlled Release of Bioactive Molecules

Oleogels have also been studied for their potential to facilitate controlled release of bioactives upon consumption such that the compound is released at a steady rate into the bloodstream over a certain period of time.[29] If not released in a controlled fashion, the ingestion of large amounts of bioactives causes a sharp increase in serum concentration, followed by a sharp decrease, owing to rapid inactivation or degradation of the molecules.[26,34] This can be undesirable, particularly owing to the negative side effects that can occur when the concentration of a bioactive in the bloodstream is above or below its therapeutic range. Accomplishing controlled release is not a simple task owing to the various factors involved in the digestion and gastric emptying processes. However, it has been achieved through techniques such as encapsulation.[34] Hydrogels have been extensively studied in this regard and used in applications for oral delivery. Polyacrylic acid hydrogels are of particular interest owing to their ability to withstand acidic conditions and therefore protect bioactive compounds from degradation through digestion.[35] These materials are also capable of adhering to the mucosal lining and slowly releasing their contents over time.[35] However, there are numerous hydrophobic bioactive compounds that cannot be easily incorporated into hydrogels, making oleogels a good alternative.[25] Oleogels also possess the potential for increasing bioavailability as during digestion TAG molecules from oils will form micelles with the hydrophobic bioactives within.[29,36] Micelles then serve to transport the bioactive molecules and disperse them throughout the body. This delivery system is ideal for oral delivery of pharmaceuticals or dietary supplements. A specific advantage of the use of EC oleogels in this application is that the melting temperature of the gels is much higher than body temperature, meaning release owing to melting will not occur.

12.4.3 *In Vitro* Bioaccessibility of β-carotene in Ethylcellulose Oleogels

One family of hydrophobic bioactive compounds that is of particular interest for this delivery method is the carotenoids, specifically β-carotene. Aside from providing yellow, orange and red colouring in many fruits and vegetables, these compounds are known to be potent antioxidants, and some provide dietary vitamin A. They are also associated with reducing the risk of chronic diseases, such as cardiovascular diseases and cancer.[37–39] Being fat soluble, the digestion of carotenoids involves their incorporation into micelles, along with dietary fat and bile acid. The micelles are then absorbed by passive diffusion into the small intestine, incorporated into chylomicrons and passed to the lymphatic system for transfer to the liver. They become bioavailable once they are incorporated into lipoproteins in the liver and subsequently released into the blood stream.[38,40] The bioavailability of carotenoids has been shown to be dependent on the efficiency of initial micellarization upon ingestion.[40]

Investigations by our group have looked into the bioavailability of β-carotene incorporated into EC oleogels using *in vitro* digestion models.[28,29] EC in canola oil oleogels were prepared using conventional methods, however, being sensitive to heat, β-carotene was added during the final minutes of the mixing process in order to reduce heat exposure. Although the composition of the gels prepared changed with the addition of β-carotene, this molecule has no potential for hydrogen bonding. This, along with the fact that the concentration was very low, meant that there was no alteration of the gel mechanical strength. Duodenal digestion was simulated and the percent transfer of β-carotene from the lipid oleogel phase to the digestate was measured. These values indicate the percentage of β-carotene that is bioaccessible to the body, where a lower percent transfer value at a certain time period indicates potential for delayed release applications. The results showed no significant difference between the transfer of gels made with 10%, 12% and 14% 10 cP EC, and the un-gelled canola oil control. In accordance with previous EC oleogel digestion conclusions, it was then important to evaluate harder gels. β-Carotene transfer in 45 cP oleogels was significantly lower than in weaker gels and un-gelled oil at time points of 60 and 180 minutes. An expected, a linear correlation between β-carotene transfer and lipolysis was found, relating the reduced transfer from harder samples to the reduced extent of breakdown previously reported during digestion. This work indicates that only sufficiently hard gels are capable of providing protection to β-carotene, and therefore have the potential for slow release delivery applications. This same observation was made in the context of increasing the strength of 12-HSA in soybean organogels used for the delivery of ibuprofen (by Iwanaga *et al.*) and policosanol in olive oil oleogels containing ferulic acid (by Lupi *et al.*).[41,42]

12.5 Considerations and Practicality of Ethylcellulose Oleogels in Food Systems

EC is of particular interest for food applications owing to being tasteless, non-caloric and physiologically inert.[14,25] Incorporation of oleogels into food systems can both improve the fatty acid profile of the food and decrease the overall fat content. Addition of EC for the gelation of oil can also improve oil migration and leakage, major issues that arise in products such as cookies, cakes, ground comminuted meats, and filled chocolates.[25,43,44] EC has been approved as a food additive by the FDA for the purposes of binding and filling dry vitamin preparations, as a protective coating for vitamin and mineral tablets and as a fixative for flavourings under regulation 21 CFR 172.868, with the limitation that the DS cannot be above 2.6 ethoxy groups per unit. It has also received Generally Regarded as Safe (GRAS) status for indirect food usage under regulation 21 CFR 182.90. Approval as a general food additive is still under review.

Heating to approximately 140 °C is required for EC gelation of oils. Some foods are capable of withstanding these high temperatures, so certain food applications may be able to employ direct dispersion of EC into an oil-containing food matrix. In most cases, this temperature exceeds that which the food materials can withstand, and gels must be prepared prior to incorporation. Due to their solid, gelled structure, incorporation of the oleogels into food products requires the use of shear to ensure that the gel will be dispersed in the matrix. Shear prior to incorporation leads to irreversible loss of structure and extensive oil leakage, and is therefore not an option. As such, incorporation of EC oleogels into foods is limited to applications that can withstand shear, such as comminuted meats, processed cheeses and as a component of shortening blends. These foods are often high in saturated fats and can greatly benefit from the fatty acid profile improvement that oleogels can provide. Stortz and Marangoni describe the potential for incorporation of glycerol mono-oleate, a food-grade surfactant, to plasticize the gels and make them thixotropic.[45] The development of a thixotropic EC oleogel, one which thins upon shearing, but thickens again once shear forces cease, could be a solution to some incorporation limitations, minimizing structure loss, and ensuring even dispersion. Afterwards, the gel could fully regenerate the solid mechanical properties over time.

The type of oil selected for gelation will depend on flavour compatibility in the desired application, and the fatty acid profile. The fatty acid profile will have an impact nutritionally and on the eventual mechanical properties. The most common oils used are canola oil, soybean oil, safflower oil and other vegetable oils. The high temperatures during production must also be monitored closely, as oil held for an excessive time at this temperature will oxidize, and the subsequent development of oxidation products will create off-flavours. Gravelle *et al.* demonstrated that after just 20 minutes of holding at 140 °C, canola oil will contain over 10 meq peroxide kg canola oil^{-1} and can no longer be termed 'fresh' oil.[19] While this seems like a short period of time,

20 minutes is sufficient for proper dissolution of EC into the oil and thus it is possible to avoid reaching this limit of freshness. There is also an option of adding antioxidants to reduce the extent of oxidation. If added, antioxidants should be food-grade, and added in concentrations ranging from 100 to 500 ppm by weight of oil.[46] Examples of food-grade antioxidants include *tert*-butyl hydroxyquinone (TBHQ), butylated hydroxytoluene (BHT), butylated hydroxyanisole (BHA), ascorbyl palmitate and tocopherols. The boiling points of these antioxidants are above 140 °C, and therefore would not be affected by the heating process. Barbut *et al.* cite the use of BHA, added to oil prior to heating, in combination with rosemary extract to reduce the off flavour associated with oil oxidation in beef frankfurters.[47] It is also possible to limit oxidation by heating oil with EC under inert atmospheric conditions, such as under vacuum or 100% nitrogen.[46]

Traditionally, oil globules are distributed evenly within food matrices and are micrometers in size, while solid fats are harder and exist as larger globules varying in size, but all over 100 μm.[43] In contrast, very small globules arise after direct replacement of solid fat with un-gelled liquid oil. The presence of very small globules has been found to cause a hard, rubbery texture, specifically in high protein and fat products, such as comminuted meats and certain cheeses. The harder texture is likely observed owing to the increased fat globule surface area, resulting in the ability of proteins to form a stronger network.[14,23] These products will also experience oil leakage with time and have an unacceptable shelf life. It is for reasons like this that the substitution of solid fat with liquid, unsaturated oils show little success. When the first trials of solid fat replacement with oleogels began, it was suspected that texture implications owing to the presence of large oleogel chunks would be unacceptable by consumers. However, this was determined not to be an issue because EC oleogels have been shown to create statistically similar microstructures in food systems, in terms of fat globule size, as the solid fats that they replace (Figure 12.3).[14,23] This adds to the body of evidence for why EC oleogels are promising candidates for successful solid fat replacement.

Other textural implications resulting from the incorporation of oleogels into foods have been documented. In general, the addition of oleogels to baked products will result in a denser or harder final product, as compared to traditional fat controls. The attempted incorporation of plant wax in sunflower oil oleogels into cookies as a replacement for traditional shortening by Mert and Demirkesen was unsuccessful owing to increased hardness in the final cookie product.[48] Blake and Marangoni determined that the replacement of laminating shortenings with oil, monoglyceride, plant wax and water emulsions in pastry products resulted in denser products when compared to the controls.[49] While these differences can be monitored instrumentally, there is a lack of sensory analysis data in this area to confirm the acceptability of the new oleogel-containing products, regardless of mechanical differences. Yilmaz and Öğütcü[50] describe the preparation of cookies with both beeswax and sunflower wax oleogels, including texture, and sensory analysis comparisons with conventional cookie controls. Although an increased bite

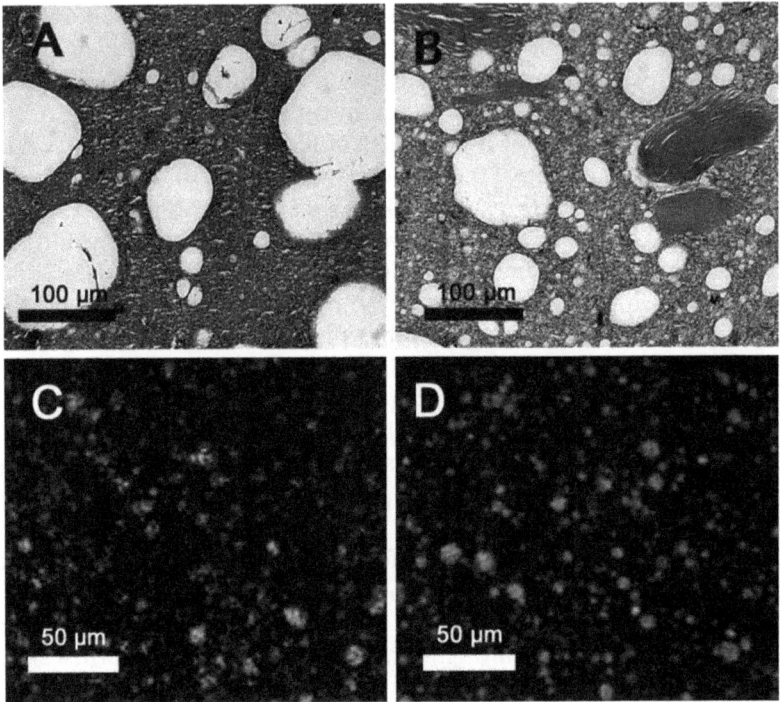

Figure 12.3   Micrographs of (A) frankfurters containing animal fat and (B) EC oleogels; (C) commercially available cream cheese and (D) cream cheese containing EC oleogels. Fat globules are white against a dark background caused by the protein network. Reprinted from ref. 21 with permission from the Royal Society of Chemistry and from *Food Research International*, Volume 85, H. L. Bemer, M. Limbaugh, E. D. Cramer, W. J. Harper and F. Maleky, Vegetable organogels incorporation in cream cheese products, 67–75, Copyright 2016, with permission from Elsevier.

hardness was noticed in oleogel-containing cookies, similar to the studies above, panelists still found the cookies to be highly acceptable. These examples illustrate the apparent concept that textural differences are likely upon solid fat replacement and must be considered when using any type of oleogel. This also indicates that beginning with only partial replacement of shortening with oleogels could produce more comparable products.[51]

Currently, most experiments using EC oleogels in food systems employ very small, non-reproducible batch processes for oleogel production. Gravelle *et al.* developed a controlled method to ensure production of oleogels that could be reproduced, involving a convection oven and an overhead stirrer.[19] For EC oleogels to be considered practical for use in commercially available products, a large-scale oleogel production process is needed. Dow Global Technologies LLC filed a patent in 2014 for the large-scale production of EC oleogels, which describes specially designed equipment.[46]

This process is described to create oleogels that are comparable or even superior in functionality to oleogels prepared using alternative methods. The process begins with combining EC and edible oils in a large extruder using feeding ports. The equipment must reach sufficient temperatures such that the EC is melted at the time of first contact with the selected oil. As the mixture travels through the extruder, the temperature is lowered such that gradual cooling begins before exiting. The exact temperature and holding time in the extruder are tailored to the oil and type of EC used in each batch. Once mixed thoroughly and heated to the necessary temperature, the mixture is extruded, giving the oleogel a smooth and homogenous texture. Not only does this create opportunity for large-scale oleogel production, another advantage to this process is that oleogels can be produced under inert atmosphere conditions, potentially reducing the extent of oxidation that will occur. However, experimentation is needed to confirm the efficacy of this technique as extreme temperature conditions still exist and maintaining a completely inert atmosphere (100% nitrogen, for example) is particularly difficult. From an efficiency standpoint, this large-scale method would also result in some oil loss or waste.

12.6 Edible Applications of Ethylcellulose Oleogels

12.6.1 Cream Cheese

Bemer *et al.* claim to have successfully reduced the overall fat content by 25% and improved the fatty acid profile in cream cheese by incorporating EC oleogels.[23] The cream cheese was produced with 45 cP EC in vegetable oil, heated to 130 °C to induce gelation, then cooled to just 75 °C at a rate of 10 °C min^{-1}, at which point milk ingredients, whey protein isolate, cultures and Rennet were added. The mixture was then blended before cooling to set the gel. This technique allowed for gelation of the oil after the mixture was completed. Replacement of milk fat with the oleogel resulted in a 120% increase in unsaturated fatty acid content and a 90% decrease in saturated fatty acid concentration. In comparison with commercially available cream cheese products, the oleogel cheese sample demonstrated similar hardness upon single penetration texture analysis. Further comparison with a non-gelled oil control confirmed the effects of ethylcellulose as a gelator. Fat-reduced cream cheese products generally demonstrate a decrease in hardness. However, it was hypothesized that the similarities in fat globule size between oleogel and full-fat samples allowed for the same extent of interactions to occur within the product, thus resulting in similar textures.[14,23,52] Texture analysis penetration tests revealed statistically comparable hardness between the samples. However, EC oleogel cream cheese showed lower adhesiveness and storage modulus (G') values as compared with the full-fat, commercially available control, limiting the potential success of these products. Another limitation is the difference in melting behaviour between milk fat and the oleogels, confirmed using temperature sweep rheology on corresponding cream cheese samples.

Full fat, commercial samples demonstrated a steep decrease in G' values as they were heated from 5 °C to 30 °C, which reflects the melting of milk fat at these temperatures. A more gradual decrease was observed with oleogel-containing samples, expected to be due to differences in the melting profile of the oleogels. These limitations mean further experimentation is required to improve oleogel-containing cream cheese, and sensory analysis is required to evaluate the acceptability of these products.

12.6.2 Frankfurters or Comminuted Meats

Animal fats are naturally solid fats, and are a component of traditional comminuted meat products, including frankfurters, bologna, mortadella or pâté. In these applications, the solid fat exists as globules dispersed throughout finely ground meat batters, which are stabilized by proteins within.[43] Animal fats are added in significant quantities, meaning these products would benefit from both total fat reduction and fatty acid profile improvement. Zetzl *et al.* investigated the production of frankfurters containing EC oleogels.[14] The oleogels were prepared with 10% EC (10 cP) in canola oil, heated initially to 145 °C, then cooled to induce gelation at a holding temperature of 22 °C. Meat batters were then prepared using 25% beef fat, canola oil or oleogel, and in all cases, the fat source was incorporated into lean ground meat by chopping at high speeds. TPA of these frankfurters revealed data on the hardness and chewiness. The textures achieved when using gelled canola oil were not found to be statistically different from using beef fat, while the canola oil samples had increased hardness and chewiness, as expected from previous investigations. Since fat globule size has been closely related to texture, light microscopy was performed to examine the samples' microstructure. Incorporation of gelled canola oil results in a substantial increase in globule size, as compared to the use of un-gelled canola oil (Figure 12.4).[14,43] Oleogel globules ranged from 20 μm to 140 μm; however, the median value for globule size was just 7 μm. This size distribution suggests that the presence of fewer large globules is sufficient to affect the hardness of frankfurters.[14]

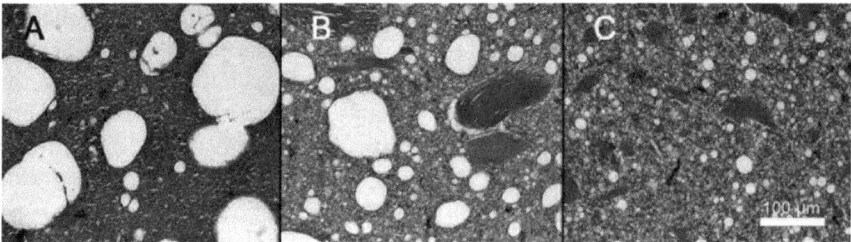

Figure 12.4 Micrographs of frankfurters containing (A) animal fat, (B) EC oleogels and (C) un-gelled oil. The fat globules are white against the dark protein network. Reprinted from ref. 21 with permission from the Royal Society of Chemistry.

However, the distribution of globules smaller than 20 μm is quite similar between beef fat and canola oil oleogel samples. Considering that meat batters containing un-gelled canola oil contain only globules under 20 μm, the greatest impact on texture may actually be owing to the absence of many small globules, rather than the presence of larger (>100 μm) globules. The results of this study are promising for the use of EC in canola oil oleogels to mimic animal fat in comminuted meat products as it allows for a particularly significant reduction in saturated fat.

In order for incorporation of EC oleogels into comminuted meats to be considered realistic, sensory analysis must also be performed. Barbut *et al.* performed sensory analysis on beef frankfurters to investigate the potential for complete replacement of beef fat with oleogels.[53] A direct comparison between the measured hardness of beef frankfurters containing beef fat, ungelled canola oil, and EC oleogels and the reported hardness determined by sensory evaluation determined that all oleogel-containing samples were harder than the beef fat control, increasing in hardness as the EC concentration was increased (Figure 12.5). The authors determined that oleogels prepared with a combination of 8–10% EC and 1.5–3% SMS were perceived by panelists to have a hardness very similar to beef fat controls, but above these concentrations perceived hardness then increased greatly once again. These EC and SMS samples were also found to have improved juiciness scores when compared to EC-only samples; however, the juiciness of oleogel samples was never found to match that of the beef fat controls. Additionally, two thirds of the panelists reported chemical off-flavours in the oleogel samples; however, the authors later cite the use of BHA as an antioxidant and rosemary extract to improve this aspect.[47] Partial replacement (20–80% oleogel) was investigated in another study, which determined again that oleogels containing 8% EC and 1.5–3% SMS were the most successful beef fat replacement. However, panelists were only able to distinguish minimal differences between the samples containing different amounts of oleogel.[47] Overall, the results indicate that frankfurters containing 8–10% EC + 1.5–3% SMS oleogels provide the closest resemblance to the beef fat controls in terms of perceived hardness and juiciness.

12.6.3 Sausages

Similar to comminuted meats, sausages are also a potential application for solid fat replacement using EC oleogels. However, sausages differ from comminuted meats, which are ground very finely, while sausages are coarsely chopped and therefore have a very different final texture. This creates new challenges as oleogels incorporated into sausages would also be present as larger pieces, the consequences of which are unknown. Barbut *et al.* formulated breakfast sausages containing EC oleogels.[54] The gels were prepared using 8%, 10%, 12% or 14% of 10 cP EC in canola oil. Some oleogels also had 1% SMS added, increasing their mechanical strength. For the control of oil oxidation, the antioxidants BHT and rosemary oleoresin were added

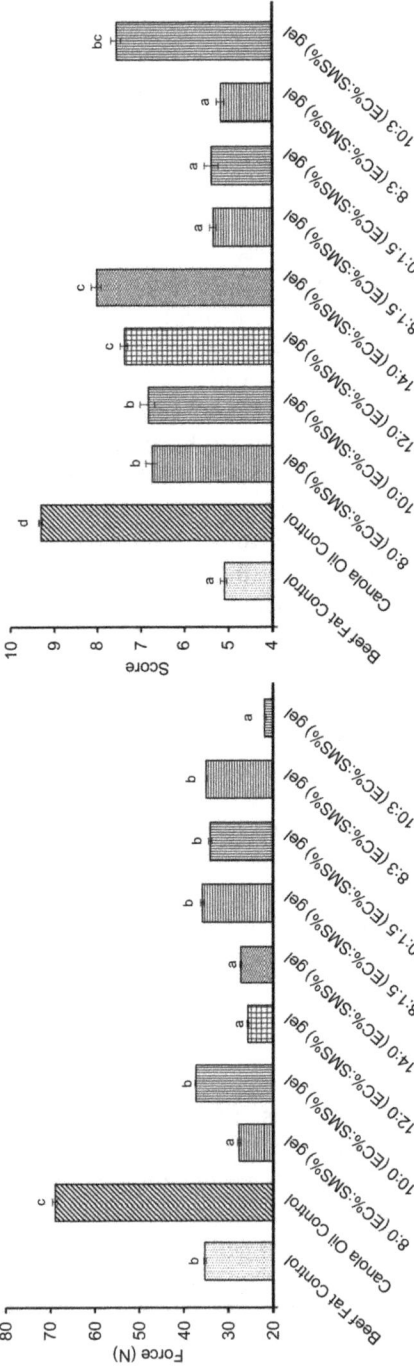

Figure 12.5 A comparison of hardness determined by instrumental TPA and sensory analysis scores (0 = very soft; 10 = very hard) of cooked frankfurters containing beef fat, un-gelled canola oil, EC gelled canola oil, and EC + SMS gelled canola oil. Presented with SEM error bars and significance represented by the letters above. Reprinted from *Meat Science*, **122**, Barbut, J. Wood and A. Marangoni, Potential use of organogels to replace animal fat in comminuted meat products, 155–162, Copyright 2016, with permission from Elsevier.

at low concentrations. Gelation was achieved by heating to 140 °C, holding for 10 min and then cooling to 20 °C. Sausage batter was prepared using coarsely ground meat and 20.8% of either pork fat, oleogel, or liquid canola oil. Each fat source was incorporated by chopping at low speeds for 1 minute, ensuring a similar grind size as the meat itself. The hardness of the sausages was then evaluated using TPA, after cooking to an internal temperature of 72 °C. Contrasting previous investigations, the hardest of the sausage samples were those made with pork fat, while samples containing liquid oil or soft oleogels demonstrated significantly lower hardness values.[14,43,55] This result is likely owing to the coarse texture of these sausages, meaning that fat was present in larger pockets throughout the sausage. Liquid oil or soft oleogels present in large pools are capable of decreasing the overall hardness of the sausages. Only harder oleogel samples containing SMS demonstrated comparable hardness to the pork fat controls, found to be not significantly different when the EC concentration was 10% or greater. Sensory analysis of these sausages found the hardness ranking by TPA accurate, but acceptance of the products did not follow the same trend. Even with SMS added for additional hardness, the 12 panelists still found that all oleogel-containing samples were softer than the pork fat control, demonstrating that instrumental hardness measurements are not indicative of sensory hardness perception. Of the samples, un-gelled canola oil was the most accepted fat replacement treatment by panelists. While results at this stage indicate that sensory acceptance of these sausages is limited, the instrumental results of sausages containing harder oleogels indicate opportunity to improve upon this application by increasing the hardness of the oleogels further. It is also encouraging to note that the addition of antioxidants in the oil successfully minimized off flavours after heating.

12.6.4 Laminating Shortenings

Laminating shortenings are solid fats used to make high-fat (on average 40% fat by weight) dough and pastry products, including yeast-leavened Danishes and croissants, as well as unleavened puff pastry. The shortenings used are commonly high in saturated fat content and are therefore of interest for saturated fat reduction.[15,56] Saturated fat, solid at room temperature, is necessary for the rising or expansion phenomenon upon baking. In laminated applications, thin layers of shortening sit between dough layers and contain numerous air bubbles stabilized by the solid fat. The air bubbles expand with heat and create the visible rise, separating the dough layers and yielding an airy, layered appearance. The conversion of water to steam during baking also contributes to the rise. Another requirement is that the fat must be solid at room temperature to prevent fat leaking out, but also plastic enough to allow for rolling without tearing the dough. These factors make it particularly challenging to eliminate the saturated fatty acid content in laminating shortenings and therefore only partial fat replacement with oleogels can be considered. Dow Global Technologies LLC has now patented a laminating shortening containing a proportion of EC oleogels.[56] In addition, a portion of the traditional fat was replaced with an alternative lower fat component, meaning the shortening has a lower overall fat content, in addition to the reduced saturated fatty acid content in favour of unsaturated fatty acids. Different formulations contain different varieties of EC depending on the laminated application, ranging from 20 to 45 cP, each with a DS of 2.46–2.57. Each shortening blend contained at least one traditional solid fat component (butter, margarine or shortening) and an EC oleogel prepared with either canola oil or sunflower oil. Additionally, flour was added, and in some cases carboxymethyl cellulose, to increase the viscosity in order to eliminate shortening squeezing out during rolling. In some cases, water was added to facilitate adequate expansion upon baking. The exact composition depended on the laminated application. Danish pastries were produced containing a roll-in fat component with 23% oleogel (12% of 45 cP EC in canola oil) and 50% butter by weight, with the remaining portion being flour. This was found to create a Danish pastry with an acceptable flaky texture. Oleogels prepared for puff pastry contained 12% of 45 cP EC in canola oil by weight. This oleogel was added at about 5% into the dough itself, which replaced butter entirely. The puff pastry roll-in shortening was composed of 35% butter and 40.5% oleogel. The remainder consisted of a solution of carboxymethyl cellulose in water, and flour. The resultant texture was also deemed acceptable, and further testing determined that CMC could be omitted while still obtaining a successful product. Finally, croissants were prepared using 7.1% of 45 cP EC in sunflower oil by weight. This decrease in EC concentration reflects the softer texture of croissant dough as compared to Danish and puff pastry doughs, and therefore the need for softer shortenings that would not tear the dough when rolling. The roll-in contained 40.3% laminating margarine, 52.7% oleogel, 1.7% water and 5.3% flour. The resultant croissants were reported to be indistinguishable from croissants made entirely with laminating margarine. This mixture allowed for a 60% reduction in saturated fat.

12.6.5   Ethylcellulose for the Reduction of Oil Migration

Oil migration, the movement of liquid oil by diffusion or capillary action, is a concern in products such as creams, chocolate confectionaries, baked goods and other-oil containing products. The structuring of oil by EC is considered a potential method to reduce this oil migration. As reviewed by Hughes *et al.*, not all oleogelators are effective at restricting oil migration in all types of food products. However, some specific examples have found EC to be successful for inhibition.[25] EC can also be added in oil systems employing alternative gelators in concentrations below what is required for gel formation to reduce the oil loss experienced over time.[57] Cookies and other high-fat baked goods are commonly susceptible to oil migration and leakage. These products are an interesting application for EC owing to the fact that the high temperatures of baking are sufficient to dissolve EC into the oil contained in the product. This means separate oleogel production could be avoided,

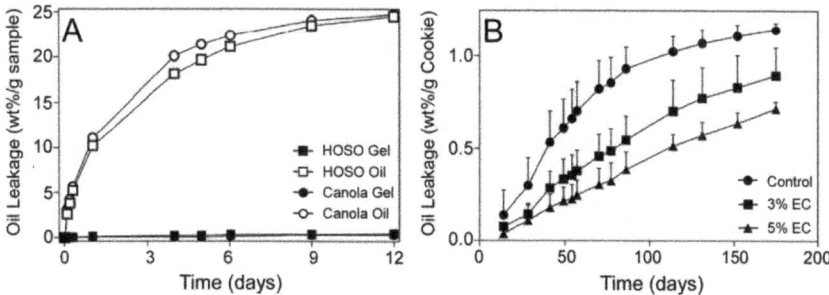

Figure 12.6 (A) Oil leakage from model cream fillings made with high oleic sunflower oil (HOSO) or canola oil, and the respective EC oleogels. (B) Oil leakage from cookies made with an overall low fat and low saturated fat shortening with added EC. Adapted with permission from T. A. Stortz, A. K. Zetzl, S. Barbut, A. Cattaruzza and A. G. Marangoni, *Lipid Technology*, 2012, **24**, 151–154. Copyright © 2012 WILEY-VCH Verlag GmbH & Co. KGaA, Weinheim.

reducing processing and energy costs. To test this procedure, Stortz *et al.* added 3 or 5% EC by weight directly into a traditional cookie recipe.[6] Oil leakage of the cookies was monitored over 200 days, and was shown to decrease with increasing concentrations of EC (Figure 12.6). Upon cooling these baked products, EC formed a network and entrapped the liquid oil, slowing migration and leakage. Since this time, a patent has been filed by Cattaruzza *et al.* for the use of EC to reduce oil migration in cookies and products of the like.[58] This outlines the use of EC ranging from 0.5 to 5% by weight.

Cream fillings, common in products such as sandwich cookies, are made using primarily saturated solid fats in combination with liquid oils in order to obtain the cream's solid, yet soft texture. EC oleogels can be employed in these cream fillings in the place of liquid oils to prevent oil leakage. In addition, incorporating EC oleogels into the creams means that the amount of solid fat used can decrease, greatly improving the fatty acid profile of the creams without compromising mechanical properties. Stortz *et al.* were able to create creams using 60% oleogel, prepared with either canola oil or high oleic sunflower oil, and 40% solid fat.[6] These oleogel-containing creams demonstrated significantly reduced oil leakage over 12 days compared to conventional controls, almost to the point of complete elimination (Figure 12.6).

Filled chocolates, and some chocolate covered confectionaries are susceptible to fat bloom due to oil migration from the inner filling to the outer surface, where visible white defects occur. As it migrates, oil will dissolve higher melting cocoa butter TAGs within the chocolate coating. Once at the surface, TAGs will recrystallize into large, stable crystals capable of scattering light and giving the chocolate an unacceptable dull, white appearance, termed fat bloom.[59] This issue has plagued confectionaries for some time and no developed methods have provided a complete solution; however, the gelation of oil does pose potential to immobilize oil. Hughes *et al.* described experimentation using cream fillings comprising 12-HSA in mixture of canola oil and interesterified hydrogenated palm oil.[25] However, this investigation actually uncovered an increase in oil migration. The authors hypothesized that the immobilization was not significant enough to affect the overall migration of oil. Further trials continue to investigate the use of other oleogelators, one of which focusses on partial replacement of the lipid source in the filling, and employed plant waxes as the gelator.[44] However, this testing is still in early stages and no conclusions have been made as of yet. The use of EC should be considered in these continuing trials, owing to its success in other oil migration reduction applications, and because of its existing uses in chocolates, described in the following section.

12.6.6 Heat-resistant Chocolate

In molten chocolate, cocoa butter acts as the continuous oil phase, so the incorporation of EC in the molten stage leads to the formation of a polymer network upon cooling. When EC is incorporated during chocolate manufacture, a significant increase in melting point is observed and the resultant chocolate has been termed heat-resistant chocolate.[60,61] This chocolate is regarded for its ability to maintain its shape at temperatures above 40 °C (Figure 12.7). For this application, an EC mix must first be prepared with

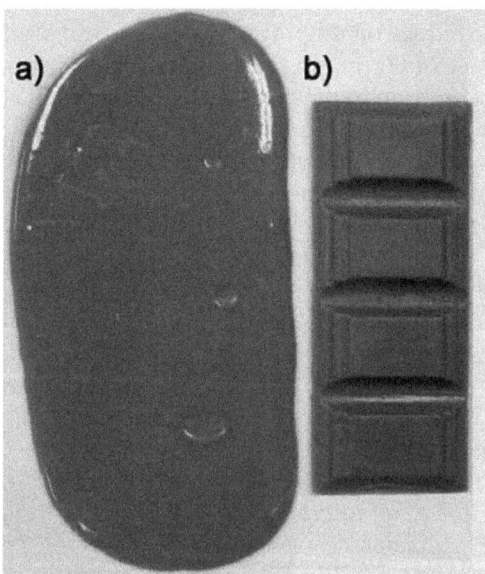

Figure 12.7   Images of (a) control chocolate and (b) heat-resistant chocolate after 2 hours in a 40 °C incubator.[66] Reproduced with permission from M. A. Rogers, T. Strober, A. Bot, J. F. Toro-Vazquez, T. Stortz and A. G. Marangoni, *Int. J. Gastron. Food Sci.*, 2014, **2**, 22–31.

20–25% EC in ethanol. The mixture is then added directly into melted, tempered chocolate at temperatures ranging from 28 °C to 40 °C depending on the type of chocolate. High temperatures are not needed here as EC is highly soluble in ethanol at room temperatures. After incorporation, volatile ethanol will evaporate, leaving the EC behind. Similar to conventional oleogels, the concentration of EC in the chocolate was found to impact the hardness at higher temperatures, though the final heat resistance of the chocolate was determined to be not dependent on the viscosity of EC used.[61] This effect is suspected to be a result of the interactions between EC and sugar molecules, caused by hydrogen bonding between sucrose and the polar groups on the EC backbone.[62] EC and sugar act to create a network with lower melting cocoa butter TAGs structured within, resulting in an increased melting point. The involvement of sugar is thought to be why milk and white chocolates exhibited greater heat resistance than dark chocolates as dark chocolate contains significantly less sugar.[61] At this point, heat-resistant chocolate is not a market ready product owing to the higher cost and lower quality of the final products. It does, however, form a good base for research to continue as heat-resistant chocolate has great potential to be a desirable commodity, specifically for hot seasons and tropical countries. A patent for this technology was filed by Marangoni in 2010.[63] A more recent study has also identified the incorporation of EC into dark chocolate as a method of modifying the rheological behaviour of molten chocolate, which did not interfere with the tempering process or chocolate stability.[64] The authors noted an increase in viscosity with increasing concentrations of EC owing to interactions between the polymer and the solid particles within chocolate.

12.7 Conclusion

Until recently, many researchers have focused on the reduction of fat as a whole to improve the nutrition of foods containing large amounts of saturated and *trans* fatty acids. Oleogels present an innovative reduction technique, where solid fat does not need to be removed entirely, but rather replaced with unsaturated sources with similar mechanical properties. Based on the evidence presented in this chapter, the use of EC oleogels is a promising route for eventual significant saturated and *trans* fatty acid reduction in the diet. EC oleogels are noteworthy owing to the ease with which the strength of the gel can be manipulated. This is advantageous owing to the impact that strength has on the feasibility of oleogel use in specific food applications, as well as their dependence on gel strength for their ability to be used as a vehicle for the controlled release of lipophilic nutraceuticals. At this point, the use of EC oleogels is most promising in certain food applications, specifically for ground meat products, soft cheeses and as a component in shortening mixtures for partial saturated fat replacement. While texture analysis and rheology provide a good sense of how the oleogels perform, the component of research lacking at this point is sensory analysis to determine the overall acceptance of these products. Additionally, taste considerations must be acknowledged to ensure that oxidization that occurs during oleogel production does not create overwhelming off-flavours. It is also worthwhile to further investigate other aspects of commercializing oleogel-containing products, including storage conditions and shelf life. Current research indicates that many EC oleogels will behave like liquid oils upon consumption. While more research is needed for strong conclusions in this regard, the potential for substantial reduction of saturated and *trans* fatty acid in foods, as well as the potential for EC oleogels to facilitate controlled release of lipophilic bioactives, make EC oleogel research an important avenue for greatly improving the nutrition of high-fat foods.

References

1. A. G. Marangoni and N. Garti, in *Edible Oleogels*, ed. A. G. Marangoni and N. Garti, AOCS Press, Urbana, IL, 2011, pp. 1–17.
2. R. P. Mensink, P. L. Zock, A. D. M. Kester and M. B. Katan, *Am. J. Clin. Nutr.*, 2003, **77**, 1146–1155.
3. D. Mozaffarian, M. B. Katan, A. Ascherio, M. J. Stampfer and W. C. Willett, *N. Engl. J. Med.*, 2006, **354**, 1601–1613.
4. J. Lunn and H. E. Theobald, *Nutr. Bull.*, 2006, **31**, 178–224.
5. M. Davidovich-Pinhas, S. Barbut and A. G. Marangoni, *Carbohydr. Polym.*, 2015, **117**, 869–878.
6. T. A. Stortz, A. K. Zetzl, S. Barbut, A. Cattaruzza and A. G. Marangoni, *Lipid Technol.*, 2012, **24**, 151–154.
7. W. Koch, *Ind. Eng. Chem.*, 1937, **29**, 687–690.
8. M. Davidovich-Pinhas, A. J. Gravelle, S. Barbut and A. G. Marangoni, *Food Hydrocolloids*, 2015, **46**, 76–83.
9. D. Roy, M. Semsarilar, J. T. Guthrie and S. Perrier, *Chem. Soc. Rev.*, 2009, **38**, 2046–2064.
10. R. H. Atalla and A. Isogai, in *Polysaccharides: Structural Diversity and Functional Versatility*, ed. S. Dumitriu, Marcel Dekker, New York, 2nd edn, 1998, pp. 123–157.
11. *EthocelTM: Ethylcellulose Polymers Technical Handbook*, Dow Cellulosics, Midland, MI, 2005.
12. A. J. Gravelle, S. Barbut, M. Quinton and A. G. Marangoni, *J. Food Eng.*, 2014, **143**, 114–122.
13. M. Davidovich-Pinhas, S. Barbut and A. G. Marangoni, *Cellulose*, 2014, **21**, 3243–3255.
14. A. K. Zetzl, A. G. Marangoni and S. Barbut, *Food Funct.*, 2012, **3**, 327–337.
15. H. M. Lai and T. C. Lin, in *Bakery Products: Science and Technology*, ed. Y. H. Hui, Blackwell Publishing, Ames, IA, 2006, pp. 3–65.
16. A. J. Gravelle, M. Davidovich-Pinhas, A. K. Zetzl, S. Barbut and A. G. Marangoni, *Carbohydr. Polym.*, 2016, **135**, 169–179.
17. T. Laredo, S. Barbut and A. G. Marangoni, *Soft Matter*, 2011, **7**, 2734–2743.
18. A. J. Gravelle, S. Barbut and A. G. Marangoni, *Food Funct.*, 2013, **4**, 153–161.

19. A. J. Gravelle, S. Barbut and A. G. Marangoni, *Food Res. Int.*, 2012, **48**, 578–583.
20. A. K. Zetzl, A. J. Gravelle, M. Kurylowicz, J. Dutcher, S. Barbut and A. G. Marangoni, *Food Struct.*, 2014, **2**, 27–40.
21. M. Davidovich-Pinhas, S. Barbut and A. G. Marangoni, *Carbohydr. Polym.*, 2015, **127**, 355–362.
22. T. Dey, D. A. Kim and A. G. Marangoni, in *Edible Oleogels*, ed. A. G. Marangoni and N. Garti, AOCS Press, Urbana, IL, 2011, pp. 295–309.
23. H. L. Bemer, M. Limbaugh, E. D. Cramer, W. J. Harper and F. Maleky, *Food Res. Int.*, 2016, **85**, 67–75.
24. N. Duffy, H. C. G. Blonk, C. M. Beindorff, M. Cazade, A. Bot and G. S. M. J. E. Duchateau, *JAOCS, J. Am. Oil Chem. Soc.*, 2009, **86**, 733–741.
25. N. Hughes, A. G. Marangoni, A. J. Wright, M. A. Rogers and J. W. E. Rush, *Trends Food Sci. Technol.*, 2009, **20**, 470–480.
26. N. Hughes, J. W. Rush and A. G. Marangoni, in *Edible Oleogels*, ed. A. G. Marangoni and N. Garti, AOCS Press, Urbana, IL, 2011, pp. 313–330.
27. H. Yu, K. Shi, D. Liu and Q. Huang, *Food Chem.*, 2012, **131**, 48–54.
28. C. M. O'Sullivan, MSc Thesis, University of Guelph, 2016.
29. C. M. O'Sullivan, S. Barbut and A. G. Marangoni, *Trends Food Sci. Technol.*, 2016, **57**, 59–73.
30. Q. Guo, A. Ye, M. Lad, D. Dalgleish and H. Singh, *Food Hydrocolloids*, 2013, **33**, 215–224.
31. Q. Guo, A. Ye, M. Lad, D. Dalgleish and H. Singh, *Food Hydrocolloids*, 2016, **54**, 255–265.
32. D. J. McClements, E. A. Decker, Y. Park and J. Weiss, *Food Biophys.*, 2008, **3**, 219–228.
33. S.-Y. Tan, E. Wan-Yi Peh, A. G. Marangoni and C. J. Henry, *Food Funct.*, 2017, **8**, 241–249.
34. P. de Vos, M. M. Faas, M. Spasojevic and J. Sikkema, *Int. Dairy J.*, 2010, **20**, 292–302.
35. C. T. Vogelson, *Mod. Drug Discovery*, 2001, **4**, 49–50.
36. S. Marze, *Crit. Rev. Food Sci. Nutr.*, 2013, **53**, 76–108.
37. A. V. Rao and L. G. Rao, *Pharmacol. Res.*, 2007, **55**, 207–216.
38. N. I. Krinsky and E. J. Johnson, *Mol. Aspects Med.*, 2005, **26**, 459–516.
39. D. A. Cooper, A. L. Eldridge and J. C. Peters, *Nutr. Rev.*, 1999, **57**, 201–214.
40. L. Ryan, O. O'Connell, L. O'Sullivan, S. A. Aherne and N. M. O'Brien, *Plant Foods Hum. Nutr.*, 2008, **63**, 127–133.
41. K. Iwanaga, T. Sumizawa, M. Miyazaki and M. Kakemi, *Int. J. Pharm.*, 2010, **388**, 123–128.
42. F. R. Lupi, D. Gabriele, N. Baldino, P. Mijovic, O. I. Parisi and F. Puoci, *Food Funct.*, 2013, **4**, 1512–1520.
43. M. K. Youssef and S. Barbut, *Meat Sci.*, 2009, **82**, 228–233.
44. A. R. Patel and K. Dewettinck, *Food Funct.*, 2016, 20–29.
45. T. A. Stortz and A. G. Marangoni, *Green Chem.*, 2014, **16**, 3064.
46. R. Ergun, R. B. Appell and D. L. Malotky, Dow Global Technologies LLC, US Pat., 20160081374A1, 2016.

47. S. Barbut, J. Wood and A. G. Marangoni, *J. Food Sci.*, 2016, **81**, 2183–2188.
48. B. Mert and I. Demirkesen, *LWT–Food Sci. Technol.*, 2016, **68**, 477–484.
49. A. I. Blake and A. G. Marangoni, *Food Res. Int.*, 2015, **74**, 284–293.
50. E. Yilmaz and M. Öğütcü, *Food Funct.*, 2015, **6**, 1194–1204.
51. B. Mert and I. Demirkesen, *Food Chem.*, 2016, **199**, 809–816.
52. M. Brighenti, S. Govindasamy-Lucey, K. Lim, K. Nelson and J. A. Lucey, *J. Dairy Sci.*, 2008, **91**, 4501–4517.
53. S. Barbut, J. Wood and A. Marangoni, *Meat Sci.*, 2016, **122**, 155–162.
54. S. Barbut, J. Wood and A. Marangoni, *Meat Sci.*, 2016, **122**, 84–89.
55. M. K. Youssef and S. Barbut, *Meat Sci.*, 2011, **87**, 356–360.
56. R. Ergun, B. S. Thomson and B. Huebner-Keese, Dow Global Technologies LLC, US Pat., 20160021898A1, 2016.
57. A. Lopez-Martinez, M. A. Charo-Alonso, A. G. Marangoni and J. F. Toro-Vazquez, *Food Res. Int.*, 2015, **72**, 37–46.
58. A. Cattaruzza, S. Radford and A. G. Marangoni, US Pat., 20140044839, 2013.
59. V. Ghosh, G. R. Ziegler and R. C. Anantheswaran, *Crit. Rev. Food Sci. Nutr.*, 2002, **42**, 583–626.
60. T. A. Stortz and A. G. Marangoni, *Trends Food Sci. Technol.*, 2011, **22**, 201–214.
61. T. A. Stortz and A. G. Marangoni, *Food Res. Int.*, 2013, **51**, 797–803.
62. T. A. Stortz, D. C. De Moura, T. Laredo and A. G. Marangoni, *RSC Adv.*, 2014, **4**, 55048–55061.
63. A. G. Marangoni, US Pat., 20120183651, 2012.
64. M. R. Ceballos, K. L. Bierbrauer, S. N. Faudone, S. L. Cuffini, D. M. Beltramo and I. D. Bianco, *Food Struct.*, 2016, **10**, 1–9.
65. M. Davidovich-Pinhas, S. Barbut and A. G. Marangoni, *Annu. Rev. Food Sci. Technol.*, 2016, **7**, 65–91.
66. M. A. Rogers, T. Strober, A. Bot, J. F. Toro-Vazquez, T. Stortz and A. G. Marangoni, *Int. J. Gastron. Food Sci.*, 2014, **2**, 22–31.

Section VI

Functional Colloids from Structured Oils

CHAPTER 13

Non-aqueous Foams Based on Edible Oils

ANNE-LAURE FAMEAU

L'Oréal, Research & Innovation, International Physical-Chemistry
Department, Saint-Ouen, 93400, France
*E-mail: afameau@rd.loreal.com

13.1 Introduction

Foams can be seen almost everywhere in our surrounding environment.
They are found in many food items providing texture to whipped cream, aerated desserts, bread, cakes, cappuccino, *etc*. Foams can exist in the solid or
liquid state. Liquid foam is a two-phase media where gas is dispersed into
a continuous fluid. There are two categories of liquid foam: aqueous and
non-aqueous foams. When the continuous fluid is water, foams are called
aqueous foams. They are formed of gas bubbles dispersed in an aqueous
phase. They are the most common types of foams and are widely used in
industrial applications, such as cosmetics, detergents, food, fire-fighting and
flotation of minerals. When the continuous phase is not water, foams are
called non-aqueous foams. In this case, foams are composed of gas bubbles
dispersed in a non-aqueous phase. These systems have an important role in
petroleum, lubrication oil systems and manufacturing industries. In comparison to aqueous foams, they are scarcely studied.[1] However, the addition
of gas bubbles to create low-calorie food products is a good way to reduce the
total fat content while providing a light and pleasant texture.[2,3] For example,

Food Chemistry, Function and Analysis No. 3
Edible Oil Structuring: Concepts, Methods and Applications
Edited by Ashok R. Patel
© The Royal Society of Chemistry 2018
Published by the Royal Society of Chemistry, www.rsc.org

it has been demonstrated that the characteristic chewy mouth-feel and low fat content of chocolate including bubbles is of crucial importance for consumers.[4] Therefore, an understanding of the key parameters leading to foam formation and stability is crucial to develop oil foams for low-calorie food products.

13.2 Aqueous Foam

The properties of aqueous foams are a very broad subject and have been described in detail in many books and reviews.[5-7] The current state of knowledge is therefore summarized briefly in order to compare with non-aqueous foams. The two most important factors in foaming a liquid are how easily it foams (its foamability) and its stability.

13.2.1 Formation of Aqueous Foam

Aqueous foam is a biphasic dispersion of air bubbles into a water phase. Aqueous foams are thermodynamically metastable systems that tend to separate with time into their individual components: gas and water. The surface of bubbles dispersed within a pure liquid possesses a strong tendency to minimize the surface area. Therefore stable foam cannot be produced in pure liquids unless the liquid possesses very high viscosity. To produce and stabilize foams, the use of stabilizing components dispersed in the aqueous phase is required. To be efficient, the foam-stabilizing component has to get rapidly to the gas–liquid interface in order to produce foam. Adsorption of surface-active components to the bubble surfaces reduces the surface tension of the liquid phase. The stabilizing components can be surfactants, polymers, proteins or particles. They are responsible for both the foamability and the stability of the resulting foams. The stabilizing components help to stabilize the foam by slowing down the three main mechanisms of foam decay: drainage, coarsening and coalescence. For food products, the foam stability is crucial. The foam food products need to be stable both on the supermarket shelves and in the consumer's home. Despite their simple composition of gas bubbles dispersed in a liquid phase, considerable effort by the scientific community has been made over a century to reach the current level of understanding.

13.2.2 Classification of Aqueous Foams

Foams may be classified as dry or wet foams according to their liquid content, *i.e.* contain a greater or lesser amount of liquid. The liquid volume fraction Φ represents this liquid content. The liquid volume fraction Φ ranges from much less than 1% to about 30%. The gas fraction simply corresponds to $(1 - \Phi)$. It determines some essential properties, such as rheological ones. Depending on the bubble size and the liquid volume fraction, the foam self-organizes into different structures. Wet foam corresponds to foam with a liquid volume fraction larger than about 20%. Wet foams consist of spherical gas bubbles surrounded by the liquid phase. When the amount of gas is small so Φ is high, the systems are sometimes called "aerated", rather than foams. When the gas fraction in foam exceeds 74%, the gas bubbles deform and transit from spherical shape to polyhedral shape. This value of 74% corresponds to the close packing of monodisperse sphere. The flattened regions become the foam liquid films. They are formed between the air bubbles and stabilized by the foam-stabilizing component.

13.2.3 Aqueous Foam: A Multiscale System

To study liquid foams, a multi-scale approach is needed starting at the molecular scale and going up to the macroscopic scale (Figure 13.1). The properties of a foam result from a complex coupling between them. At the nanometer scale, it is important to study the foam-stabilizing components adsorbed at the gas–liquid interfaces and the mechanisms responsible for the stabilization of thin films separating the individual bubbles where two interfaces are in close proximity. Zooming further out, at the microscopic scale, three adjacent thin films meet to create Plateau borders. Nodes are made by the connections of four Plateau borders. They form the main liquid skeleton of interconnected channels. The liquid flows through these channels and controls the foam properties. Looking at individual bubbles, their diameter can vary from tens of micrometers to centimeters. The macroscopic scale corresponds to the whole foam, whose properties are dictated by those in all the length-scales below.

13.2.4 Methods of Aqueous Foam Production

There is a large variety of foaming methods: shaking, whipping, turbulent mixing of liquid and gas jets, decompressing gas, blowing gas through porous membranes, *etc.*[8] The bubble size and the liquid volume fraction can be varied by the foaming methods used to produce foams. A parameter often used to compare different foaming systems is the overrun, *i.e.*, the

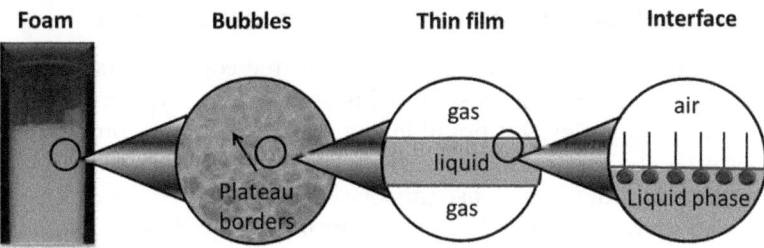

Figure 13.1 Foam structure at different length scales from the macroscopic scale (foam) to nanometric scale (interface).

percentage increase in volume owing to incorporation of gas. The foam over-run can be obtained by using eqn (13.1), for which the volume fraction of the air corresponds to $V_t - V_0$:

$$\text{Overrun(\%)} = \left[\frac{V_t - V_0}{V_0} + 100 \right] \tag{13.1}$$

V_t corresponds to the volume of foam after a time t and V_0 corresponds to the initial volume of liquid used to produce the foam.

13.2.5 Mechanisms of Aqueous Foam Destabilization

All foams are thermodynamically metastable dispersions of a gas in a liq-uid that are evolving with time.[9] Foam evolves owing to the redistribution of liquid inside the foam channels or/and gas between the bubbles. The term *foam drainage* describes the process by which fluid flows out of a foam. It is caused by both gravity and capillary pressure from the Plateau borders.[10] As a result of the drainage, the liquid volume fraction Φ decreases, leading to drier foams with a higher gas volume fraction. At low gas volume fractions, the creaming mechanisms of foam destabilisation predominate, whereas at higher gas fractions the liquid drainage prevails. If the film between two bub-bles is unstable it can break, leading to the merging of the two bubbles. This mechanism is called *coalescence*. This destabilization mechanism is also responsible for foam collapse as the bubbles coalesce at the top of the foam, leading to the disappearance of the foam.[11] The diffusion of gas between bubbles is a destabilization phenomenon called *coarsening*. It involves the transport of gas between bubbles of different radii owing to differences in the Laplace pressure. Bubbles smaller than the average size shrink, whereas larger bubbles grow, leading to the disappearance of the smaller bubbles[10,12]. Coalescence and coarsening lead to the growth of average bubble sizes with time. These two processes are strongly coupled with drainage. The drainage of the fluid between bubbles leads to a closer approach of the bubble sur-faces and can lead to the rupture of the thin liquid film when a critical thick-ness is reached.[13]

13.2.6 Mechanisms of Aqueous Foam Stabilization

Aqueous liquid foams always have at least three components: an aqueous liquid phase, a gas and a foam-stabilizing component.[14] To control the foam stability and properties, it is possible to modify these three components.[15-17] The nature of the gas and the viscosity of the liquid phase are simple param-eters to tune in order to increase the foam stability.

One important parameter is the viscosity of the liquid phase, which modi-fies the drainage strongly, as well as influences coarsening and coalescence. An easy method to slow down the foam drainage is to increase the aque-ous liquid phase viscosity. For example, the addition of a small amount of glycerol in the liquid phase leads to an increase of the bulk viscosity, which results in a decrease of the drainage rate.[18] Most of the aerated food products are wet foams, since bubbles remain spherical and are separated from each other. In these systems, creaming of the bubbles need to be prevented. To achieve this, the continuous aqueous phase should have a yield stress. For example, it is possible to use a gelling polymer such as gelatin. Above 30 °C, gelatin makes a viscous liquid and can then be beaten to incorporate bubbles. By cooling down the system to below 20 °C, the gelling occurs and the foam is mechanically stabilized. In the same way, in whipped cream, the fat globules cover the air bubbles and also form a continuous network in the aqueous phase forming a bubble network.[19]

The nature of the gas is another important parameter since it diffuses through the liquid to lead to coarsening of the bubbles. For example, by using a gas that is very poorly soluble in the aqueous phase, the gas transport across the water phase can be slowed down in comparison to a more soluble gas. Therefore, foam stability is higher in the presence of N_2 gas than CO_2 owing to the poorer solubility of N_2 compared with that of CO_2 in water. To drastically slow down the coarsening, gases of very low solubility, such as fluorinated gases, can be used.[20] Another possibility to improve foam stability by slowing down coarsening is to use a small amount of a poorly soluble gas. The transfer of the more soluble gas will be faster, and therefore the large and the small bubbles will have different mixture compositions of gas. This leads to a difference in osmotic pressures, which can help balance the Laplace pressure.[21] This will slow down the coarsening.

13.2.7 Importance of the Surface Tension and Viscoelastic Properties

The presence of a foam-stabilizing component is essential for foam generation and stabilization. They adsorb at the gas–liquid interface and reduce the surface tension of the liquid phase. Foaming is related to surface tension and the surface tension of the foaming solution gives a measure of the surface activity. Although a low dynamic surface tension usually corresponds to a high foamability, the reduction of the surface tension is not sufficient to lead to a stable foam. For example, ethanol–water mixtures have a low surface tension (40% ethanol in water leads to a surface tension of 30 mN m^{-1}), but make very unstable foams.[22] Foam stability is usually more dependent on the viscoelasticity and repulsive interactions within the film lamellae. The viscoelasticity can arise from steric or electrostatic interactions between the stabilising components or through the formation of a mechanically rigid layer.[23] In the case of surfactants, the mechanisms of stabilization of bubble interfaces are linked to their ability to resist surface tension gradients. The resistance to changes in the surface area or surface concentration is described by the interfacial viscoelasticity. In the case of the common surfactants as foam stabilizers, they adsorb rather rapidly to the interface and populate the gas–liquid interfaces, leading to the formation of surfactant monolayers.

The presence of this surfactant monolayer influences the coarsening process depending on the mechanical properties of the surface. The requirement to decrease the coarsening process is that the surfactant monolayers exhibit a surface elastic modulus and a resistance to compression. The elastic modulus is defined as:

$$E = A\frac{d\gamma}{dA} \tag{13.2}$$

Where A is the bubble area and γ is the surface tension. A high compressional surface elasticity means that small changes in surface area lead to large changes in γ.

In the case of surfactant monolayers, an exchange of surfactant monomers can occur between the bulk and the interface due owing the free desorption and adsorption of the monomers at the interface for soluble surfactants. This phenomenon leads to the low resistance of the monolayer to compression. The resulting foam resistance to coarsening is low. When the surfactant monomers can desorb from the interface, the surface coverage can in some cases be too low to avoid the coalescence phenomenon. The presence of a surfactant monolayer influences the coarsening and coalescing processes depending on the mechanical properties of the surface.[24,25] In the case of polymers and proteins, the adsorption energies are typically larger than those for surfactants. The molecular exchange at the interface is slowed down and the surfaces are often more viscoelastic than those with surfactants. The particles used for foam stabilisation increase the adsorption energy even further and can lead to even more elastic interfaces. Indeed, it is possible to completely arrest the ageing of the foams with suitably chosen particles.[26] There is a large choice of particles leading to aqueous foam stabilisation.

13.2.8 Main Differences Between Aqueous and Non-aqueous Foams

Contrary to aqueous foams, few studies have been devoted to the understanding of oil foams.[1,27] It was only 70 years after the first publication by McBain and Perry that this topic really started to gain interest in the foam community. Nowadays, this topic is intensively studied, especially over the last two years. Indeed, this topic is important for various industries: the petroleum industry, lubrication oil systems, the personal care industry and the food industry. For the food industry, the addition of gas bubbles to create low-calorie food products seems a good way to reduce the total fat content while providing a light and pleasant texture.[2,3]

In aqueous foams, the liquid phase is always based on the same solvent: water. In non-aqueous foams, the liquid phase can be any solvent except water. In a first step to describe and understand non-aqueous foams, all the information and concepts described in the literature for aqueous foam formation and stability are useful, but specific considerations must be taken into account by replacing water with nonaqueous solvents in the continuous phase. Indeed, water is unique since this low-viscosity liquid has a high surface tension owing to extensive hydrogen bonding. In the case of nonaqueous solvents, the surface tension is relatively low. The main difference between non-aqueous and aqueous foaming systems comes from the surface tension at the liquid–gas interface. In aqueous foams, surfactants strongly adsorb at the interface reducing the high water–air surface tension ($\approx$72 mN m^{-1} at 25 °C) to lower values ranging from 20 to 40 mN m^{-1}. For non-aqueous foam systems, the typical non-aqueous liquids already have rather low surface tension between 15 to 30 mN m^{-1}, making the adsorption of hydrocarbon-based surfactants energetically unfavorable. For this reason, the surface tension is not the main criterion to understand the foamability and foam stability in the case of non-aqueous foams.[1] Moreover, in the case of oil systems, the dielectric constant is low and the electrostatic double layer repulsion is minimal, contrary to aqueous foams. The electrostatic stabilization of foam films is prevented. The non-aqueous foam properties are mainly driven by the physical properties of the non-aqueous liquid phase: polarity, viscosity, density, conductivity, dielectric constant and Hansen solubility parameter values. Three different foam stabilizer categories can be used to stabilize non-aqueous foams: specialty surfactants, solid particles and crystalline particles.[27]

13.3 Non-aqueous Foams Based on Surfactants

In the literature, some surfactants have been shown to produce oil foams. The efficiency of a surfactant to produce and stabilize oil foams is linked to its molecular structure.

13.3.1 Non-aqueous Foams Based on Hydrocarbon-type Surfactants

The first report dealing with the reduction of surface tension of oils by surfactants was published in the 1940s by McBain and Perry.[28] They showed that lauryl sulfonic acid can reduce the surface tension of mineral oils, tetraisobutylene and hydrogenated tetra-isobutylene from about 3 to 5 mN m^{-1}, but it has a less pronounced effect on the surface tension of benzene, xylene and heptane. This report highlighted the influence of the liquid phase on the surfactant's ability to decrease the surface tension. In the 1980s, Ross and coworkers reported that sorbitan monolaurate (Span 20) decreased the equilibrium surface tension of mineral oil by 2 mN m^{-1}.[7] They also showed weak foaming in lubricating oils.[7] Zisman showed that packed methyl groups yield a surface tension of 23 mN m^{-1} while methylene groups lower the surface tension by a value of 28 mN m^{-1} [29]. These results suggest that the lowering of surface tension is linked to the surface concentration of oriented methyl groups.

In the 1970s, Sanders *et al.* demonstrated that certain types of hydrocarbon surfactant with specific functional groups (*i.e.* alcohols, glycols, soaps, acids, amines) in mineral oil could lead to non-aqueous foam formation.[6,30] For example, mineral oil foams can be stabilized by ethoxylated stearyl alcohol or polyethylene-based surfactants. In the 1980s, Friberg *et al.* reinforced this hypothesis by studying xylene foams stabilized by triethanolammonium oleate surfactant.[6] No foam was produced when the surfactant formed an isotropic liquid phase, but foams were produced in the region where the surfactant became insoluble and formed crystalline particles. These papers highlight that long-chain hydrocarbon molecules with particular functional groups (acids, alcohol or amine groups) can decrease the surface tension. Moreover, the solubility of the hydrocarbon-based surfactant in the oil phase seems to be the critical parameter to produce oil foams.

In general, hydrocarbon-based surfactants produce relatively unstable foams with a non-aqueous liquid phase with low viscosities. However, they can lead to very stable non-aqueous foams when the hydrocarbon-based surfactant self-assembles to form crystalline particles in the non-aqueous phase (see Section 13.5 about crystalline particles).

13.3.2 Non-aqueous Foams Based on Polymethylsiloxane-type Surfactants

Contrary to the non-aqueous foams obtained by using hydrocarbon surfactant, polymethylsiloxane-based surfactants are able to produce relatively stable non-aqueous foams. The surface activity of polymethylsiloxane-based surfactants is due to the presence of hydrophilic Si–O groups, which are shielded by the lyophobic methyl groups, which orient and pack at the surface. Silicone surfactants are obtained by adding a lyophilic moiety to the lyophobic polysiloxane. Owing to the incompatibility of the polysiloxane moieties of the silicone surfactant with themselves and other molecules, they are highly effective in lowering surface tension. For example, below the solubility limit, some polydimethylsiloxane $[Si-(OCH_3)_2]n$ are able to decrease the surface tension of synthetic lubricant oils leading to foaming.[7] However, above the solubility limit, they are efficient foam breakers for both aqueous and non-aqueous foams. These silicone-based surfactants increase surface viscosity owing to the polyether portion of the surfactant. Thus, the surface shear viscosity increases with the amount of polyether attached to the polysiloxane. Branched polysilane block copolymers have been shown to yield higher foam stability than linear surfactants.[31] Based on results described in the literature, silicone surfactants seem to stabilize non-aqueous foams by both setting up a surface tension gradient and by increasing the surface viscosity.[7]

13.3.3 Non-aqueous Foams Based on Fluoroalkyl-type Surfactants

Fluoroalkyl surfactants can reduce the surface tension of liquids to low values (<20 mN m^{-1}) owing to their fluorocarbon moiety. These fluoroalkyl-type surfactants decrease the surface tension of non-aqueous liquids lower than any other known solute.[32] The surface activity of this fluoro-carbon moiety is owing to the fluorine atoms, which have the proper covalent radii to shield the carbon chain sterically. Fluoroalkyls $(C_3[CF_2]n)$ have the lowest interactions both with themselves and other molecules than any other type of groups. In the literature, fluorine-containing compounds with different types of polar groups have been described. Nonionic fluoroalkyl-type are the most efficient to decrease the surface tension of organic solvents. For example, some of these surfactants reduce the surface tension of heptane, leading to the production of relatively stable foams with diesel fuels and kerosene.[7]

13.3.4 Non-aqueous Foams Based on Asphaltenes and Resins

One of the most abundant types of non-aqueous foam is petroleum-based foam. The continuous phase is made of crude oil. Crude oil is a very complex natural liquid composed of variety of organic compounds, mainly saturated and aromatic hydrocarbons and components with amphiphilic properties, such as resins and asphaltenes. The composition of crude oil varies as a function of the location and age. Crude oil can foam upon depressurization during gas separation processes. These foams can be stable for several hours or days but they are unwanted in the oil industry, because they reduce oil production capacity, damage equipment and lead to poor oil–gas separation efficiency. In the 1980s, several studies were published showing that foaming in several different types of crude oils was due to the presence of surface-active organic materials.[33,34] The composition of the crude oil controls the eventual foaming properties. Several constituents can lead to foam formation and its stabilization: short-chain carboxylic acids and phenols, asphaltenes and resins.[27] Several parameters have been identified to control the foaming behavior of crude oils: bulk viscosity and density, oil–gas surface tension, asphaltene and resin content, and their molecular weight.[35]

13.4 Non-aqueous Foams Based on Solid Particles

13.4.1 Wettability of Solid Particles

Aqueous and non-aqueous foams can be obtained from specific colloidal particles, which adsorb at fluid–fluid interfaces.[36] If these solid particles are adsorbed irreversibly at the interface, they can lead to bubbles protected against coalescence and disproportionation. The driving force of the particle adsorption at the fluid interfaces is the free energy gain by losing an area of interface. Particle adsorption is thermodynamically favorable in the case of solid–liquid–air systems, if the sum of the solid–air tension and the solid–liquid tension is less than the original liquid–air tension. To adsorb, the solid particles need to exhibit suitable wettability. The wettability of the particles is quantified by the contact angle (θ) the particles make with the air–liquid interface measured through the liquid phase (Figure 13.2). The contact angle is given by Young equation:

$$\cos\theta = \frac{\gamma_{\text{solid/air}} - \gamma_{\text{solid/liquid}}}{\gamma_{\text{liquid/air}}} \tag{13.3}$$

Young's equation suggests that the particle contact angle could be changed by altering the respective interfacial tensions. To produce aqueous foams, there are two common approaches for changing the particle contact angle and tuning particle wettability: chemical modification of the particle surface and surfactant adsorption to the particle surface.[36] When the liquid phase is oil, the spherical particle is completely wetted by this liquid, when θ is 0°. The result is the dispersion of particles inside the oil phase. When θ is 180°, the particle is completely non-wetted by the liquid. In these two cases, no air entrainment even after vigorous mixing is observed, since no particle adsorption occurs at the air–liquid surface (Figure 13.2). For intermediate θ values between 0° and 180°, particles are more wetted by one of the two phases: liquid or air. They can adsorb at the interface. For $\theta < 90°$, particles are partially oleophobic and stable oil foams should result after aeration of the

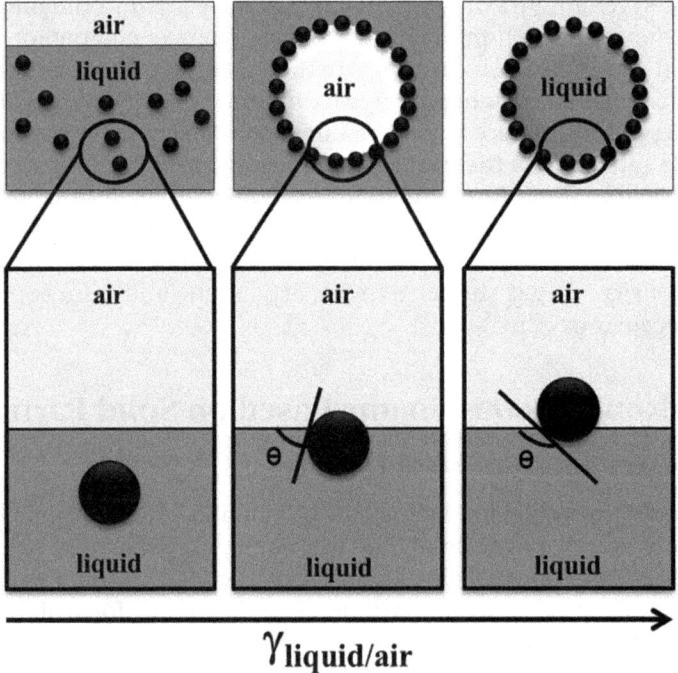

Figure 13.2   Upper: oil dispersion (left), air-in-oil foam (center), and liquid-in-air powder (right). Lower: position of particle in mixtures of liquid and air as a function of contact angle and surface tension. Contact angle increases from zero to medium to high values from left to right.

mixture (Figure 13.3a). For $\theta > 90°$, particles are extremely oleophobic and the inverted case of oil-in-air droplets should be stabilized. Particles must be partially oleophobic in order to exhibit contact angles between 0 and 180°.

The difficulty is generating repellency against oils on particle surfaces to obtain these partially oleophobic particles. Contrary to aqueous foams, it is not easy to tune particle wettability. As for aqueous foams, the particle contact angle can be modified by chemical modification of the surface particles. Indeed, in the case of particles coated by hydrocarbon-containing groups, they are wetted by many oils and no oil foam can be produced. However, when they are coated by fluoro groups they can be oleophobic to certain oils. Oil foams are obtained. The difference comes from the lower surface energy of fluorocarbons than that of hydrocarbons. Therefore, the preparation of particle-stabilized oil foams is much more challenging than for aqueous foams.[37] Moreover, the industrial application of oil foams stabilized by particles coated by fluoro groups is limited to specific applications and cannot be used for food applications. In the case of oil foams, the wettability of the particles can be changed not only by chemical modification of the surface particles but also by the surface tension of the liquid phase.

13.4.2 Formation and Properties of Non-aqueous Foams Obtained from Solid Particles

Oil foams can be produced by hand-shaking or by introducing air bubbles by mechanical mixing (blender, rotor-stator homogeniser, *etc.*). In terms of foam stability, the non-aqueous foams obtained from solid particles are called ultrastable since the foam volume remains constant for months (Figure 13.3.a). In some cases, the oil phase drains and the bubbles slowly cream upwards. Altough their volume fraction increases, they remain stable to disproportionation and coalescence. The high stability of these foams

(a) (b)

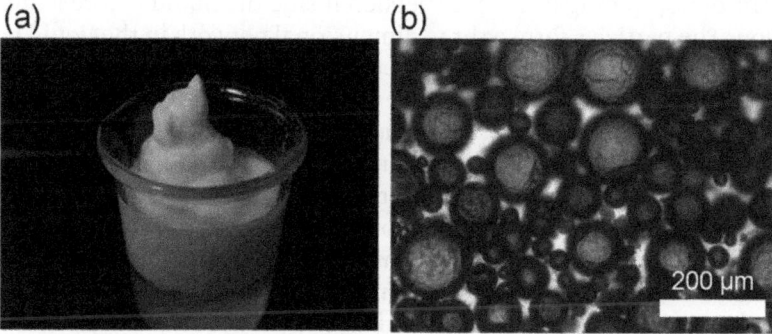

Figure 13.3 (a) Photograph of non-aqueous foam prepared from a mixture of silicone oil and ethyl benzoate with OTFE particles. (b) Optical micrograph of air bubbles stabilized by OTFE particles. Reproduced from ref. 37 by permission of John Wiley and Sons. Copyright 2010 WILEY-VCH Verlag GmbH & Co. KGaA, Weinheim.

comes from the particle layer at the bubble surface acting as a physical barrier against coalescence and disproportionation.[36] This behaviour is similar to particle-stabilised aqueous foams in which the close-packed particle layer also acts as a physical barrier to these destabilisation phenomena. The presence of a dense and rigid layer of jammed particles leads to non-spherical bubbles (Figure 13.3.b). This rigid layer prevents the relaxation to a spherical shape. The bubble surface also appears textured as a result of packing of irregularly shaped particles. Oil foams can show a slow foam collapse over days. This phenomenon comes from the progressive wetting of the particles by the oil that desorbs slowly from the bubble surface. The bubble surface is no longer protected and coalescence occurs.[38]

In terms of foamability, three parameters are important to take into account: the oil viscosity, the concentration of particles and the aggregation state of the particles. During the foam formation, the viscosity of oil can either help to stabilize the foam or prevent the foam formation. If the oil viscosity is high, the oil phase can act to retard oil drainage from the films between bubbles and reduce the frequency of coalescence between bubbles that are not coated or are only partially coated by particles.[37] However, when the oil is too viscous and the method to produce foam is a low-energy method, such as hand-shaking, the oil viscosity may prevent bubble formation.

The foamability is linked to the concentration of particles since a larger area of air–oil surface can be stabilised by increasing the concentration. For example, in the case of polytetrafluoroethylene (PTFE) particles, Binks *et al.* showed that the foamability increases progressively with particle concentration.[39] An important parameter to take into account is that the foamability not only depends on the concentration of particles but also on the aggregation state. For example, fluoro-coated fumed silica particles lead to oil foams when they are produced by hand-shaking. However, if ultrasound is applied before foam production, no more oil foams can be produced and viscous dispersions are obtained.[40] The aggregates are broken into smaller aggregates by ultrasound, leading to their dispersion inside the liquid before foam production. The particles form a three-dimensional network in the liquid owing to the formation of a siloxane bond between free silanol groups on adjacent aggregates. This network increases the viscosity of the oil phase and prevents the foam formation. However, when the particle powder and liquid are hand-shaken to produce an oil foam without applying ultrasound, the aggregates are bigger and attach directly to the air bubbles before entering the continuous phase, preventing the formation of siloxane bonds between aggregates. All the results described in the literature highlight that the production of oil foams stabilized by particles is very sensitive to the shearing process.

An alternative process to easily obtain oil foams is to use the catastrophic phase inversion phenomenon from dry oil powders. This catastrophic phase inversion is driven by the oil:particle ratio. This is the same catastrophic phase inversion process that occurs in aqueous systems for dry water.[38,41] At a fixed mass of particles, when the oil content increases, the size of oil droplets or their number can increase and the amount of non-adsorbed particles decreases. Above the critical oil:particle ratio (COPR), the amount of particles to stabilize the total area of oil droplets surfaces is not sufficient in comparison to the air surfaces within the container. The dry powder is metastable and upon gentle shaking it converts to an oil foam. The COPR depends on both the particles and the surface tension of oil. For example, COPR increases with an increase in the degree of fluorination on clay platelet particles.[40]

13.4.3 Modification of the Contact Angle by the Non-aqueous Liquid Surface Tension

From Young's equation [eqn (13.3)], one simple way to tune the particle contact angle is to modify the liquid surface tension. In the case of aqueous foams, it is the hydrophobicity of the particles themselves that is varied by either chemisorbed or physisorbed molecules and not the liquid phase.[42,43] The liquid surface tension for a given surface can be easily modified just by changing the nature of the liquid phase. In the 1950s, Fox and Zisman established a simple relation between θ and $\gamma_{liquid/air}$ for the wetting of smooth solid surfaces of PTFE.[29] They showed that $\cos\theta$ increases more or less linearly with a decrease in $\gamma_{liquid/air}$. They defined the critical surface tension of the solid $\gamma_{critical}$ such that liquids with $\gamma_{liquid/air} \leq \gamma_{critical}$ wet the solid completely, whereas those with $\gamma_{liquid/air} > \gamma_{critical}$ do not wet the solid. For PTFE, $\gamma_{critical}$ is 18 mN m^{-1} at 20 °C. On smooth PTFE surfaces, $\theta = 0$ ° for pentane, 46 ° for hexadecane and 108° for water. θ increases as the cohesive energy density of the liquid increases as the interactions between molecules in the solid and those in the liquid change from purely dispersion to those including hydrogen bonds.[44]

This simple concept based on the nature of the liquid phase has been used by Binks and coworkers in order to produce oil foams from PTFE and oligomers of tetrafluoroethylene (OTFE) particles. First, they studied 30 oils with various $\gamma_{liquid/air}$.[44] They used an easy method to obtain qualitative information on the wetting behavior of the particles. To determine if the liquid wetted the particles or not, they placed the particles on the liquid surface contained in a glass bottle. Since the powdered particles contain air, which must be expelled before the solid is in contact with the liquid, the bottles must be gently agitated. After agitation, the wetting behavior depends on if the particles are immersed in the liquid. There are three cases: particle layer enters the liquid completely, is partially wet or remains completely non-wetted.[38] A dispersion of particles in oil was obtained for oils that wet the powder, that is to say with surface tensions below 30 mN m^{-1}, such as squalane, toluene and linalol. For oils with intermediate tension (45 mN m^{-1} > $\gamma_{liquid/air}$ > 30 mN m^{-1}), such as sunflower oil or eugenol, the particles were partially wetted and oil foams were obtained. For polar liquids with a much higher tension, like glycerol and water, the curvature of the interface is inverted and a powder-like material is formed containing droplets of water or glycerol coated by particles. This powder-like material is similar to dry oil or dry water but it does not flow and the application of shear causes the rupture of the droplets and the release of liquid. Then, to obtain quantitative data on these systems, the main parameter to determine is the contact angle. In the studies decribed in the literature, the size of the primary particles varies from 0.5 to around 10 µm. Most of the time, these kinds of particles are strongly agglomerated in air owing to surface forces leading to particle agglomerates ranging in size from tens to hundred of micrometers. Moreover, even if the particles can be separated, it is difficult to properly measure the contact angle of a small particle with a liquid surface. The best way to measure the contact angle in this case is to prepare a flat substrate pellet composed of compressed particles.[39] The surface of the resulting pellet is not perfectly smooth and the value of θ will depend on whether the liquid advances or recedes on the surface. Therefore, both contact angles need to be measured. For example, the cosine of contact angle (advancing and receding) has been measured in the case of PTFE particles. Both contact angles increase linearly with a decrease of $\gamma_{liquid/air}$ confirming the relation established by Fox and Zisman (Figure 13.4). Oil foams are obtained for $\cos\theta$ below 0.8, that is to say for contact angles above 35°. The wetting behavior of OTFE and PTFE particles significantly depends on surface tension of oils.

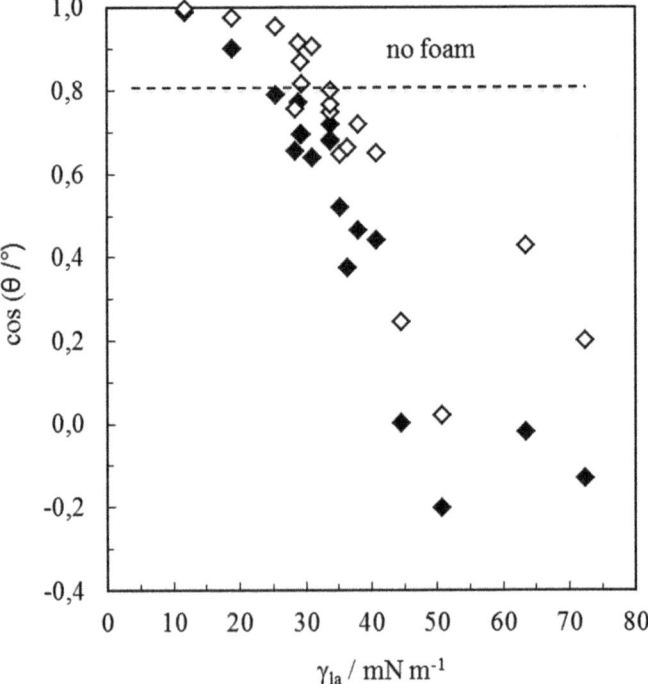

Figure 13.4   Cosine of (♦) advancing and (◊) receding contact angle θ of liquid in air on a compressed powder tablet of Zonyl MP1100 particles as a function of the liquid surface tension. Reproduced from ref. 39 by permission of the Royal Society of Chemistry.

Another way to prepare oil foams is to change the wetting behavior of particles by tuning the oil properties. To achieve this modification, oil mixtures can be used. For example, Murakami and Bismarck studied OTFE particles in the presence of a mixture of silicone oil and ethyl benzoate. They varied the mole fractions of ethyl benzoate in order to modify the contact angle.[37] The advancing contact angle and the wetting behaviour have been determined for the various mixtures between silicone oil and ethyl benzoate. A wetting transition between non-wetted and wetted state occurs around an advancing contact angle value of 46°, which corresponds to a mixture of 70% ethyl benzoate and 30% silicone oil. Stable oil foams can only be obtained for contact angles ranging from 32.8° to 49.6°. The maximal foam volume is reached around the wetting transition. The appropriate oil or oil mixture needs to be chosen in order to adjust the surface activity of the particles at the liquid–air surface to obtain oil foams.

13.4.4 Modification of the Contact Angle by the Surface Chemistry

As for aqueous foams, the surface of the particles can be tuned by surface chemical modification in order to generate repellency against oils.[42,43] The preparation of particle-stabilized oil foams is much more challenging than for aqueous foams. It is not easy to tune the particle surface to obtain these partially oleophobic particles. Indeed, in the case of particles coated by hydrocarbon-containing groups, they are completely wetted by many oils. In order to obtain particles that are partially oleophobic to certain oils, the surface needs to be coated by suitable fluoro groups with lower surface energy, making the particles oleophobic to certain oils. The degree of fluoro-modification can also be tuned.

The first sudy was done by Binks *et al.* on fluoro-coated fumed silica particles. Three degrees of fluorination were studied: low amount of fluorine content (75% SiOH), intermediate (59% SiOH) and high content (50% SiOH). Oil foams can be obtained both in systems of oils of high surface tension (>32 mN m^{-1}) and particles of intermediate fluorine content (59% SiOH) and with oil of low surface tension (<28 mN m^{-1}) and particles of high fluorine content (50% SiOH). To explain these results, they measured the resulting contact angle for each degree of fluorination and for various oils.[40] The surface energy of the particles, calculated from contact angle data, decreases upon increasing the degree of fluorination. For particles with different degrees of fluorination, a nonlinear dependence between $\cos \theta$ and $\gamma_{liquid/air}$ is observed. At high values of $\gamma_{liquid/air}$, θ is more or less constant over a range of $\gamma_{liquid/air}$ (Figure 13.5). However, at a given surface tension, a transition between the wetted and non-wetted state occurs abruptly. This transition is observed for all particles. This critical surface tension depends on the degree of fluorination. For particles with the lowest amount of fluorine (75% SiOH), the transition occurs at a surface tension of around 65 mN m^{-1}. For particles with intermediate fluorine content (59% SiOH), this transition occurs

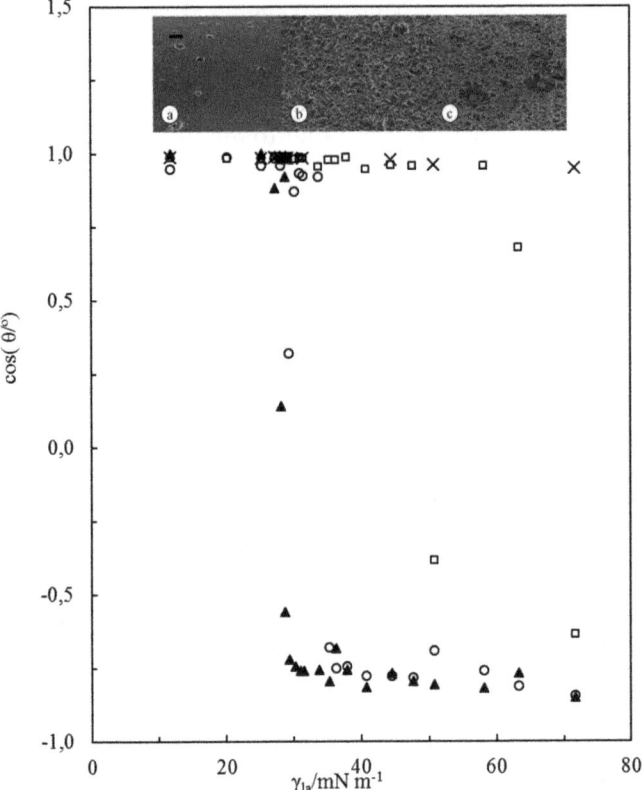

Figure 13.5   Cosine of advancing contact angle (measured through the liquid) of 0.1 cm³ of various liquids in air on glass slides spin coated with 100% SiOH silica (x), 75% SiOH fluorosilica (□), 59% SiOH fluorosilica (o) and 50% SiOH fluorosilica (▲) particles as a function of the surface tension of the liquid at 20 °C. Inset: SEM images of particle-coated slides for (a) 75% SiOH, (b) 59% SiOH and (c) 50% SiOH fluorosilica particles; scale bar = 200 μm. Reproduced from ref. 40 by permission of the Royal Society of Chemistry.

at a lower surface tension of around 33 mN m⁻¹. For particles with the highest fluorine content (50% SiOH), the transition occurs for surface tension of around 28 mN m⁻¹. These particles are oleophobic to many oils and only oils of low surface tension (<28 mN m⁻¹) can lead to a complete wetting of those particles. Oil foams can be produced by choosing the appropriate fluorine content of the particles and the oil in order to obtain a contact angle suitable for oil foam production. The same conclusions were obtained for fluorinated clay platelets and for TiO₂-based particles coated to different extents with fluoroalkylsilanes.[38,41] The hydrophobicity of silica particles has also been tuned by coating the surface to different extents with dichlorodimethylsilane instead of perfluoro-alkoxysilane.[41,45] These particles led to oil foams in ethylene glycol.

13.5 Non-aqueous Foams Based on Oleogels

The structure of many lipid phases of food products is based on networks formed by crystalline triacylglycerol. The properties of the triacylglycerol depend on the fatty acid composition and distribution on the glycerol of each single lipid. They generally contain high levels of saturated fatty acids, which is one factor contributing to cardiovascular diseases. Therefore, within the food industry, there is a trend to develop new and healthier food products. From a technical point of view, saturated fatty acids are ideal ingredients for structuring oils. An alternative for saturated fatty acids to structure the lipid phase needs to be found in order to reduce their levels inside food products. The structuring of the oil phase is also known as oleogel systems. An oleogel can be defined as a liquid oil entrapped within a thermo-reversible, three-dimensional gel network. This gel network is formed by the self-assembly of materials at relatively low concentrations, such as low molecular weight gelators (LMW), polymers, waxes, steroids and plant sterols. The research in the area of oil gelation for food applications is a hot topic. It is difficult to find oleogel systems that can fully replace all the requirements of texture and mouthfeel normally encountered with food products based on crystalline triacylglycerol. That is why another way has recently appeared in the literature to reduce the fat content of food products. It consists of incorporating gas bubbles inside an oleogel to create low-calorie food products. The presence of gas bubbles has been shown to provide a light and pleasant texture.[2,3] These systems are known as oil foams and are based on oleogels.

13.5.1 Formation of Oleogel Systems to Produce Non-aqueous Foams

In the 1970s and 1980s, several groups demonstrated that some hydrocarbon surfactants with specific functional groups (*i.e.* alcohols, glycols, soaps, acids, amines) in oil could lead to non-aqueous foam formation.[6,30] When the surfactant formed an isotropic liquid phase, no foam was produced. However, when the surfactant became insoluble and formed crystalline particles, non-aqueous foams were produced. They showed that foamability and foam stability dramatically increased close to the condition where the foam stabilizer becomes insoluble. This condition is likely to maximize the tendency of the foam stabilizer to adsorb at the liquid–air surface and therefore promote the foam formation. These papers highlighted that the solubility of the surfactant in the oil phase seems to be the critical parameter to produce oil foams. The formation of crystalline particles is the main parameter to produce non aqueous foams from these systems based on hydrocarbon surfactants. These crystalline particles are also very well known in the literature to be involved in the formation of oleogel. Based on this concept, oil foams have been designed recently from various crystalline particles and oleogel systems: mixtures of mono and diglycerides,[46] fatty alcohols,[47] fatty acids,[2] triacylglycerols[3] or sucrose esters and sunflower lecithin.[48] Myristic acid and fatty alcohols crystallize into two-dimensional plate-like crystals.[2,47] Tria-cylglycerol forms three-dimensional spherulitic crystals. The size and poly-morphs (α, β and β') of triacylglycerol are linked to the process.

The presence of crystalline particles is essential to obtain foams from these systems. Therefore, it is important to determine the solubility limit for each chemical substance in a given solvent. For an ideal solution, the varia-tion of solubility as a function of the temperature is given by:

$$\ln x = \frac{\Delta_{fus}H}{R}\left(\frac{1}{T_f} - \frac{1}{T}\right) \tag{13.4}$$

where x is the mole fraction of solute in the saturated solution at absolute temperature T, R is the gas constant, $\Delta_{fus}H$ is the enthalpy of fusion and T_f is the temperature of fusion.

For example, to better understand the foaming properties of myristic acid in high oleic sunflower oil, Binks *et al.* used this equation to calculate the variation of myristic acid solubility temperature in oil as a function of myristic acid concentration (Figure 13.6).[2] First, the myristic acid must be dissolved in the oil by heating. Then, upon cooling, the solution becomes supersaturated. The saturation concentration corresponds to the concentra-tion at which the addition of more solute does not increase the molecular concentration in solution and the excess amount of solute begins to self-as-semble. Following or accompanying supersaturation is crystal formation occurring by nucleation and growth. Above a certain crystal concentration, the critical gelation concentration can be reached, leading to formation of a gel called an oleogel. Upon heating, the crystalline particles melt above the solubility boundary leading to a gel–sol transition. For the myristic acid in high oleic sunflower oil, the formation of plate-like crystals occurs by cool-ing the molecular solution below the solubility boundary.[2] To produce oil foams from these systems, the appropriate concentration of myristic acid needs to be chosen. Above the solubility boundary, no foam can be produced from the molecular solution, but below the solubility boundary foam can be obtained. The solubility boundary is the main parameter to determine for these systems before studying the foaming properties. Differential scanning calorimetry (DSC) is a useful technique to determine both the temperature at which the crystal formation occurs (precipitation) and the temperature at which the crystals dissolve to form a molecular solution.[47] In this system, the solubility limit can be approximated by the DSC curves and corresponds to the temperature at which the crystal melting process is over (dissolution).

13.5.2   Production of Non-aqueous Foams Based on Oleogels

Two main foaming techniques have been described in the literature to pro-duce non-aqueous foams from crystalline particles. The reader needs to keep in mind that whatever the foaming process, foam formation is only possible when crystals are present in solution, *i.e.* below the solubility boundary.

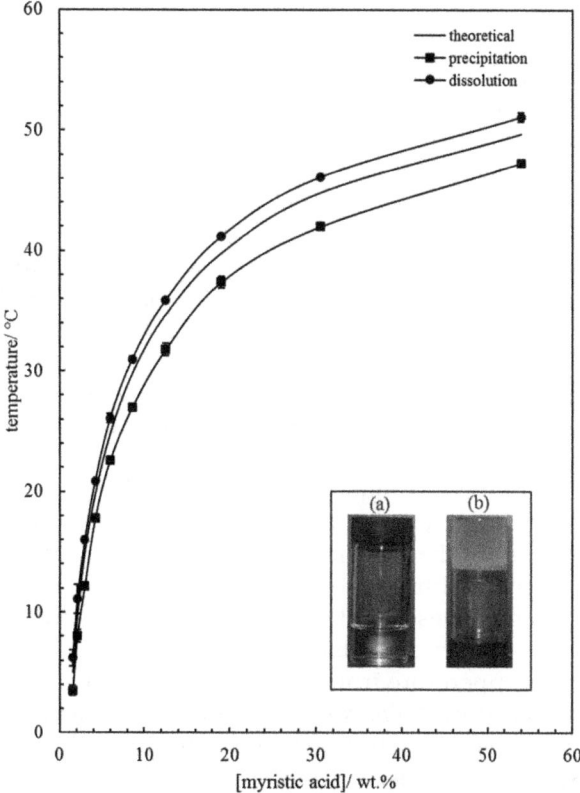

Figure 13.6 Variation of myristic acid solubility in high oleic sunflower oil with temperature, with points obtained from cooling at 0.1 °C min⁻¹ (squares) and heating at 0.1 °C min⁻¹ (circles). The full line without points is that calculated using eqn (13.4). Inset: photographs of (a) oil solution, upright at 40 °C, and (b) oil gel, inverted at 22 °C, for 10 wt% myristic acid. Reproduced from ref. 2 by permission of the Royal Society of Chemistry.

Foaming can be achieved through phase transition. This method was used by Shrestha and coworkers.[49–52] They used liquid petroleum gas (LPG) since it can be liquefied under pressure at normal temperature. LPG is easily available and commonly used for hair care cosmetics or shaving foams. In their studies, LPG was partially soluble in the chosen hydrocarbon oil. LPG is dissolved under high pressure and then released to create the bubbles in the oil phase. In this foaming device, the solution containing oil and crystals is poured in an air-tight bottle fitted with a nozzle and then the LPG is added into this solution. The LPG content is around 20 wt% in total. Before foaming, the mixture is shaken and the gas is released through the nozzle. For some systems at high concentration, the solubility boundary is above 25 °C so crystal particles are present inside the bottle in high amounts, leading to the formation of a gel network, which is difficult to release through the nozzle. Thus, to avoid the formation of gel, it is important to adjust the temperature closer to the solubility boundary before foaming with this technique. Moreover, if this foaming device is kept at a temperature above the solubility boundary, no crystals are present inside the oil. So, no foam can be produced. The same observations have been made with a depressurization process used to produce oil foams.[53] Around 250 g of the solution containing oil and crystals is placed at high temperature inside a 0.5 L cream dispenser. The dispenser is pressurized with N_2O gas at a pressure of approximately 60 bars. The foam is formed at the exit nozzle when the dispenser valve is opened. The depressurization leads to the cooling of the oil mixture. The two techniques do not seem to be conceivable for an industrial application since the consumer would need to keep the foaming device at a precise temperature close to the boundary solubility. It is the main drawback of these techniques for producing non-aqueous foams.

The second foaming technique is the kitchen blender. This is a highly popular foaming device especially in the food industry. Various kitchen mixers have been described in the literature to produce oil foams. Whisks of different size and at different beating speeds have been used. The main difference between all these studies is the temperature at which the foam is produced: above or below the solubility boundary. Foams can be produced directly at room temperature from the oleogel[2,46] or during the crystallization process induced by decreasing the temperature below the solubility boundary during shearing.[47] In this foaming technique, the production of foam is based on air entrainment and systematic bubble-breakup under shear. Air is entrained at the free surface of the blended liquid, which creates large bubbles, which are then broken under the continuous shearing action of the blender.[8] During shearing, the gas fraction increases and the average bubble size decreases. Over time, this process at the macroscopic scale leads to oil foam, which is more and more white and solid-like. When the equilibrium state is reached, the gas fraction and the bubble size cannot be changed anymore by the beating device. Both the rheological properties of the liquid and the beating speed control the characteristic gas fraction and bubble size of a foaming solution.[8] The reader should keep in mind: the higher the viscosity of the foaming liquid, the smaller the bubble size and the lower the gas fraction that can be obtained. To do a systematic study on the effect of the size and shape of the crystals on the foaming properties, the foam needs to be produced at a temperature below the solubility boundary after the formation of the oleogel in a control way. Indeed, when the foam is produced during the crystallization process by shearing, it is more difficult to control the crystallization kinetic.

Recent experiments have been performed by using a high-shear rotor-stator mixer called Mondomix, which is frequently used in the food industry.[53] With this technique, higher aeration levels can be reached in comparison to those with a kitchen blender. This study shows the importance of the foaming technique on the foaming properties.

13.5.3 Properties of Non-aqueous Foams Based on Oleogels

In oil foams based on oleogel systems, most of the bubbles are non-spherical and possess textured surfaces, such as described previously for non-aqueous foams obtained with solid particles (Figure 13.7a). Each foam bubble is covered by adsorbed crystalline particles. This solid layer prevents the relaxation of the bubble to a spherical shape. The crystals can be clearly observed on the bubble surface by using microscopy techniques. The crystals are permanently attached at the interface since even after diluting the foam with pure oil, crystals remain attached to the bubbles.[53] Proof of the adsorption of the crystalline particles has been made by Gunes and coworkers.[53] They studied a planar air–oil surface. The density of crystals is higher than that of the surrounding melt. However, crystals are observed at the planar interface, which unambiguously proves their adsorption. As previously described for solid particles, the crystalline particles must be partially oleophobic in order to exhibit suitable contact angles and adsorb to the interface. Binks *et al.* measured the contact angle through the high oleic sunflower oil by using discs of compressed myristic acid powder.[2] The contact angle is around 40°. The result is consistent with the adsorption of myristic acid crystals to the air bubble surfaces in oil [eqn (13.3)]. The same contact angle has been measured by Gunes and coworkers for monoglycerides and high oleic sunflower oil.[53] In their study, the measurements were made on a film of pure monoglycerides prepared by spin coating. Synchrotron radiation microbeam X-ray diffraction has been recently used to understand how the crystalline particles stabilised

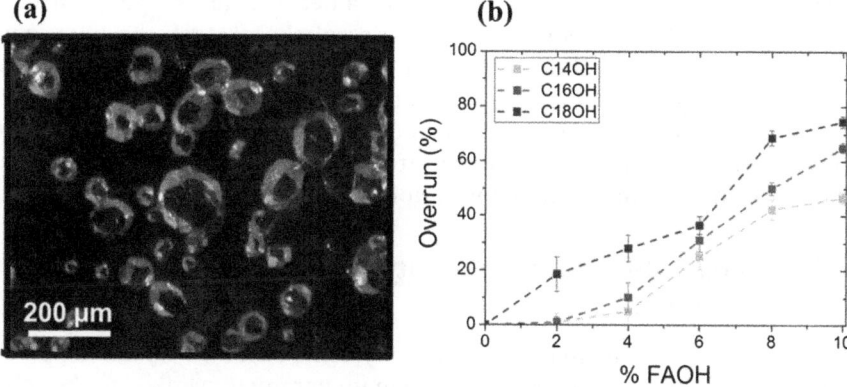

(a) **(b)**

Figure 13.7 (a) Micrograph of foam taken 1 minute after foam formation from an oleogel containing 10 wt% 1-octadecanol in sunflower oil under polarized light. Foam bubbles are stabilized by the fatty alcohol particles, which adsorb to the bubble surface, leading to the formation of non-spherical bubbles with textured surfaces. (b) Foam overrun for samples produced from sunflower oil and containing 1-tetradecanol ($C_{14}OH$), 1-hexadecanol ($C_{16}OH$) and 1-octadecanol ($C_{18}OH$) as a function of FAOH concentration. Reprinted with permission from ref. 47. Copyright 2015 American Chemical Society.

the interface. Mishima and coworkers produced salad oil foams stabilized by triacylglycerol crystals (fully hydrogenated rapessed oil rich in behenic acid).[3] By using the X-ray diffraction method, they showed that the lamellar planes of the crystals near the air–oil surface are arranged almost parallel to the surface. From this result, we can conclude that the lamellar planes composed of methyl end groups are facing the air phase, whereas the lateral planes composed of glycerol groups are connected to each other through the crystals adsorbed at the air–oil surface. We suppose that the same hypothesis can be made for plate-like crystals based on fatty acids or fatty alcohols. The faces expose methyl groups, which are in contact with air, whereas the edges expose mainly methylene and carboxylic or hydroxyl groups, interacting with each other through these groups within the air–oil surface.[2]

Another specific feature of these non-aqueous foams based on crystalline particles is related to the maximum gas volume fraction. In all the results described in the literature, the maximal gas volume fraction is around 55%, whatever the nature of the crystals and their concentration, when the foam is produced with a kitchen blender. For aqueous foams, the maximal air fraction can be higher than 99%. The random close packing fraction of monodisperse spheres is 64%, above which randomly distributed spheres are compressed against each other owing to packing constraints. The value of 55% is close to the packing fraction of monodisperse spheres. In many studies, the authors used the overrun to compare their results between various foaming systems [eqn (13.1)].

Another important parameter to take into account in these non-aqueous foam systems is the particle concentration. When the excess of crystalline particles in the continuous oil phase forms a network, an oleogel is created. As a function of the rheological properties of the oleogel, the buoyancy-driven creaming of air bubbles within the foam can be slowed down or almost stopped [11]. When the foam is turned upside down and no flow is observed, the oil foam is therefore in a gelled state. Rheological measurements have shown that these foams behave like a strong gel [35]. They sustain their own weight without any apparent deformation.

13.5.4   Foamability and Solubility Boundary of Oleogel Systems

In the literature, the foamability of these systems based on crystalline particles has been shown to be driven by two main parameters linked to the solubility limit: the concentration of crystalline particles and the temperature at which the foam is produced. Foamability can be tuned either by modifying the particle concentration at a fixed temperature or producing the foam at different temperatures at a fixed particle concentration.

The foamability of a given system increases until reaching the maximum overrun by increasing the concentration of the component that crystallizes, *i.e.* the crystalline particle concentration. When the concentration of the crystallizing component increases, it leads both to an increase of the boundary solubility temperature and to an increase of the particle concentration below the boundary solubility. For example, at 20 °C below the solubility boundary in the 1-octadecanol fatty alcohol in sunflower oil system, the overrun increases from 20% at 2 wt% of fatty alcohol to 75% for 10 wt% (Figure 13.7b).[47,54] The foamability increases owing to the presence of more particles and a larger air–oil surface area can be stabilized. The same observations have been obtained for various systems. For a fixed concentration of crystallizing component, the temperature at which the foam is formed is another crucial parameter in the foamability of these systems. For example, in the case of 8 wt% myristic acid dispersed in high oleic sunflower oil, the foam volume decreases by increasing the temperature from 20 to 35 °C[2] at 20 °C for 8 wt% myristic acid, the system is below the solubility boundary leading to the presence of large amount of crystalline particles, which stabilize the air bubbles to reach an air fraction of 50 vol% (Figure 13.6). Above 25 °C, the boundary solubility is crossed and the crystalline particles amount decreases owing to progressive crystals melting. At 30 °C, the air fraction decreases to reach 20 vol% since the amount of crystalline particles is lower than that at 20 °C. From 35 °C, above the solubility boundary, it is only a molecular solution; no crystalline particles are present in the system. Therefore, no air bubbles can be stabilized and no foam can be produced. This foamability decrease is linked directly to the solubility boundary of the system.

Only one study has focused on the effect of crystal size and shape on the foamability. For aqueous foams, it is known that these two parameters affect the foamability. Mishima *et al.* recently produced oil foams based on fully hydrogenated rapeseed oil rich in behenic acid (triacylglycerol) in salad oil.[3] They modified the crystallization and tempering process in order to change the size and polymorphs of triacylglycerols crystals. They showed that the overrun is modified by the tempering process, highlighting the influence of crystal size and morphology. The best foaming agent was tiny β-fat crystals. More studies are needed to better understand how the crystal shape and size modify the foamability.

Moreover, the reader needs to keep in mind that an increase of the crystalline particle concentration in such systems can lead to the formation of an oleogel, which is a highly viscoelastic mixture, since large amounts of liquid oil are immobilized by the presence of crystalline particles. The rheological properties of the oleogel likely have a great importance on the foam formation and overrun, but this specific point still needs to be studied.

13.5.5 Foam Stability and Solubility Boundary of Oleogel Systems

The oil foam stability depends on the three main mechanisms of foam destabilization: drainage, coarsening and coalescence. In the same way as for the foamability of these systems based on crystalline particles, two main parameters linked to the solubility limit govern the foam stability: the concentration of crystalline particles and the temperature at which the foam is kept.

Foam stability can be tuned from stable to unstable either by modifying the particle concentration at a fixed temperature or keeping the foam at different temperatures at a fixed particle concentration.

Foam stability is linked to the particle concentration at a fixed temperature: the higher the crystalline particle concentration, the higher the foam stability. For example, in the case of foams generated by myristic acid dispersed in high oleic sunflower oil, various foam stabilities as a function of the myristic acid concentration have been observed from low to high stability.[2] In this study, the foams were kept at 22 °C for months. After one day, for low myristic acid concentration (<6 wt%), around 30% of the initial oil volume drains. The foam stability is relatively low. For low myristic acid concentration (<6 wt%), the solubility boundary is almost reached at 22 °C, the temperature at which the foam is produced and kept (Figure 13.6). A large proportion of the crystals have melted to form the molecular solution. Therefore, the foaming systems do not contain a large amount of crystalline particles to stabilize the foam. For high myristic acid concentration (>8 wt%), no oil drainage occurs and no signs of bubble coalescence or disproportionation were observed. The bubble size distribution remains constant. For high myristic acid concentration, foams are stable for months without any sign of destabilization. These non-aqueous foams are ultrastable since no drainage, no coalescence and no coarsening occur. For high myristic acid concentration (>8 wt%), the solubility boundary is much higher than 22 °C. A large quantity of crystals is present to stabilize the foam. This increase of foam stability by increasing the myristic acid concentration at a fixed temperature (22 °C) is linked to the solubility boundary. In the same way, foams obtained from diglycerol mono-myristate in olive oil at 25 °C have been shown to be stable for months at 10 wt%, but only stable for 1 hour at 1 wt% of surfactant.[52] This difference between low and high foam stability is directly linked to the solubility boundary of the system.

Foam stability is also linked to the temperature at which the foam is kept at a fixed particle concentration owing to the boundary solubility temperature of each system. To illustrate the effect of the boundary solubility temperature on foam stability, the same foam needs to be kept at different temperatures: below the boundary solubility and close to it. For example, the system containing 10 wt% of 1-tetradecanol in sunflower oil has a boundary solubility between 13.1 and 26.6 °C, which corresponds to the beginning and the end of the crystal melting process determined by DSC.[47] The foam produced from this system is kept at 10 °C, below the boundary solubility, and the foam is ultrastable, whereas at 20 °C, close to the solubility boundary, drainage occurs. Some of the crystals already melted at 20 °C and the crystal amount is not sufficient to avoid drainage. The foam is less stable. Below the solubility boundary, the foam stability is high, whereas close to it the foam stability is low. Above the solubility boundary, the foam cannot be produced.

Shrestha and coworkers showed that foam stability increases with monoglyceride chain length at a fixed concentration of monoglycerides and at a fixed temperature. Foam stability at 25 °C was in the order of minutes for systems based on glycerol monocaprylin (11 carbons) and hours for systems based on glycerol monocaprin (13 carbons).[50] For glycerol monocaprylin, the solubility boundary was almost reached at 25 °C, whereas for glycerol monocaprin the solubility boundary was much higher than 25 °C. The amount of crystalline particles needed to stabilize the foam at 25 °C is higher for glycerol monocaprin than monocaprylin, leading to higher foam stability. The same conclusions have been obtained for oil foams stabilized by fatty alcohol crystals: the higher the fatty alcohol chain length, the higher the foam stability at a fixed temperature. This effect of the chain length on the foam stability is directly linked to the solubility of the crystallizing component into the oil. The solubility decreases by increasing the chain length. As a rule of thumb, close to the boundary solubility, the foam stability is low, but below it, the foam stability is high.

13.5.6 Rheological Properties of Non-aqueous Foams Based on Oleogels

These oil foam systems are stabilized efficiently due to the presence of a dense layer of irreversibly adsorbed crystals around the bubbles that considerably reduces gas permeability and coalescence. Moreover, the non-adsorbed crystals in excess in oil can form a gel network in the continuous oil phase; *i.e.* an oleogel. The presence of this network impends buoyancy-driven creaming of air bubbles within the foam.[2] The oil drainage is directly linked to the rheological properties of the continuous phase, which depend on the crystal network. However, there are no studies in the literature regarding the rheological properties of the oleogel and the resulting foams. These ultrastable non-aqueous foams are resistant to the three instability mechanisms owing to the combination of solid crystals both at the interface and in the continuous oil phase. Some of these ultrastable oil foams behave like strong gels, as shown by rheological measurements, and no flow is observed when the foam is turned upside down (Figure 13.8a).[47] They sustain their own weight without any apparent deformation when they are kept below the solubility boundary temperature. However, when they are kept at a temperature close to the solubility boundary temperature, the foam flows (Figure 13.8b). Moreover, this ultrastable characteristic has been emphasized by Brun and coworkers, who studied the shear resistance of oil foams by processing them in a Couette's cell at a shear rate of 1000 s^{-1}.[46] They showed that foams are not destabilized by the applied shear, the air fraction remains the same and the bubble size is decreased by a factor two.

13.5.7 Responsive Non-aqueous Foams Based on Oleogels

The link between foam stability and solubility boundary has been used in the literature to show that these non-aqueous foams can be responsive to temperature changes.[2,47] When ultrastable non-aqueous foams are stored at

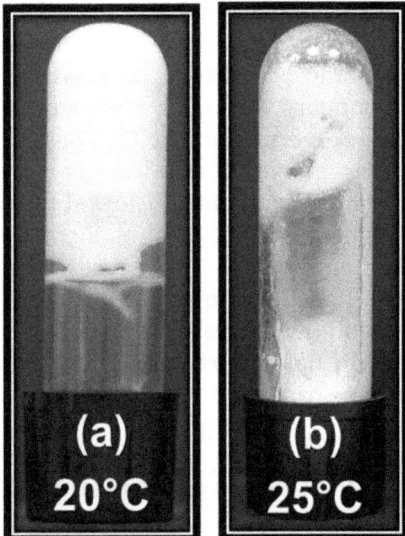

Figure 13.8   Photographs of foams taken 30 minutes after foam generation at 20 °C
(below the solubility boundary of the system) (a) and at 25 °C (close to
the solubility boundary of the system) (b). Foam does not flow when
the vial is inverted at 20 °C but flows at 25 °C. Reprinted with permission from ref. 47. Copyright 2015 American Chemical Society.

temperatures at which crystals dissolve, they completely collapse in a few
minutes owing to the disappearance of the crystals both at the interface and
in the continuous oil phase.[2,46] No foam remains and only a clear molecular
solution is present. Since the crystals re-form from the molecular solutions
on cooling below the solubility boundary, the foam stability can be easily
alternated between high and low by tuning the temperature above or below
the solubility boundary through multiple temperature cycles. For example,
for the system based on 10 wt% 1-octadecanol in sunflower oil, the temperature at which the crystals completely dissolve is around 48 °C and they form
at around 31 °C as determined by DSC experiments.[47] Figure 13.9 shows that
the oil foam is stable at 20 °C since the temperature is below the solubility
boundary. However, when the temperature is increased to 50 °C (above the
solubility boundary), the crystals begin to melt, leading to foam destabilization in a few minutes. By cooling the foam to 20 °C, the destabilization
process is halted and the foam becomes stable again owing to the recrystallization process both in the continuous phase and around the bubbles. By
again increasing the temperature to 50 °C, foam destabilization mechanisms
are reactivated and the foam completely collapses in a few minutes. Moreover, an approach used for aqueous photothermoresponsive foams has been
extended to develop non-aqueous photoresponsive foams.[55] Light has many
advantages as a stimulus in comparison to temperature. Instead of heating the foam externally, light avoids the physical contact with the foam.[17,56]

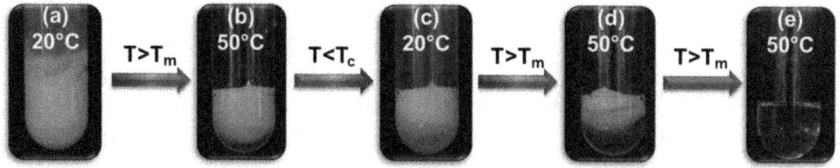

Figure 13.9 Photographs showing the stabilization/destabilization phenomena with temperature for a foam produced from oleogel containing 10 wt% 1-octadecanol in sunflower oil. (a) Stable foam at 20 °C. (b) At 50 °C (above the solubility boundary temperature), foam destabilization started occurring in less than 1 minute. (c) By decreasing the temperature back to 20 °C, the foam became stable again. (d) Upon increasing the temperature to 50 °C for a second time, the onset of foam destabilization could again be observed in less than 1 minute. (e) After 2 minutes at 50 °C, the foam was completely destroyed. Reprinted with permission from ref. 47. Copyright 2015 American Chemical Society.

It can be directed at a precise location of interest with high resolution. The concept is based on the use of internal heat sources incorporated into the foam matrix to generate the heat. In this study, carbon black particles were used as internal heat sources. They are known to absorb UV light and dissipate the adsorbed energy as heat. The carbon black particles were dispersed and entrapped inside the continuous oily phase. Without UV, the oil foams are ultrastable, but under UV illumination, the foam destabilization process begins and bubbles disappear.[47] Under UV illumination, the carbon black particles absorb the light and act as photothermal heat generators, leading to a foam temperature increase above the solubility boundary. The crystals melt inside the foam, leading to foam destabilization. By removing the UV light, the foam destabilization process is stopped. This example highlights that non-aqueous foams based on crystalline particles can be similarly responsive to aqueous foams.

13.6 Conclusion and Outlook

Research in the area of oil foams for food applications is a new topic that has just begun. It is a new way to create low-calorie food products by incorporating gas bubbles inside an edible oil phase. Non-aqueous foams have been obtained from different systems: specialty surfactants, solid particles and crystalline particles. The specialty surfactants, such as polymethylsiloxane-based surfactants, fluoroalkyl surfactants, asphaltenes and resins, cannot be applied for food applications. In the same way, all the solid particles coated by fluoro groups are unsuitable for food applications. One strategy to obtain edible oil foams from solid particles would be to use edible particles of biological origin.[57] The only current way to obtain edible oil foam is to use oleogel systems based on edible oil and crystalline particles from fatty acid, mono and diglycerides, sucrose esters, *etc.* The crystalline particles are the only substance needed to stabilize the air bubbles in the oil phase. There is no need to add another foaming agent. The stability of the foam is linked to the presence of the crystalline particles, which depends on the solubility boundary (Figure 13.10). It is a simple strategy to obtain oil foams based on the aeration process of oleogel systems and could be applied to various oleogel and organogel systems. However, a lot of questions remain open on these systems.

For example, it is known that the properties of oleogels are determined by both thermodynamic considerations (solubility) and kinetic aspects (crystallization kinetics), as crystal size and shape depend strongly on the kinetic properties of the crystallization process of gelator crystallization. However, there are almost no data about the effect of the size and shape of crystalline particles on the foaming properties, both foam stability and foamability. Only Mishima and coworkers highlighted that the triacylglycerol crystallization process tunes the foamability.[3] In all the other studies, the crystallization kinetics are not controlled. In aqueous foams, the shape of the particles is known to have an effect on the foaming properties.[15,36] This parameter needs to be studied in the case of non-aqueous foams in order to understand the stabilization mechanisms of crystalline particles at the bubble surface.

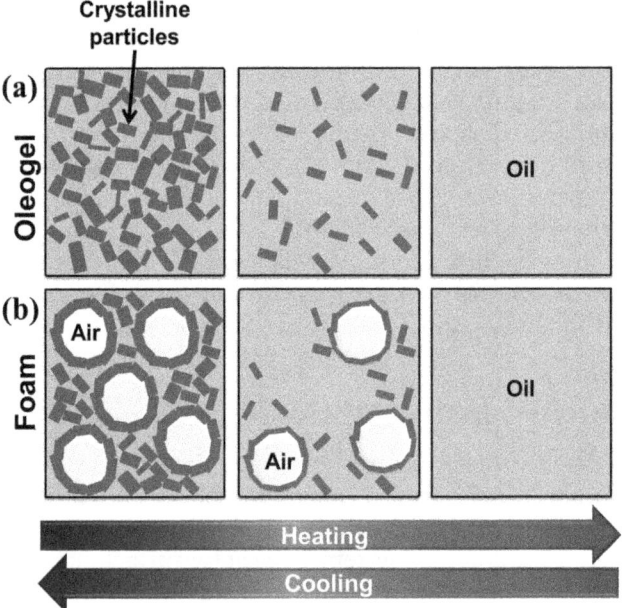

Figure 13.10   Schematic illustrating that crystalline particles lead to the formation of an oleogel and an oil foam. (a) The oleogel begins to transit from solid to liquid state owing to crystal dissolution above the solubility boundary temperature. (b) The foam stabilization is owing to the presence of crystalline particles at the oil–air interface below the sol–gel transition. By increasing the temperature above the solubility boundary temperature, crystalline particles located in the bulk as well at the interface dissolve, resulting in foam destabilization.

It could be interesting to determine the rheological properties of these interfaces stabilized by adsorbed particles.

In the state of art, the foam is obtained by shearing the oleogel. However, the rheological properties of the oleogel depend on the nature of the oil as well as the crystal size and shape. The link between the rheological properties of the oleogel and the resulting foaming properties needs to be established in order to control the formation and stability of the foam. It would be interesting to vary the foaming process and to study its impact on the resulting foam and on the bubble size distribution.[53] Moreover, it is still not known if all the oleogel and organogel systems can lead to non-aqueous foam formation.

Oil foam systems are similar to Pickering emulsions. The abundant literature about Pickering emulsions could be helpful to better understand the foaming properties of non-aqueous foams.[36,58-60]

It is just the beginning of studies on oil foams for food applications. Two recent patents on oil foams highlight the interest in this subject, which will grow in the coming years in parallel with the research on oleogel systems.[61,62]

References

1. S. E. Friberg, *Curr. Opin. Colloid Interface Sci.*, 2010, **15**, 359.
2. B. P. Binks, E. J. Garvey and J. Vieira, *Chem. Sci.*, 2016, 7, 2621.
3. S. Mishima, A. Suzuki, K. Sato and S. Ueno, *J. Am. Oil Chem. Soc.*, 2016, **93**, 1453.
4. J. Haedelt, S. Beckett and K. Niranjan, *J. Food Sci.*, 2007, **72**, E138.
5. I. Cantat, S. Cohen-Addad, F. Elias, F. Graner, R. Höhler, O. Pitois, F. Rouyer and A. Saint-Jalmes, *Foams: Structure and Dynamics*, OUP Oxford, 2013.
6. R. J. Pugh, *Bubble and Foam Chemistry*, Cambridge University Press, 2016.
7. R. K. Prud'homme and S. A. Khan, *Foams-Theory, Measurements and Applications*, New York, 1996.
8. W. Drenckhan and A. Saint-Jalmes, *Adv. Colloid Interface Sci.*, 2015, **222**, 228.
9. D. Weaire and S. Hutzler, *The Physics of Foams*, Oxford, 1999, Oxford University edn.
10. A. Saint-Jalmes, *Soft Matter*, 2006, **2**, 836.
11. D. Langevin and E. Rio, in *Encyclopedia of Surface and Colloid and Science*, ed. P. Somasundaran, Taylor and Francis, New York, 2012, 2nd edn, pp. 1–15.
12. S. Hilgenfeldt, S. A. Koehler and H. A. Stone, *Phys. Rev. Lett.*, 2001, **86**, 4704.
13. V. Carrier and A. Colin, *Langmuir*, 2003, **19**, 4535.
14. P. Garrett, *Chem. Eng. Sci.*, 1993, **48**, 367.
15. A.-L. Fameau and A. Salonen, *C. R. Phys.*, 2014, **15**, 748.
16. E. Rio, W. Drenckhan, A. Salonen and D. Langevin, *Adv. Colloid Interface Sci.*, 2014, **205**, 74.

17. A. L. Fameau, A. Carl, A. Saint-Jalmes and R. Von Klitzing, *ChemPhysChem*, 2015, **16**, 66–75.
18. M. Safouane, M. Durand, A. Saint Jalmes, D. Langevin and V. Bergeron, *J. Phys. IV*, 2001, **11**, 280.
19. P. Walstra, *Physical Chemistry of Foods*, CRC Press, 2002.
20. D. Weaire and V. Pageron, *Philos. Mag. Lett.*, 1990, **62**, 417.
21. A. Webster and M. Cates, *Langmuir*, 2001, **17**, 595.
22. R. Tuinier, C. G. Bisperink, C. van den Berg and A. Prins, *J. Colloid Interface Sci.*, 1996, **179**, 327.
23. D. Langevin, *Adv. Colloid Interface Sci.*, 2000, **88**, 209.
24. N. D. Denkov, S. Tcholakova, K. Golemanov, K. Ananthpadmanabhan and A. Lips, *Soft Matter*, 2009, **5**, 3389.
25. S. Tcholakova, Z. Mitrinova, K. Golemanov, N. D. Denkov, M. Vethamuthu and K. Ananthapadmanabhan, *Langmuir*, 2011, **27**, 14807.
26. A. Stocco, E. Rio, B. P. Binks and D. Langevin, *Soft Matter*, 2011, **7**, 1260.
27. C. Blázquez, E. Emond, S. Schneider, C. Dalmazzone and V. Bergeron, *Oil Gas Sci. Technol.*, 2014, **69**, 467.
28. M. McBain and L. Perry, *J. Am. Chem. Soc.*, 1940, **62**, 989.
29. H. Fox and W. Zisman, *J. Colloid Sci.*, 1950, **5**, 514.
30. P. A. Sanders, *J. Soc. Cosmet. Chem.*, 1970, **21**, 377.
31. T. Kendrick, B. Kingston, N. Lloyd and M. Owen, *J. Colloid Interface Sci.*, 1967, **24**, 135.
32. A. Ellison and W. Zisman, *J. Phys. Chem.*, 1959, **63**, 1121.
33. I. Callaghan, C. Gould, R. Hamilton and E. Neustadter, *Colloids Surf.*, 1983, **8**, 17.
34. I. Callaghan, A. McKechnie, J. Ray and J. Wainwright, *Soc. Pet. Eng. J.*, 1985, **25**, 171.
35. M. K. Poindexter, N. N. Zaki, P. K. Kilpatrick, S. C. Marsh and D. H. Emmons, *Energy Fuels*, 2002, **16**, 700.
36. B. P. Binks, *Curr. Opin. Colloid Interface Sci.*, 2002, **7**, 21.
37. R. Murakami and A. Bismarck, *Adv. Funct. Mater.*, 2010, **20**, 732.
38. B. P. Binks, S. K. Johnston, T. Sekine and A. T. Tyowua, *ACS Appl. Mater. Interfaces*, 2015, **7**, 14328.
39. B. P. Binks, A. Rocher and M. Kirkland, *Soft Matter*, 2011, **7**, 1800.
40. B. P. Binks and A. T. Tyowua, *Soft Matter*, 2013, **9**, 834.
41. B. P. Binks, T. Sekine and A. T. Tyowua, *Soft Matter*, 2014, **10**, 578.
42. B. P. Binks and T. S. Horozov, *Angew. Chem.*, 2005, **117**, 3788.
43. U. T. Gonzenbach, A. R. Studart, E. Tervoort and L. J. Gauckler, *Angew. Chem., Int. Ed.*, 2006, **45**, 3526.
44. B. P. Binks and A. Rocher, *Phys. Chem. Chem. Phys.*, 2010, **12**, 9169.
45. A. K. Dyab and H. N. Al-Haque, *RSC Adv.*, 2013, **3**, 13101.
46. M. Brun, M. Delample, E. Harte, S. Lecomte and F. Leal-Calderon, *Food Res. Int.*, 2015, **67**, 366.
47. A.-L. Fameau, S. Lam, A. Arnould, C. d. Gaillard, O. D. Velev and A. Saint-Jalmes, *Langmuir*, 2015, **31**, 13501.
48. K. Karp, Master thesis, Lodz University of Technology, 2016.

49. H. Kunieda, L. K. Shrestha, D. P. Acharya, H. Kato, Y. Takase and J. M. Gutiérrez, *J. Dispersion Sci. Technol.*, 2007, **28**, 133.
50. L. K. Shrestha, K. Aramaki, H. Kato, Y. Takase and H. Kunieda, *Langmuir*, 2006, **22**, 8337.
51. L. K. Shrestha, R. G. Shrestha, S. C. Sharma and K. Aramaki, *J. Colloid Interface Sci.*, 2008, **328**, 172.
52. R. G. Shrestha, L. K. Shrestha, C. Solans, C. Gonzalez and K. Aramaki, *Colloids Surf., A*, 2010, **353**, 157.
53. D. Gunes, M. Murith, J. Godefroid, C. Pelloux, H. Deyber, O. Schafer and O. Breton, *Langmuir*, 2017, **33**, 1563.
54. A. G. Marangoni, *J. Am. Oil Chem. Soc.*, 2012, **89**, 749.
55. A.-L. Fameau, S. Lam and O. D. Velev, *Chem. Sci.*, 2013, **4**, 3874.
56. A.-L. Fameau, A. Arnould, M. Lehmann and R. von Klitzing, *Chem. Commun.*, 2015, **51**, 2907.
57. S. Lam, K. P. Velikov and O. D. Velev, *Curr. Opin. Colloid Interface Sci.*, 2014, **19**, 490.
58. V. Schmitt, M. Destribats and R. Backov, *C. R. Phys.*, 2014, **15**, 761.
59. D. Rousseau, *Curr. Opin. Colloid Interface Sci.*, 2013, **18**, 283.
60. R. Aveyard, B. P. Binks and J. H. Clint, *Adv. Colloid Interface Sci.*, 2003, **100**, 503.
61. D. Z. Gunes, H. Chisholm, O. Shafer, H. Deyber and C. Pelloux. Pat. 14558-EP-EPA, 2015.
62. H. Chisholm, D. Z. Gunes, C. Gehin-Delval, C. Appolonia Nouzille, E. Garvey, M. Destribats, S. Nalur, J. Vieira and J. German, Pat. 14560-EP-EPA, 2015.

CHAPTER 14

Innovative Dispersion Strategies for Creating Structured Oil Systems

ASHOK R. PATEL*

Sci Five Consulting Services, Coupure 164F, 9000 Gent, Belgium
*E-mail: ashok2510@gmail.com

14.1　Introduction

One of the main formulation challenge that is currently faced by the food industry constitutes the minimization of solid fats (*trans* and saturated fats) in development of lipid-based food products. Oleogelation, which by definition is gelation of liquid oil by components that form specific molecular assemblies that are capable of creating a continuous 3D network to physically immobilize the liquid oil, is considered by many researchers to provide a solution for partial or complete replacement of solid fats in food products. Although the oleogelation approach has a lot of potential, there are still some nagging issues that stand in the way of their commercial acceptance, such as the regulatory status of some efficient gelators (waxes, silica particles, ethylcellulose, shellac resin, hydroxylated fatty acids *etc.*), the cost-effectiveness of some gelators (phytosterols + oryzanol) and the organoleptic properties of the formed gels (monoacylglycerols and wax-based gels).

Food Chemistry, Function and Analysis No. 3
Edible Oil Structuring: Concepts, Methods and Applications
Edited by Ashok R. Patel

As alternatives to oleogelation, several smart dispersion strategies have also been tried to create structured systems using raw materials that are commonly used in food formulations. These strategies include: (a) formulating structured biphasic systems such as Coasun SE (a concentrated oil-in-water emulsion stabilized by interfacial accumulation of hydrated multilayers of saturated monoacylglycerols),[1] (b) creating a gelled water-in-oil emulsion through jamming of a large volume of gelled water droplets (volume fraction of water > 0.7),[2] (c) using indirect ways to exploit structuring abilities of water soluble proteins[3] and polysaccharides,[4] and (d) transforming water continuous emulsions into oil gels through water removal; the oil–water interfaces in these emulsions are made elastic through the use of cross-linked protein layers,[5] discrete protein particles,[6] surface-active triterpenoids such as saponins,[7] or a synergistic combination of two food polymers.[8-11] Some of these strategies have a favourable profile with respect to the ingredients used and may thus get a positive nod for commercial exploration by the food industry.

New opportunities can be created by exploring other research ideas in the area of formulating structured systems through the use of dispersion strategies that are feasible for scale-up.

In the following section, three dispersion strategies are reported that are currently being investigated for creating structured systems that may contribute to solid fat reduction in lipid-based food products.

14.2 Structured O/W/O Double Emulsions

In recent years, research on multiple/double emulsions has received considerable interest from industrial scientists working in the field of food structuring. They can be used for a number of applications including compartmentalization of incompatible bioactives, controlled delivery of micronutrients and fat and salt reduction in reformulated food products.[12,13] For fat reduction, mainly water continuous double emulsions (water-in-oil-in-water, W/O/W) have been explored where the goal is to decrease the energy density of the formulation by replacing a part of oil with water for applications in products such as mayonnaise, salad dressings, sauces and dips.[12-14] On the other hand, oil continuous double emulsions (oil-in-water-in-oil, O/W/O) have not received the same level of attention as WOW emulsions, mainly because of the processing difficulties encountered in the preparation of oil continuous double emulsions[15-17] and/or the requirement of high levels of surfactant(s).[17] While W/O/W emulsions are applicable for reducing the fat content of food products, O/W/O emulsions may find use in reduction of the amount of solid fat required for structuring of food products that are based on indirect emulsions (W/O) (such as table spreads, baking margarines and cooking fats).

The lowering of solid fats in indirect emulsions through use of O/W/O emulsions is explained with the help of a hypothesis shown in Figure 14.1a. Basically, the partitioning of a portion of liquid oil in the inner water phase results in the lowering of the amount of liquid oil left to be structured in the

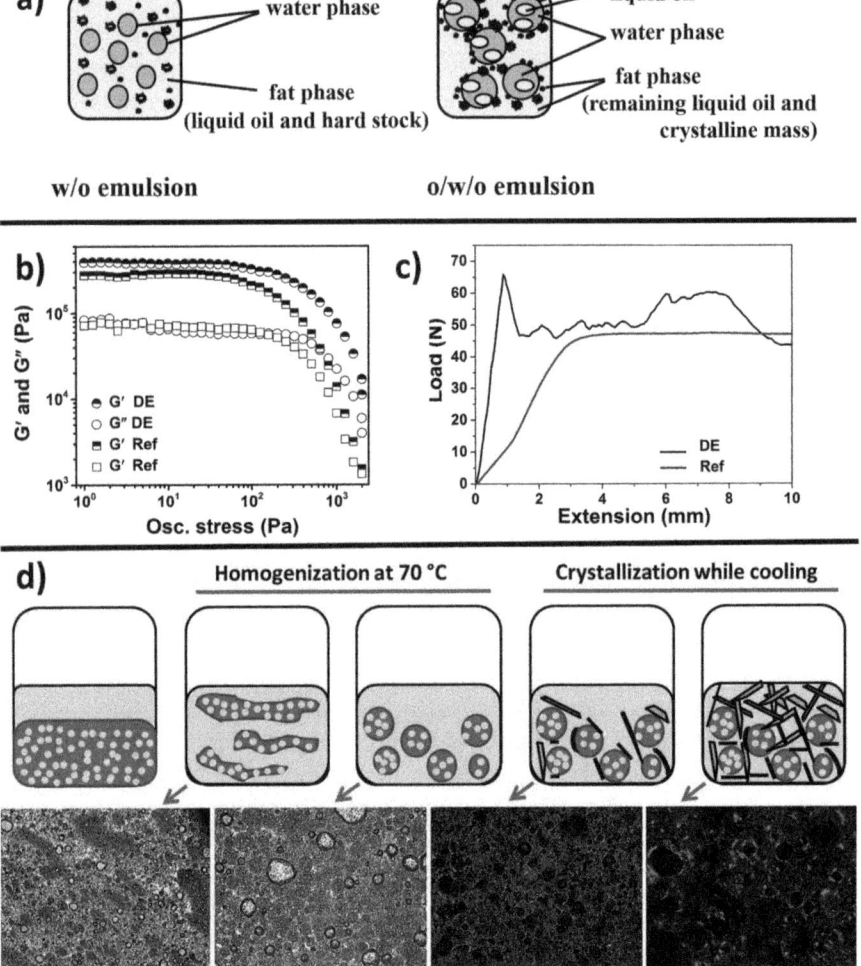

Figure 14.1 (a) Schematic representation of hypothesis to explain the use of O/W/O for structuring indirect emulsion based products. As a part of liquid oil is partitioned into the inner water phase, the amount of crystalline mass (solid fat) required for structuring the external oil phase can be decreased. (b) and (c) Amplitude stress sweep and load-extension curves comparing the firmness and deformation properties of DE and the corresponding indirect emulsion (Ref) containing the same levels of components. Both DE and Ref contained 30 wt% sunflower oil, 20 wt% water (with 1.6 wt% biopolymers) and 50 wt% palm stearin. However, the phase distribution in DE was 30:20:50 (sunflower oil:water:palm stearin) whereas in Ref it was 20:80 (water:sunflower oil + palm stearin). This is a proof-of-concept to prove the hypothesis that by encapsulating liquid oil in the inner water droplet, we can get an increase in the firmness and consistency of DE over Ref owing to the relatively higher proportion of crystalline mass in the external oil phase. (d) Representation of stages involved in the formation of O/W/O. The microstructure changes are explained with the help of microscopy images (image widths = 400 μm). Modified from A. R. Patel, Surfactant-free oil-in-water-in-oil emulsions stabilized solely by natural components-biopolymers and vegetable fat crystals, *MRS Adv.*, **2**, 1095–1102, reproduced with permission.

external oil phase and thus a better texture may be obtained at a relatively lower mass fraction of crystalline fat. With this hypothesis, O/W/O emulsions were first prepared and the rheological and the textural properties of the double emulsion were compared with an indirect emulsion containing the same level of hard stock (Figures 14.1b and c) as a proof-of-concept. As can be seen from the figure, by distributing the same phases differently, a higher firmness is achieved in the double emulsion compared to in the indirect emulsion.

Please note that in the rest of this section, the description is related to O/W/O emulsions prepared using palm oil (PO) as hard stock instead of palm stearin.

In order to make it acceptable for eventual use in the food industry, the primary focus was to first develop a method to prepare stable O/W/O emulsions using only food-grade components and preferably without using any emulsifiers. Double emulsions are most commonly created using a two-step emulsification technique by first preparing the primary emulsion followed by homogenization in the external phase.[18] There are some reports where a single-step emulsification has been successfully used for preparing double emulsions,[19-21] but a two-step technique is generally considered to provide better control over the microstructure and encapsulation yield. One of the main concerns while preparing O/W/O emulsions, especially in the absence of small molecular weight emulsifiers, is the susceptibility of phase inversion, leading to the formation of O/W or W/O emulsions. A recent paper reports the formation of surfactant-free, O/W/O emulsions when the oil phase containing a rather high concentration of carnauba wax (15% wt of oil phase) was homogenized with the water phase with simultaneous cooling.[22] The emulsion droplets stabilized by wax particles were non-spherical in nature but showed exceptional stability against phase separation on storage. However, using waxes at such high concentrations is unacceptable in food products owing to the regulatory constraints and the undesirable greasy taste caused by the high melting waxes.

In the current work, the concept of Pickering stabilization was exploited by using crystallizing vegetable oil (PO) as the hard stock in the external oil phase. O/W/O emulsions were prepared using a two-step emulsification technique by first preparing O_1/W primary emulsions with varying water contents followed by homogenization with the external oil phase containing PO and sunflower oil (SFO) in varying proportions. In order to create fine crystals of PO that could accumulate at the water–oil interfaces, homogenization was carried out simultaneously with the lowering of temperature (to 5 °C). The stages involved in emulsion formation are shown in Figure 14.1d, microscopy images included in the figure provides further details about microstructure changes at different stages.

The primary emulsion (Figure 14.2a) was prepared using a specific combination of biopolymers—gelatin (GL) and xanthan gum (XG)—that has previously been proven to be very effective at stabilizing concentrated emulsions in the absence of any added emulsifiers.[10] In addition to improving the elasticity of the interfacial membrane formed by GL, XG is also responsible for

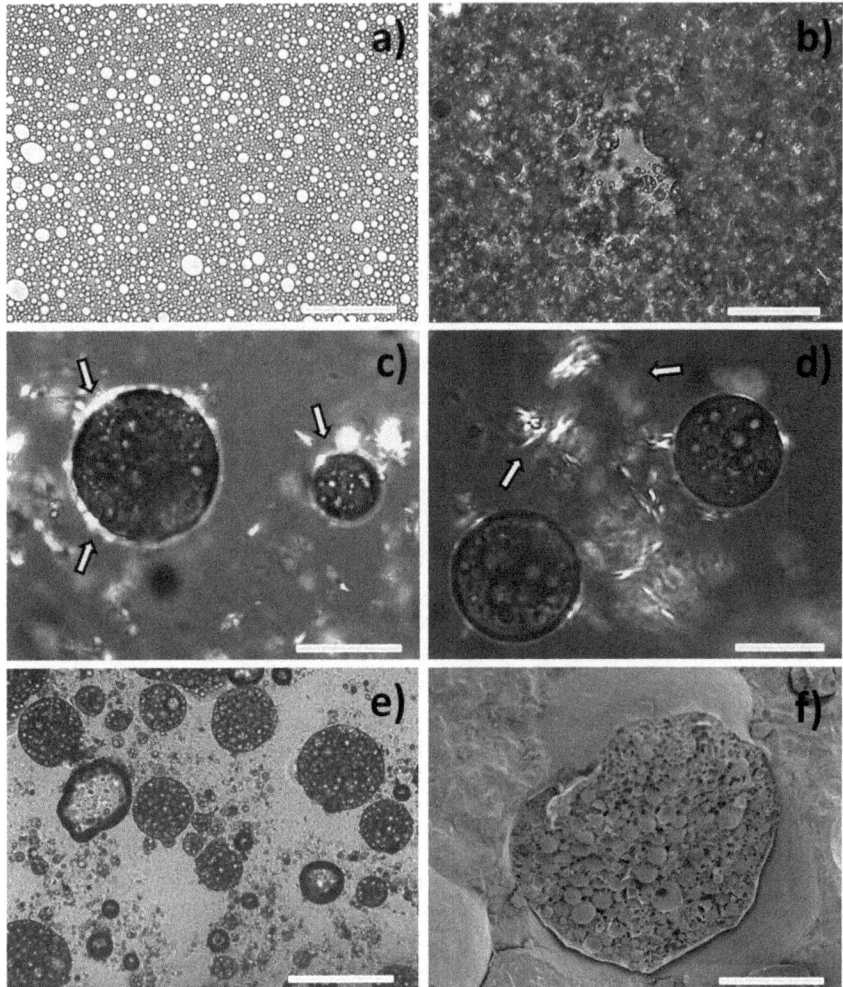

Figure 14.2   Microscopy images of (a) primary O/W emulsion (oil content = 60 wt%),
scale bar = 100 μm, and (b) O₁/W/O₂ emulsion (30:20:50), scale bar =
50 μm. The emulsion droplets are stabilized by interfacial crystallization (c) and physical entrapment in the network formed by bulk
crystallization (d), scale bars = 25 μm. (e) Optical microscopy image
of diluted double emulsion, scale bars = 20 μm. (f) Cryo-SEM image
of freeze-fractured double emulsion droplet showing a high encapsulation of inner oil droplets, scale bar = 15 μm. Modified from A. R.
Patel, Surfactant-free oil-in-water-in-oil emulsions stabilized solely by
natural components-biopolymers and vegetable fat crystals, *MRS Adv.*,
2, 1095–1102, reproduced with permission.

structuring of the bulk water phase and this results in making the emulsion
viscoelastic and have a solid-like deformation behaviour. The response of
the primary emulsion to shear is apparent from the way it fragments (elon-
gation followed by breaking into discrete spherical units) during the initial
homogenization steps, as shown in Figure 14.1d. This viscoelastic property

of the primary emulsion was also responsible for preventing phase inversion and formation of either O/W or W/O emulsions. The homogenization and crystallization were carried out simultaneously to stabilize primary emulsion droplets in double emulsion through interfacial accumulation of fat crystals as well as physical entrapment in the solidified crystalline network in the bulk oil phase (Figures 14.2b, c and d). Use of fat crystals for stabilizing water droplets in oil continuous phase has been studied extensively over the last decade.[23-25] It has been observed that a combination of interfacial crystal accumulation and bulk crystallization is quite effective in stabilizing oil continuous emulsions against coalescence. Most of these studies have been done in the presence of emulsifiers that can provide a templating effect by adsorbing and crystallizing at the interface prior to the crystallization of the fat phase. Crystals of emulsifiers formed at the interface act as impurities seeding nucleation for fat crystallization.[26] As expected, *in situ* crystallization of fat provides a better stabilization compared to the pre-formed crystals.[24,27]

In the current work, fat crystals were found to accumulate around water droplets (even in the absence of an emulsifier), providing a robust coating that prevents coalescence of water droplets. Fat crystals by themselves are not considered to be highly surface active but owing to the *in situ* crystallization there is a possibility that droplet boundaries may act as seeding surfaces where fat can crystallize (as seen in this case, Figure 14.2c). Additional stabilization of emulsion droplets was achieved through structuring of the bulk phase by crystalline fat network (Figure 14.2d).

Selection of the proportion of GL and XG used for stabilization of primary (O₁/W) emulsions was based on the prior knowledge obtained from our study on the emulsion-templated approach.[10] An optimized combination of 1 wt% GL and 0.6 wt% XG was found to stabilize concentrated O/W (60 wt% oil) by providing desirable elasticity to the oil–water interface and viscoelasticity to the bulk phase. It was also found that the concentration of the surface-active component (GL) had a stronger influence on the fineness of water droplets compared to the viscosity-building component (XG). Although XG did not contribute substantially to the colloidal stability of the primary emulsion, it was still required for the improvement of the viscoelastic properties of the primary emulsion, thereby stabilizing it against gravitational instability (creaming). Moreover, primary emulsions stabilized only with GL (in the absence of XG) resulted in the formation of water continuous emulsions (O/W) when mixed with the molten oil phase before showing a complete phase separation. Therefore, the presence of XG was necessary both for the stability of the primary emulsion and for ensuring the formation of oil continuous double emulsions.

The viscoelastic properties of the primary emulsion are responsible for the initial formation of islands of macrophase consisting of emulsion droplets, which are broken down into discrete double emulsion droplets under homogenization (as shown in Figure 14.1d). Because of this process (which simply involves breakdown of the macrophase of primary emulsion into discrete droplets), a high oil encapsulation (100% yield of inner oil phase) was obtained in the double emulsion, as confirmed from the optical microscopy image of the diluted emulsion and the cryo-SEM image of a freeze-fractured emulsion droplet (Figures 14.2e and f).

The possibility of preparing surfactant-free double emulsions with such high loading of the inner oil phase could be exploited in spread-like products where a part of the liquid oil can be encapsulated in the inner water phase so a relatively lower proportion of mobile oil phase is left in the external phase to be structured by the crystalline mass. In addition, this kind of formulation could also be used in spreads to partition incompatible phases, such as fats (like coconut and palm oil) that show different crystallization behaviours.

Spreads constitute a broad range of products differing in the content of water and fat phases as well as the fat types, and these phases together influence the bulk properties of the systems, such as texture and rheology. To explore the flexibility of double emulsions for applications in spread-like products, the amount of solid fat content in the external oil phase (O_2) as well as the ratio of the primary emulsion to the external oil phase $(O_1/W):O_2$ were varied and the influence of this variation on the bulk properties of double emulsions was characterized using a combination of microstructure studies, rheology tests and texture analysis.

The solid fat content of the external oil phase was altered (Figure 14.3a) by partially replacing PO rich in saturated fatty acids (~9 wt%) with liquid SFO containing a high amount of unsaturated fatty acids (saturated fatty acids <10 wt%). Stable double emulsions were prepared by replacing as much as 50% of PO with SFO; any further replacement of PO resulted in phase separation (within 24 hours), thus suggesting that a minimum 10 wt% of crystalline fat (at 5 °C) in the O_2 phase is required to stabilize the double emulsion against coalescence and eventual phase separation. The microscopy images of emulsions prepared at PO:SFO ratios of 100:0, 85:15, 65:35 and 50:50

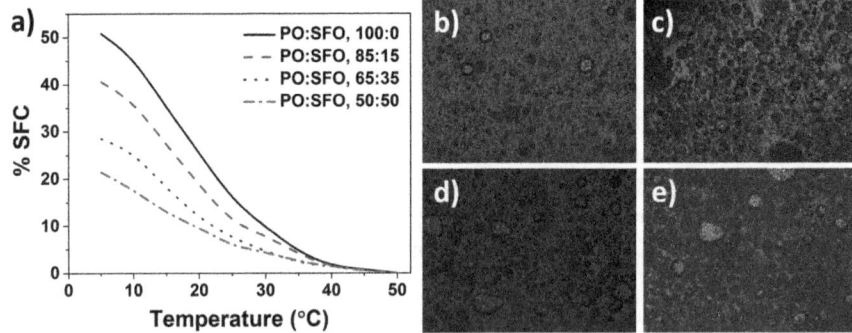

Figure 14.3   (a) Solid fat content profiles of different fat phases used for the preparation of double emulsions. (b)–(e) Polarized light microscopy images of emulsions prepared using fat phases containing different proportions of PO:SFO, 100:0; 85:15; 65:35 and 50:50 respectively, image widths = 400 µm. Modified from A. R. Patel, Surfactant-free oil-in-water-in-oil emulsions stabilized solely by natural components-biopolymers and vegetable fat crystals, *MRS Adv.*, **2**, 1095–1102, reproduced with permission.

are shown in Figures 14.3b–e. It was noticed that a decrease in the PO level resulted in a higher population of double emulsion droplets owing to the decrease in the viscosity of the continuous phase.

For possible applications in low-fat spreads, the ratio of $(O_1/W):O_2$ was also changed to investigate the possibility of formulating double emulsions with high water content. Double emulsions with up to 45 wt% water (and a minimum of 15% inner oil phase) could be prepared without addition of any emulsifier. The levels of all three phases were altered in order to obtain emulsions with high water content while making sure that the inner oil phase is at a level sufficient enough for future encapsulation-related applications.

Small amplitude oscillatory shear rheology and texture analysis studies (Figures 14.4a and b) were used to understand the effect of PO replacement and different levels of water incorporation on the rheology and texture of double emulsions. As expected, the resultant decrease in SFC owing to the

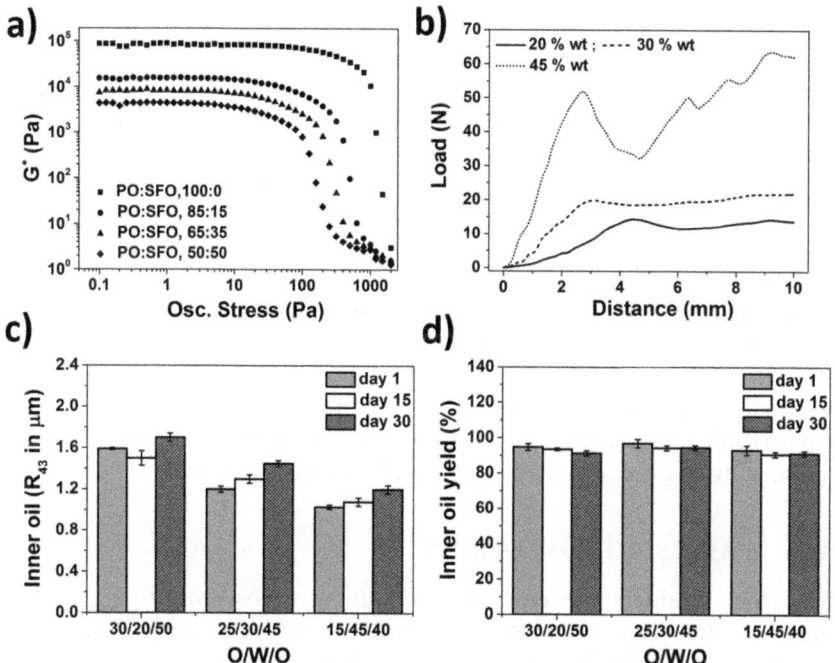

Figure 14.4 (a) Data from amplitude sweeps conducted on $O_1/W/O_2$ (30:20:50) emulsions prepared using different proportion of PO and SFO in the external oil phase (O_2). (b) Load–distance curves showing the large deformation behaviour of double emulsions prepared at different water contents. (c) and (d) Mean size (R_{43}) and yield of inner oil droplets measured over the storage period of 30 days for double emulsions prepared at different water contents. Modified from A. R. Patel, Surfactant-free oil-in-water-in-oil emulsions stabilized solely by natural components-biopolymers and vegetable fat crystals, *MRS Adv.*, 2, 1095–1102, reproduced with permission.

replacement of PO with SFO led to a softer structure, as confirmed from comparative curves of amplitude sweeps. Both the 'at-rest' plateau and oscillatory yield stress showed a substantial decrease with the decrease in SFC. The plateau value of the complex modulus is a measure of 'at-rest' resistance to deformation or stiffness, while the yield stress is a measure of structure strength or breaking strength. The load distance curves obtained from large deformation studies suggested that water incorporation led to a significant increase in the structural cohesive properties of the samples. The peak load value representing the hardness of the sample showed an increase from 14 N for double emulsion with 20 wt% water to 52 N for 45 wt% water. In addition, the slope of the linear region leading to the peak value as well as the area under the curve (energy/work required to disrupt the internal structure) showed a substantial increase, thus confirming the effect of water incorporation on the stiffness and internal strength of the emulsion structure. However, with the increase in the water content, the elasticity of the sample was decreased, which led to a more 'brittle-type' failure, as confirmed from smaller deformation distances.

The mean size $(R_{4,3})$ and yield of inner oil droplets (Figure 14.4c and d) were obtained by using Murday and Cott's model whereby the hindered diffusion behaviour in the oil droplets was obtained by measuring the echo intensity of the NMR signal as a function of the gradient strength. As can be seen from figure, the inner oil droplet size showed a slight increase over 30 days of storage but the inner oil yield remained more or less unaffected. This indicates that although the inner oil droplets did show some degree of coalescence, the diffusion of oil droplets into the continuous oil phase was prevented. The stability of encapsulated oil droplets against diffusion could be viewed as a positive attribute of these surfactant-free double emulsions. These emulsions could find interesting applications in developing spreads (with low levels of saturated fats), as a controlled release formulation for compartmentalizing incompatible fats with different crystallization behaviours, and controlling the release of hydrophobic bioactives loaded in the inner oil phase.

14.3 'Arrested' Oil Foams

Most food products are complex colloidal systems comprising multiple phases and interfaces that are created by close contact of dispersed and continuous phases.[28] Foams are one such system where dispersed air bubbles are incorporated and stabilized in a continuous phase that is usually fluid in nature.[29,30] In recent years, the food manufacturing industry has shown an increased interest in edible foams because of their enormous potential for reformulating edible products with reduced calories and the possibility of creating systems with unique textures and mouth feel for molecular gastronomy purposes.[31-34] Unfortunately, research in the area of food foams has not received the same level of attention as other edible colloids, such as emulsions and gels. Moreover, most work done in this field to date is mainly restricted to aqueous foams with a main focus on the stabilization of air–water interfaces.[35-38] Some interesting results have been published recently in the area of non-aqueous foams where stabilization was achieved using pre-formed crystals of fatty acids,[39] fatty alcohols[40] and partial glycerides[41,42] cre-ated by cooling their solutions in liquid oil below the solubility limit. Some of these gelled foams are also referred to as oleofoams or foamed oleogels because the crystalline components used in formulating these foams are oil-structuring agents that are capable of gelling the liquid oil through for-mation of a continuous crystalline network.

In this current work, the aim was to utilize crystal-coated air bubbles as structuring units to transform liquid oil into a viscoelastic system using an emulsifier (sucrose esters, SE), which is otherwise unable to gel the oil. Sucrose esters of fatty acids are non-ionic emulsifiers based on sucrose and edible fatty acids. Depending on the degree of esterification of the sucrose molecule, it is possible to obtain emulsifiers with hydrophilic lipophilic balance (HLB) rang-ing from 1 up to 16. They are approved for food use in a number of countries including Japan, USA, China and Republic of Korea. In the European Commu-nity, sucrose esters are included in annex 4 of Directive 95/2 under E473.[43]

The oil structuring properties of SE were first studied in the preliminary trials. It was found that all grades of hydrophobic SE (HLB values < 10) were unable to gel liquid oil (even at concentrations as high as 10 wt%) showing a clear phase separation of liquid oil and sediments of crystalline mass. On view-ing the samples under the microscope, oil dispersion of SE (HLB 2) showed the presence of crystal-coated air bubbles scattered sparsely in the liquid oil phase with an aggregated crystalline mass in the sediment (Figures 14.5a–c). This suggested that (a) SE can accumulate at the low free energy interface of two hydrophobic phases (air–oil) and (b) the crystalline structures formed by SE undergo a strong aggregation leading to contraction and consequent phase separation. Based on these observations, it was hypothesized that incor-porating a large number of air bubbles in the oil dispersion could promote

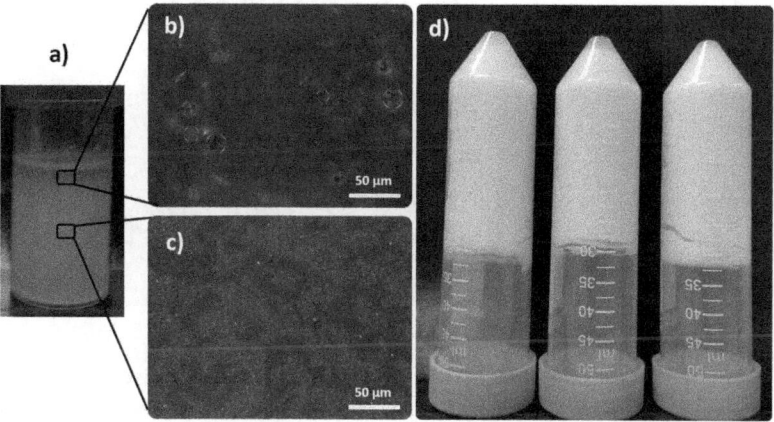

Figure 14.5   (a) Photograph showing clear phase separation in oil dispersion con-taining 10 wt% SE. (b)–(c) PLM images of the clear liquid oil layer and sediment respectively. (d) Photograph of oil foams prepared from 5, 7.5 and 10 wt% SE.

interfacial crystallization of SE crystals and thereby decrease the aggregation of crystalline mass. In addition, the crystal-coated air bubbles could act as robust structuring units that gels the liquid oil much like the gelled water droplets in oil continuous emulsions.[2] To prove this hypothesis, foams were prepared by aerating oil dispersions of SE at high temperature followed by cooling to 5 °C to trigger crystal formation at the air–oil interfaces. By incorporating a large amount of air bubbles ($\varphi_{air} \gg 0.5$), a jammed structure was created in the system that displayed solid-like elastic properties and resistance to flow under the influence of gravity (Figure 14.5d).

The viscoelastic behaviour of these 'arrested' foams is further evident from the rheological data presented in Figures 14.6a and b. When subjected

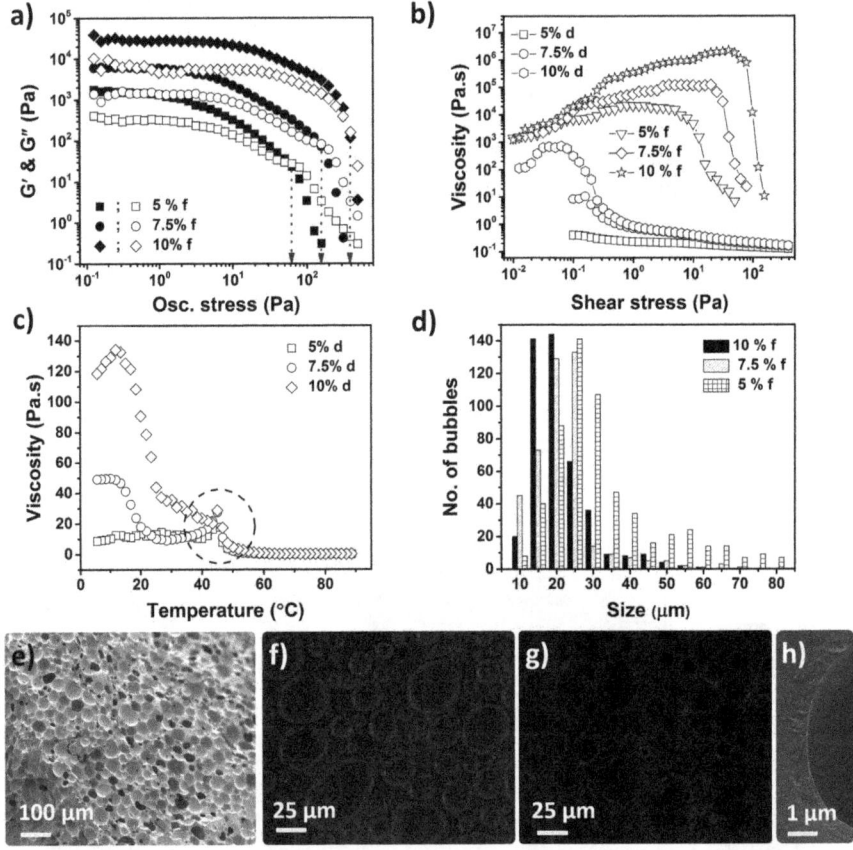

Figure 14.6   Plots from (a) amplitude sweeps done on foams (f); (b) stress ramp done on foams and oil dispersions (d); and (c) viscosity changes of 5–10% (d) as a function of temperature. (d) Histogram showing the bubble distribution in 5–10% (f); (e)–(g) Cryo-SEM, PLM and confocal microscopy images of diluted 10% (f), and (h) cryo-SEM image showing the presence of crystalline coating around the air bubble. Adapted from Ref. 59 with permission from The Royal Society of Chemistry.

to amplitude sweeps, foams prepared at all three concentrations of SE (5, 7.5 and 10 wt%) displayed gel-like behaviour with the elastic moduli being higher than the viscous moduli in the linear response region. Elastic moduli reached well above 20 000 Pa for 10% foam (Figure 14.6a). The crossover point, which is an indication of gel–sol transformation, was seen to shift towards the right with the following values: 63, 158 and 394 Pa for the 5, 7.5 and 10% foams respectively. Dispersions and foams prepared at the three concentrations were subjected to the stress growth test (Figure 14.6b) to understand their flow behaviour. The oil dispersions at 7.5 and 10 wt% SE did show some sort of structure build-up at very low stress values (<0.1 Pa), which could be attributed to the loose network of crystalline aggregates. In comparison, the three-dimensional stress-bearing structures formed by closely packed air bubbles imparted a yielding behaviour to the foams. When a high force was applied to cause local rearrangements of bubble packing, the complex structural changes resulted in a 1000-fold increase in the yield stress value of foam compared to the oil dispersion prepared at the same concentration of SE.

To understand the effect of temperature on the foam microstructure, samples were heated on a hot-stage microscope and images were captured periodically (Figure 14.7). At 70 °C, which is above the melting range of SE (melting range of SE in 5–10 wt% dispersions was found to be around 65 °C, DSC data not shown), foams prepared at all three concentrations showed a significant collapse, suggesting that the bubbles were stabilized by crystalline particles of SE. However, this is in contradiction to the preparation process of foams as the aeration was carried out at 80 °C. Based on the results observed in this study, it can be speculated that SE had the right molecular features to be adsorbed at air–oil interfaces, which was why a large amount of air bubbles could be incorporated in heated oil dispersions. Further, on cooling, the crystallization of adsorbed SE molecules resulted in the formation of a robust crystalline coat that prevented coalescence. When these foams are heated above the melting range of SE, the molecular layers left at the air–oil interfaces are probably not rigid enough to resist the expansion of gas, which leads to the coalescence and eventual collapse of the foams. From Figure 14.7, it can also be observed that the foams prepared at higher concentrations of SE were relatively more stable than the 5% foam at 50 °C, which is below the melting range.

In summation, the results obtained in this study confirm that air bubbles encased in a robust coating of fine crystalline particles could act as structuring units to transform liquid oil into a viscoelastic gel that displays a jammed and glassy state over a range of applied shear. Some of the applications of these foams could be explored in formulating whipped shortenings (with reduced hard stock) for bakery applications and for developing table spreads with high volume (and low fat content). These foams could also be exploited for developing new product format as an alternative to cooking oil.

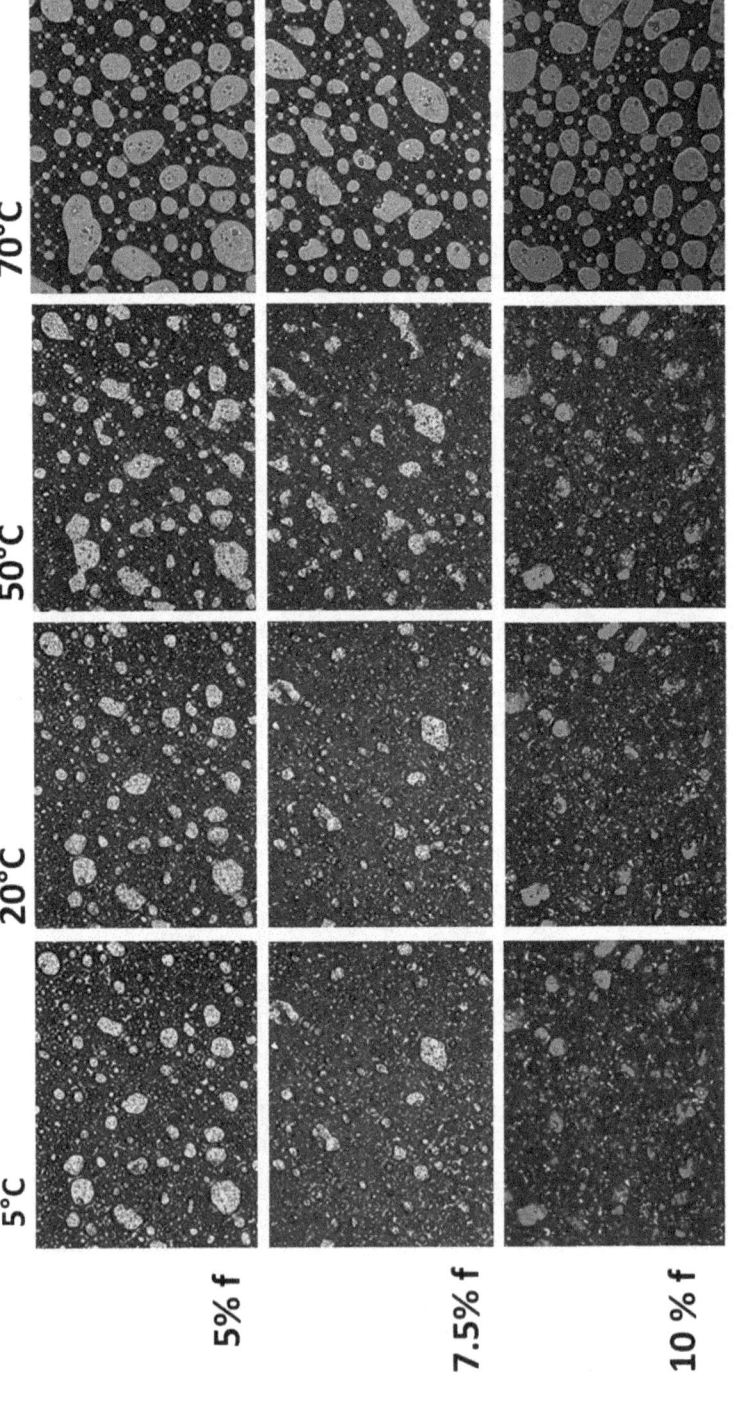

Figure 14.7   Changes in microstructure of foams (5–10% f) as a function of temperature. Image width = 400 μm.

14.4 Polymer-coated Crystalline Microcapsules

Structuring agents explored so far for the gelation of liquid oil are broadly classified as either low molecular weight organogelators (LMOGs) or polymeric organogelators. LMOGs (such as waxes, partial glycerides *etc.*) that form crystalline units can provide fat-related features such thermoreversible processing and 'melt-in-mouth' effect, on the other hand, polymeric organogelators can provide textural properties such as firmness and consistency. Therefore, the possibility of creating the ideal oleogel exists if we find a way to combine these two categories of gelators together. Such (polymer–crystal forming component) hybrid oleogels have been studied with promising results in the past, for example, combining ethyl cellulose (a hydrophobic polymer) with a liquid crystal-forming component (glycerol monooleate) to create a thixotropic oil gel that recovers all of its viscosity on removal of shear,[44] using binary mixtures of stearyl alcohol:stearic acid to influence plasticity of ethyl cellulose resulting in a hybrid oleogel,[45] incorporating palm stearin (PS) to improve the thixotropic structure recovery of oil gel made from freeze dried hydroxyl propyl methylcellulose,[46] and using Pickering emulsions as templates to utilize structuring capabilities of protein–stearate complexes.[6] However, the harsh or energy-intensive processing (heating at high temperatures or freeze drying) required for creating these systems makes them unattractive for commercial implementation. As an alternative, an emulsion-templated approach[8] with a crystallizing component (hard stock) added to the oil phase could be explored, but the issue of oil oxidation owing to high-temperature drying is still encountered in these systems (unpublished results). Hence, there is a need to explore other ways of combining polymers and crystalline particles together to create ideal oleogels that are feasible for scale-up and commercialization.

In the current work, a new approach was explored to create polymer-coated microcapsules that could be used as templates to introduce (per-hydrated) hydrophilic polymers in the matrix of the network formed by crystalline particles. A range of hydrophilic polymers (methylcellulose, gelatin, sodium caseinate and whey protein) were used together with either PS or a commercial crystal starter (Palsgaard® 6111) as the crystal-forming component. In the following section, only results from PS and methylcellulose (ME) are discussed.

One of the most common ways of preparing microparticles involves an emulsion-based process where dispersed droplets are transformed into solid capsules through either internal gelation (for W/O emulsions) or coacervation and spray-drying (for O/W emulsions). In the current work, ME-coated particles of PS were prepared using an emulsion-based process by first hydrating ME in aqueous phase, followed by dispersion and homogenization of melted PS to create a fine (O/W) emulsion. The hot emulsion was quickly diluted in ice-cold water to trigger solidification of droplets. The solidified droplets were separated and dried at mild temperature (~32 °C) to obtain polymer-coated crystalline microcapsules. These capsules were then used as templates to create oleogels. A schematic representation of the process involved in the formation of microcapsules and the oleogel is shown in Figure 14.8.

ME is an amphiphilic polymer and, based on the previous study, a minimum concentration of 2 wt% was found to efficiently emulsify dispersed oil droplets.[8] ME also has a unique property of high-temperature gelation owing to hydrophobic interactions above its lower critical solution temperature.[47] This property of ME is often exploited to create a coating barrier to stabilize fried food products.[48,49] The incipient gel temperature for ME is dependent on the concentration of the polymer and is usually found to be below 65 °C.[50,51]

The approach described in this work was designed based on the above-mentioned information about emulsion-based process and properties of ME; PS was melted at 70 °C (well above its melting range) followed by dispersion and homogenization in a pre-prepared aqueous solution of ME. Interfacial accumulation of ME at the oil droplets (due to its inherent surface activity) followed by a subsequent gelation of adsorbed layers of ME at the surface of the oil droplets (due to the high temperature of the oil phase) resulted in a stable emulsion formation. Emulsions were prepared at 10, 30 and 60 wt% oil while keeping the concentration of ME constant at 2 wt% of total emulsion. The hot emulsions were then quickly poured into ice-cold water (five times the volume of the emulsion) to trigger immediate solidification of oil droplets. The large size of the emulsion droplets ($\gg$1 μm) coupled with the dilution of the bulk phase resulted in a time-dependent creaming of solidified droplets. The creamed layer of solidified droplets was scooped out with

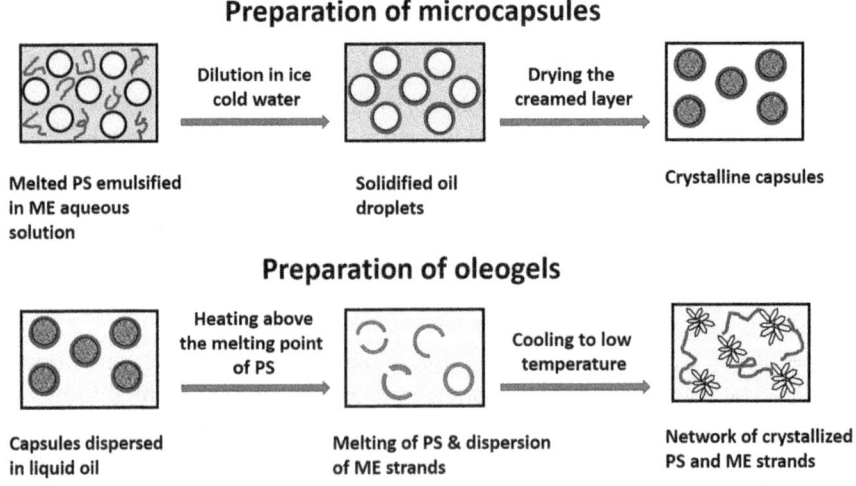

Figure 14.8 Schematic representation of steps involved in preparation of microcapsules and the use of microcapsules as templates in oleogel preparations. Modified from *Materials Chemistry and Physics*, 195, A. R. Patel, Methylcellulose-coated microcapsules of palm stearin as structuring templates for creating hybrid oleogels, 268–274, Copyright 2017, with permission from Elsevier.

ease and dried at a mild temperature (~32 °C) to obtain a free-flowing powder consisting of crystalline PS microcapsules coated with ME. Images of capsules from polarized light microscopy and cryo-SEM are shown in Figures 14.9a and b. A closer look at the surface of the capsule (Figure 14.9c) reveals the presence of adsorbed layers of polymeric sheets. Capsules showed a monomodal distribution of particle size (Figure 14.9d) with the average size ranging from 5 to 20 µm, and the volume of oil phase in the initial emulsion did not have any significant influence on the final size of the capsules.

To evaluate the enhancement in the oil structuring properties of PS in the capsule form, oleogels (70 wt% sunflower oil) were prepared with unprocessed PS and capsules obtained from emulsions with 10, 30 and 60 wt% oil phase (C10, C30 and C60, respectively). The oleogel preparation is shown in Figure 14.9e where the first step was to melt PS and capsules in liquid sunflower oil followed by cooling to room temperature and storing at 5 °C for further structure development. On melting the crystalline PS capsules at 70 °C, dispersed polymer sheets (remnants of the capsule coats) were noticeable in the oil continuous phase compared to the clear solution obtained on melting of unprocessed PS. On subsequent cooling of these hot dispersions

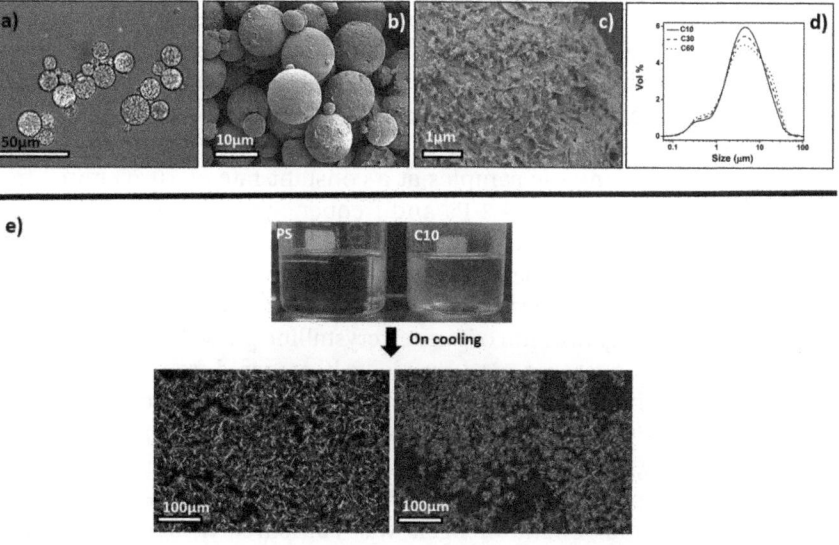

Figure 14.9 (a) and (b) PLM and cryo-SEM images of capsules (C10). (c) Cryo-SEM image showing the presence of adsorbed layers of polymer on the surface of capsules. (d) Particle size distribution of C10, C30 and C60 capsules. (e) Schematic representation of oleogel preparation depicted with the help of photographs and polarized light microscopy images. A comparison is shown between oleogels prepared from unprocessed PS and capsules (C10). Modified from *Materials Chemistry and Physics*, 195, A. R. Patel, Methylcellulose-coated microcapsules of Palm stearin as structuring templates for creating hybrid oleogels, 268–274, Copyright 2017, with permission from Elsevier.

below the crystallization temperature of PS (*i.e.* on undercooling), structured oleogels were formed. The following differences were noticed on comparing the crystalline microstructure of these oleogels: (i) lower crystalline mass fraction in C10 oleogel compared to PS oleogel and (ii) controlled crystallization of PS in discrete spherical units in C10 (probably due to the crystal memory) compared to the uncontrolled growth and aggregation in the PS oleogel. Furthermore, it can also be speculated that the dispersed polymer sheets provided additional linkages that helped strengthen the network of crystalline particles.

The oleogels were subjected to a range of rheological tests to determine their consistency and deformation properties. Plots from amplitude (stress) and frequency sweeps (Figures 14.10a and b) clearly suggest that the structuring properties of PS were enhanced when processed in capsule form. The presence of the dispersed polymer contributed to the strengthening of the crystalline framework supporting the gel structure. The highest firmness (higher complex moduli values and broader linear response region) was observed for the C10 oleogel. In addition, the slope of the yield zone (*i.e.* the region of curve between critical stress and oscillatory yield point,[52,53] data not shown), suggests that the presence of the dispersed polymer in C10 and C30 contributed to brittle-type failure of the oleogels (*i.e.* uniformity in bonds among structuring units, which are broken-off all at once).[52,54] The response of the oleogels to the increased rate of deformation (frequency) suggests a weak gel structure with slightly positive slopes of the curves highlighting the frequency-dependence of moduli values.[55]

Microstructure development as a function of temperature (Figure 14.10c) was followed by cooling the samples at a constant rate of 10 °C min^{-1} from 70 to 5 °C (oscillatory stress = 2 Pa and frequency = 1 Hz). For all samples, a surge in the complex modulus was observed close to the offset value of the crystallization peak of PS (data not shown), suggesting that microstructure development initiates following the end of crystallization. Such behavior, which involves a reorganization of formed crystalline phase for network formation, has been previously reported for oleogels prepared from waxes containing high melting components.[56] As expected, samples made from PS capsules showed comparatively higher elastic moduli values throughout the entire range of frequency, further confirming that the presence of dispersed polymer in the continuous phase contributed to the network development of the gels.

The flow behaviour of the oleogels was compared through shear ramp and three interval thixotropy tests to determine the yielding and structure recovery properties. As seen from Figure 14.10d, the PS oleogel displayed the lowest yield stress value (σ_Y = 398 ± 4.3 Pa) among the studied samples. The yielding property was substantially enhanced for capsule-based oleogels with an almost 2.5 fold increase in the yield stress value (1005 ± 7 Pa) for the C10 sample. Yield stress determined by method such as stress ramp is often defined as static yield stress (*i.e.* minimum stress required to initiate the flow) as compared against the apparent yield stress (*i.e.* minimum stress required to maintain the flow) obtained from a flow test.[52,57,58] Static yield

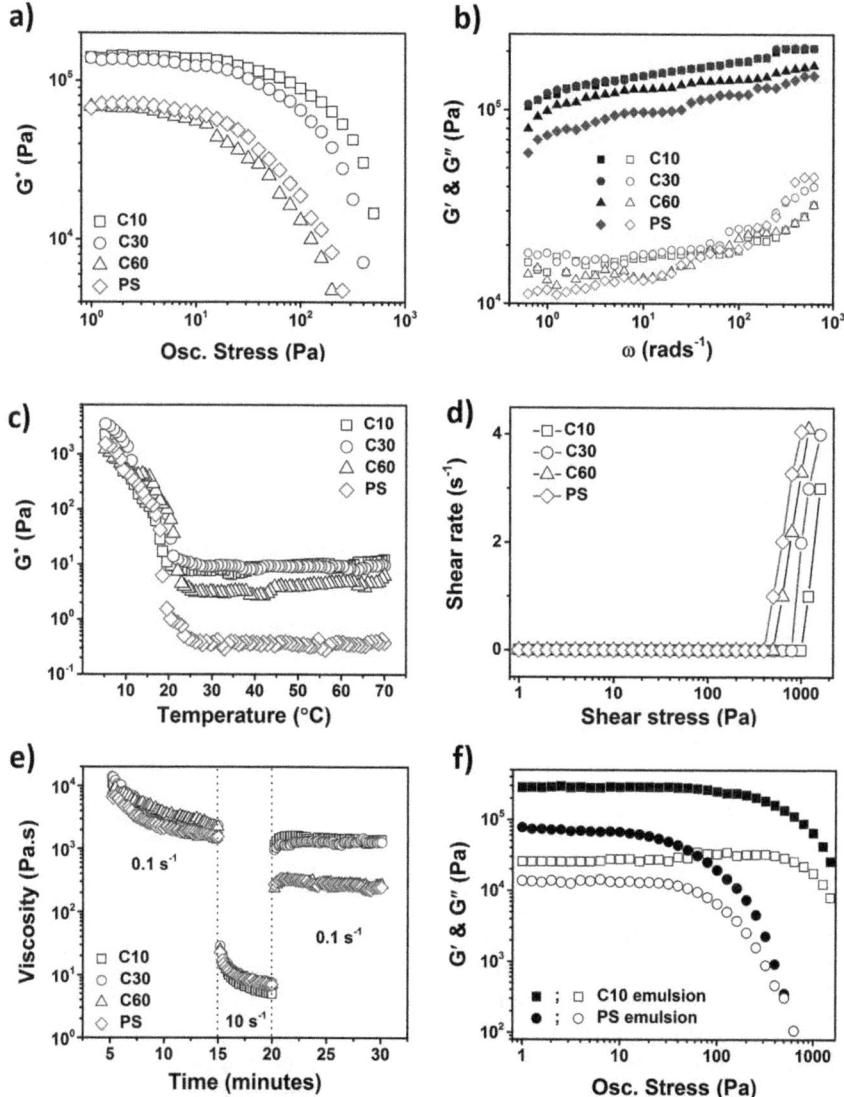

Figure 14.10 (a) and (b) Amplitude and frequency sweeps done on oleogels (70 wt% oil) prepared using unprocessed PS and the capsules. (b) Microstructure development followed during cooling of oleogels. (c), (d) and (e) Shear ramp and three interval thixotropy test done on oleogels (70 wt% oil) prepared using unprocessed PS and the capsules. (f) Data from stress sweeps performed on 10 wt% (W/O) emulsions prepared using oleogels (PS and C10) as continuous phase. Modified from *Materials Chemistry and Physics*, 195, A. R. Patel, Methylcellulose-coated microcapsules of Palm stearin as structuring templates for creating hybrid oleogels, 268–274, Copyright 2017, with permission from Elsevier.

stress is usually higher than apparent yield stress and is thus considered a more accurate measure of yielding (or permanent deformation), which in the case of fluid is represented as flow.[57,58] The relatively higher values of yield stress in capsule oleogels confirm the build-up of a robust structure at rest as compared to PS oleogel. The structure build-up was found to be dependent on the oil–polymer proportion in the emulsions used for capsule preparation, *i.e.* lower oil content = higher yield stress.

Thixotropic structure recovery is one of the most important rheological properties to consider when formulating oleogels as it can have a significant impact on the practical performance of structured oil in food products. The three interval thixotropy test is a standard test that allows tracking of material response resulting from stepwise changes in shear rate, making it the most suitable method for structure recovery studies. Results from the three interval thixotropy test (Figure 14.10e) revealed that oleogels prepared from capsules (C10 and C30) showed >90 and 70% structure recovery, respectively. In contrast, the PS oleogel performed quite poorly with a low value of 20.9%. Such high improvement in structure recovery can be attributed to the presence of two kinds of structuring units—crystalline particles and polymer sheets. The structure developed by combination of these two structuring units is responsible for reversible microstructure development on removal of mechanical force. This is in agreement with a previously reported study where enhancement in thixotropic recovery was observed when polymer and crystalline units were used together to create hybrid systems.[46]

The tolerance of these hybrid oleogels to water incorporation was also studied by fabricating 10 wt% water-in-oil emulsions using PS and C10 oleogels as oil continuous phases. Both emulsions were found to be stable without any sign of structure collapse (which is usually seen when water is incorporated in polymer oleogels). Furthermore, as expected, the structure of the C10 emulsion was comparatively better than PS as confirmed from the curves of amplitude stress sweeps shown in Figure 14.10f. Compared to the PS emulsion, the C10 sample showed higher moduli values, a broader linear response region and a higher oscillatory yield point (crossover between elastic and viscous modulus).

Promising results were also obtained when commercial crystal starter (Palsgaard® 6111) was processed in capsule form with the help of food proteins such as gelatin, caseinate and whey. It can be claimed that the results are a clear indication of the potential of this approach to enhance the structuring properties of hard stock by incorporating hydrophilic food polymers as co-structurants.

14.5 Conclusion

The preliminary results reported in this chapter suggest that innovative concepts beyond conventional oleogelation can be exploited to formulate structured oil systems. Dispersion techniques that utilize conventional ingredients (which are commonly used in formulating lipid-based food products), and are feasible for scale-up, will be of significant interest to food manufacturers that want to reformulate their existing line of products or develop new product formats.

In terms of innovation, oil continuous colloidal systems or 'oleocolloids' have unfortunately not received the same level of attention as water continuous systems, the main reason being the limited range of food products that are based on oil continuous systems. However, the recent policy changes with regards to the removal of *trans* fats from food products and the rising concerns among consumers about palm oil usage have provided new motivations for food formulators to focus on oil continuous systems.

In this chapter, preliminary data related to three different oleocolloids (O/W/O emulsions, oil foams and hybrid oleogels) were discussed with respect to the preparation methods and characterization results.

Structured O/W/O emulsions could be used as an alternative to reformulate conventional indirect emulsions such that a part of liquid oil is partitioned into the inner water droplets leaving a lower proportion of liquid oil to be structured in the external oil phase, so in principle the amount of crystalline phase could be reduced in products such as bakery margarines, cooking fats and fat spreads. In addition, owing to the possibility of high encapsulation of oil droplets in the inner water phase, these double emulsions could also be used to compartmentalize two incompatible fats (like coconut and palm oil) that show different crystallization behaviours.

Arrested oil foams work on the principle of using dispersed air bubbles with robust crystalline coats as units that structure oil dispersions of an emulsifier, which is otherwise unable to gel liquid oil on its own. By introducing a large number of air bubbles ($\varphi_{air} \gg 0.5$) in the oil dispersion at high temperature, the emulsifier can be forced to crystallize at the air–oil interfaces and thus prevent the aggregation-induced contraction of the crystalline phase. The foams show interesting viscoelastic properties and can be used to reformulate traditional lipid-based products, such as shortening, or to create a new product format, such as cooking foams instead of cooking oils.

The approach described for creating hybrid oleogels is based on the use of principles from the field of microencapsulation to facilitate incorporation of hydrophilic polymers as co-structurants together with fat hard stock in liquid oil. The work is still at a very early stage but the promising preliminary results warrant further investigation of this approach. It could be of significant interest to ingredient suppliers, who could use this strategy to modify the structuring properties of fat hard stock or other crystal promoters such that a lower amount of these components is required for providing functionalities in a range of lipid-based food products. Furthermore, the use of approved raw materials and industrial-friendly process will make this approach more appealing to the food industry.

References

1. A. G. Marangoni, S. H. J. Idziak, C. Vega, H. Batte, M. Ollivon, P. S. Jantzi and J. W. E. Rush, *Soft Matter*, 2007, **3**, 183–187.
2. A. R. Patel, Y. Rodriguez, A. Lesaffer and K. Dewettinck, *RSC Adv.*, 2014, **4**, 18136–18140.

3. A. de Vries, A. Wesseling, E. van der Linden and E. Scholten, *J. Colloid Interface Sci.*, 2017, **486**, 75–83.

4. A. R. Patel, D. Schatteman, A. Lesaffer and K. Dewettinck, *RSC Adv.*, 2013, **3**, 22900–22903.

5. A. I. Romoscanu and R. Mezzenga, *Langmuir*, 2006, **22**, 7812–7818.

6. Z.-M. Gao, X.-Q. Yang, N.-N. Wu, L.-J. Wang, J.-M. Wang, J. Guo and S.-W. Yin, *J. Agric. Food Chem.*, 2014, **62**, 2672–2678.

7. X.-W. Chen, J.-M. Wang, J. Guo, Z.-L. Wan, S.-W. Yin and X.-Q. Yang, *Food Funct.*, 2017, **8**, 823–831.

8. A. R. Patel, N. Cludts, M. D. Bin Sintang, B. Lewille, A. Lesaffer and K. Dewettinck, *ChemPhysChem*, 2014, **15**, 3435–3439.

9. A. R. Patel, N. Cludts, M. D. B. Sintang, A. Lesaffer and K. Dewettinck, *Food Funct.*, 2014, **5**, 2833–2841.

10. A. R. Patel, P. S. Rajarethinem, N. Cludts, B. Lewille, W. H. De Vos, A. Lesaffer and K. Dewettinck, *Langmuir*, 2015, **31**, 2065–2073.

11. I. Tavernier, A. R. Patel, P. Van der Meeren and K. Dewettinck, *Food Hydrocolloids*, 2017, **65**, 107–120.

12. F. Jiménez-Colmenero, *Food Res. Int.*, 2013, **52**, 64–74.

13. G. Muschiolik, *Curr. Opin. Colloid Interface Sci.*, 2007, **12**, 213–220.

14. J. Surh, G. T. Vladisavljević, S. Mun and D. J. McClements, *J. Agric. Food Chem.*, 2007, **55**, 175–184.

15. W. W. L. Altrock and J. A. Ritums, *US Pat.*, 4366180 A, 1982.

16. M. S. Miller, G. S. Buliga and R. L. Meibach, *US Pat.*, 5472728 A, 1995.

17. S. Okonogi, R. Kato, Y. Asano, H. Yuguchi, R. Kumazawa, K. Sotoyama, K. Takahashi and M. Fujimoto, *US Pat.*, 5279847 A, 1994.

18. F. Leal-Calderon, J. Bibette and V. Schmitt, in *Emulsion Science: Basic Principles*, Springer, New York, NY, 2007, pp. 173–199.

19. J. M. Morais, O. D. H. Santos and S. E. Friberg, *J. Dispersion Sci. Technol.*, 2010, **31**, 1019–1026.

20. A. R. Patel, P. Dumlu, L. Vermeir, B. Lewille, A. Lesaffer and K. Dewettinck, *Food Hydrocolloids*, 2015, **46**, 84–92.

21. P. S. Clegg, J. W. Tavacoli and P. J. Wilde, *Soft Matter*, 2016, **12**, 998–1008.

22. P. Szumała and N. Luty, *Colloids Surf., A*, 2016, **499**, 131–140.

23. D. Rousseau, *Food Res. Int.*, 2000, **33**, 3–14.

24. S. Ghosh and D. Rousseau, *Curr. Opin. Colloid Interface Sci.*, 2011, **16**, 421–431.

25. D. Rousseau, *Curr. Opin. Colloid Interface Sci.*, 2013, **18**, 283–291.

26. M. Douaire, V. di Bari, J. E. Norton, A. Sullo, P. Lillford and I. T. Norton, *Adv. Colloid Interface Sci.*, 2014, **203**, 1–10.

27. S. Ghosh, T. Tran and D. r. Rousseau, *Langmuir*, 2011, **27**, 6589–6597.

28. J. E. Norton and I. T. Norton, *Soft Matter*, 2010, **6**, 3735–3742.

29. B. P. Binks, S. Campbell, S. Mashinchi and M. P. Piatko, *Langmuir*, 2015, **31**, 2967–2978.

30. E. Dickinson, *Annu. Rev. Food Sci. Technol.*, 2015, **6**, 211–233.

31. A. J. Green, K. A. Littlejohn, P. Hooley and P. W. Cox, *Curr. Opin. Colloid Interface Sci.*, 2013, **18**, 292–301.

32. S. Roosth, *Am. Anthropol.*, 2013, **115**, 4–16.
33. P. Barham, L. H. Skibsted, W. L. P. Bredie, M. Bom Frøst, P. Møller, J. Risbo, P. Snitkjær and L. M. Mortensen, *Chem. Rev.*, 2010, **110**, 2313–2365.
34. R. N. Zúñiga and J. M. Aguilera, *Trends Food Sci. Technol.*, 2008, **19**, 176–187.
35. B. P. Binks and R. Murakami, *Nat. Mater.*, 2006, **5**, 865–869.
36. B. P. Binks, M. Kirkland and J. A. Rodrigues, *Soft Matter*, 2008, **4**, 2373–2382.
37. B. P. Binks, K. Muijlwijk, H. Koman and A. T. Poortinga, *Food Hydrocolloids*, 2017, **63**, 585–592.
38. S. Lam, K. P. Velikov and O. D. Velev, *Curr. Opin. Colloid Interface Sci.*, 2014, **19**, 490–500.
39. B. P. Binks, E. J. Garvey and J. Vieira, *Chem. Sci.*, 2016, **7**, 2621–2632.
40. A.-L. Fameau, S. Lam, A. Arnould, C. Gaillard, O. D. Velev and A. Saint-Jalmes, *Langmuir*, 2015, **31**, 13501–13510.
41. M. Brun, M. Delample, E. Harte, S. Lecomte and F. Leal-Calderon, *Food Res. Int.*, 2015, **67**, 366–375.
42. D. Z. Gunes, M. Murith, J. Godefroid, C. Pelloux, H. Deyber, O. Schafer and O. Breton, *Langmuir*, 2017, **33**, 1563–1575.
43. *EFSA J.*, 2004, **2**, 106.
44. T. A. Stortz and A. G. Marangoni, *Green Chem.*, 2014, **16**, 3064–3070.
45. A. J. Gravelle, M. Davidovich-Pinhas, S. Barbut and A. G. Marangoni, *Food Res. Int.*, 2017, **91**, 1–10.
46. A. R. Patel, in *Alternative Routes to Oil Structuring*, Springer International Publishing, 2015, ch. 3, pp. 29–39.
47. T. Chatterjee, A. I. Nakatani, R. Adden, M. Brackhagen, D. Redwine, H. Shen, Y. Li, T. Wilson and R. L. Sammler, *Biomacromolecules*, 2012, **13**, 3355–3369.
48. A. Salvador, T. Sanz and S. M. Fiszman, *Food Hydrocolloids*, 2008, **22**, 1062–1067.
49. K. I. Holownia, M. S. Chinnan, M. C. Erickson and P. Mallikarjunan, *J. Food Sci.*, 2000, **65**, 1087–1090.
50. T. Sanz, M. A. Fernández, A. Salvador, J. Muñoz and S. M. Fiszman, *Food Hydrocolloids*, 2005, **19**, 141–147.
51. J. Desbrières, M. Hirrien and S. B. Ross-Murphy, *Polymer*, 2000, **41**, 2451–2461.
52. T. Mezger, *The Rheology Handbook: For Users of Rotational and Oscillatory Rheometers*, Vincentz Network GmbH & Co. KG, Hannover, Germany, 2006.
53. A. Y. Malkin and A. I. Isayev, *Rheology: Concepts, Methods and Applications*, Chemtech Publishing, Toronto, Canada, 2006.
54. D. C. H. Cheng, *Rheol. Acta*, 1986, **25**, 542–554.
55. K. P. Menard, *Dynamic Mechanical Analysis: A Practical Introduction*, CRC Press, Florida, USA, 2008.
56. A. R. Patel, M. Babaahmadi, A. Lesaffer and K. Dewettinck, *J. Agric. Food Chem.*, 2015, **63**, 4862–4869.

57. F. J. Galindo-Rosales and F. J. Rubio-Hernandez, *Appl. Rheol.*, 2010, **20**, 22787.
58. F. J. Galindo-Rosales, F. J. Rubio-Hernández, J. F. Velázquez-Navarro and A. I. Gómez-Merino, *J. Am. Ceram. Soc.*, 2007, **90**, 1641–1643.
59. A. R. Patel, *Food Funct.*, 2017, **8**, 2115–2120.

Subject Index

aberrant crypt foci (ACF), 56
aqueous foam, edible oils
    classification of, 278–279
    destabilization, 280
    formation of, 278
    multiscale system, 279
    and non-aqueous foams,
        282–283
    production, 279–280
    stabilization, 280–281
    surface tension, 281–282
    viscoelastic properties,
        281–282
'arrested' oil foams, 316–320
attenuated transmission reflection
    infrared (ATR-IR) analysis, 141

beeswax (BSX), 42, 85
berry wax (BEX), 42, 85
biobased molecular structuring
    agents
    multifunctional molecular
        gelators, 44–47
    vegetable oil structuring
        biobased methods,
            33–44
        chemical methods, 28–33
biomimicry
    ceramides
        ceramide oleogels,
            57–61
        health aspects of, 56–57
    stratum corneum lipid
        domains, 61–65
    stratum corneum (SC), 54–55

biopolymer complexes formation
    pH and concentration,
        178–179
    properties of, 179–181
biopolymeric framework, 16
biopolymeric microcapsules, 16

candelilla wax (CW), 42, 84–85, 112,
    113, 219, 231, 232
capillary interactions, 166
caprylic acid derivatives, 47
carnauba wax (CRW), 42, 85, 232
cellulose derivatives, 43–44
ceramides, 37
    ceramide oleogels, 57–61
    health aspects of, 56–57
cryogenic scanning electron
    microscopy (cryo-SEM), 81
crystallographic mismatch
    branching (CMB), 40

degree of substitution (DS), 136
dibutyl phthalate (DBP), 139
diethyl phthalate (DEP), 139
differential scanning calorimetry
    (DSC), 78, 99
di(2-ethylhexyl)phthalate
    (DOP), 139

edible applications
    of ethylcellulose oleogels
        comminuted meats,
            264–265
        cream cheese, 263–264
        frankfurters, 264–265

edible applications (*continued*)
    heat-resistant chocolate, 270–271
    laminating shortenings, 267–268
    oil migration, 268–270
    sausages, 265–267
  of wax-based oleogels
    in bakery products, 225–232
    in chocolate and confectionery products, 233–237
    in comminuted meat products, 242–244
    in dairy products, 237–242
    margarine and spread production, 218–224
    nutraceutical, 244–246
edible oils
  aqueous foam
    classification of, 278–279
    destabilization, 280
    formation of, 278
    multiscale system, 279
    and non-aqueous foams, 282–283
    production, 279–280
    stabilization, 280–281
    surface tension, 281–282
    viscoelastic properties, 281–282
  non-aqueous foams, oleogels
    foamability, 298–299
    foam stability and solubility boundary, 299–301
    formation of, 293–294
    production of, 294–296
    properties of, 297–298
    responsive non-aqueous foams, 301–303
    rheological properties, 301
    solubility boundary of, 298–299
  non-aqueous foams, solid particles
    contact angle, 291–292
    formation and properties of, 287–289
    non-aqueous liquid surface tension, 289–291
    wettability of, 285–286
  non-aqueous foams, surfactants
    asphaltenes, 285
    fluoroalkyl-type surfactants, 284–285
    hydrocarbon-type surfactants, 283–284
    polymethylsiloxane-type surfactants, 284
    resins, 285
emulsifiers, 13
emulsions
  oil-based systems, 102
  water-in-oil and oil-in-water, 103
ethylcellulose (EC), 12, 43
ethyl-cellulose oleogels, 251–252
  characteristics, 136–138
  edible applications of
    comminuted meats, 264–265
    cream cheese, 263–264
    frankfurters, 264–265
    heat-resistant chocolate, 270–271
    laminating shortenings, 267–268
    oil migration, 268–270
    sausages, 265–267
  in food systems, 260–263
  gelation mechanism, 140–141
  gel properties
    oil type, 142–144
    oleogel fractionation, 146

polymer concentration and molecular weight, 142
surface-active molecule addition, 144–145
thermal treatment on, 145–146
health implications of
bioactive molecules, 258
β-carotene in, 259
in vitro and *in vivo* digestion, 255–258
physical properties of, 252–253
analysis, 253–254
parameters affecting, 254–255
thermo-gelation of, 138–140
evaporative light scattering detector (ELSD), 74
extra virgin oil (EVO), 161

fat blending, 30–33
fatty acids (FA), 97–98
and fatty alcohols, 98–101
fatty alcohol-rich oleogels, 101
fatty alcohols (FO), 98, 144
and fatty acids, 98–101
Fourier transform infrared spectroscopy (FTIR), 142
free fatty acids, 144
fruit wax (FRX), 42, 85, 86

gelation
mechanisms of, 6
of oil by waxes, 74–78
gelators, chemical structures of, 46
gel stiffness, 156
gel-to-sol transition temperature, 35
generally recognized as safe (GRAS), 72
gliadin, 206–207
glycerol monooleate (GMO), 144
glycerol monostearate (GMS), 144

hazelnut oil–beeswax (HBW), 224
H-bonding, 5

high internal phase emulsions (HIPEs), 151, 176–177
food applications of, 191–193
oleogels
preparation of, 186–188
properties of, 188–191
SC-ALG, 189–191
WPI–LMP and SC–LMP, 188–189
particle size distribution, 182–183
preparation and microstructure of, 181–182
properties of, 182–186
rheology, 183–186
high molecular weight gelators (HMWG), 136, 138
high-performance liquid chromatography (HPLC), 74
hydrogenated palm stearin (HPS), 245
hydrogenation, 28–29
hydrophilic lipophilic balance (HLB), 45
hydrophobicity, 162
(R)-12-hydroxyoctadecanamide (HOA), 114, 115, 116
hydroxypropyl methylcellulose (HPMC), 43, 44
12-hydroxystearic acid (12-HSA)
molecular structure and mechanism, 110–112
organogels, 116–123
polar gelator molecules, 124–129
ammonium chloride salt derivatives of, 124–129
shearing and cooling rate, 112–116
simulated molecular ordering in, 110
structures of, 41, 109
vegetable oil, 116–123

jojoba wax (JW), 42

kafirin, 205–206

lauric acid, 32
layer-by-layer (LbL) technique, 177
lecithin–sorbitan tristearate
    oleogels, 8
low-density lipoprotein (LDL), 32
low molecular weight gelators
    (LMWGs), 33–43, 107, 136, 138
low molecular weight
    organogelators (LMOGs), 321

mannitol, 46
mannitol-based gelators, 47
methylcellulose (ME), 321
microplatelet–microplatelet
    interactions, 112
minimum gelator concentration
    (MGC), 42
monoacylglycerols (MAGs), 7, 36
myristic acid, 32

natural waxes, 72–74
natural wax gelators
    candelilla wax (CLX), 84–85
    carnauba wax (CRX), 85
    rice bran wax (RBX), 82–84
    sunflower wax (SFX), 82–84
next-generation oil structuring
    agents, 44–47
non-aqueous foams, edible oils
    oleogels
        foamability, 298–299
        foam stability and
            solubility boundary,
            299–301
        formation of, 293–294
        production of, 294–296
        properties of, 297–298
        responsive non-aqueous
            foams, 301–303
        rheological properties,
            301
        solubility boundary of,
            298–299

solid particles
    contact angle, 291–292
    formation and properties
        of, 287–289
    non-aqueous liquid
        surface tension,
        289–291
    wettability of, 285–286
surfactants
    asphaltenes, 285
    fluoroalkyl-type
        surfactants, 284–285
    hydrocarbon-type
        surfactants, 283–284
    polymethylsiloxane-type
        surfactants, 284
    resins, 285
N-octadecyl chain, 121
(R)-N-octadecyl-12-hydroxyoctade-
    canamide (OHOA), 114, 115, 116

oil foams, arrested, 316–320
oil-in-water-in-oil (O/W/O), double
    emulsions, 309–316
oil oxidation, 142
oils oleogels
    oil–water interfacial tension,
        161
    viscosity, 161
oil structuring
    capillary interactions, role of,
        165–168
    oil type, effect of, 160–164
    potential applications,
        169–171
    protein oleogels
        from protein aggregates,
            156–160
        from protein hydrogels,
            153–156
    solvent exchange route, 152
    suitable protein building
        blocks, 164–165
oil–water interfacial tension, 161
oleic acid, 29

oleogelation
  from colloidal gel perspective, 5–12
  concepts, 4–5
  monocomponent gels, 5–8
  multi-component gels, 8–11
  oleogelators, 5–12
  pie charts categorizing literature, 13
  polymer gels, 11–12
  supramolecular structures, 5
oleogelators
  with crystal particles system, 39–43
  with self-assembly mechanism, 35–38
oleogel emulsion preparation process, 220
oleogels, 4. *See also* ethyl-cellulose oleogels; oils oleogels
  ceramide, 57–61
  high internal phase emulsions (HIPEs)
      preparation of, 186–188
      properties of, 188–191
γ-oryzanol, 10, 37

palmitic acid, 32
palm oil (PO), 311
partially hydrogenated oils (PHOs), 199
peroxide, 142
phytosterols, chemical structures of, 38
poly(lactic-*co*-glycolic acid) (PLGA), 136
polyethylene glycol (PEG), 136
polymer-coated crystalline microcapsules, 321–326
polymeric gelators (cellulose derivatives), 43–44
potential oil structurant, fatty acid + fatty alcohols system, 161–163
  cost, 103
  food-grade status, 103
  impact on taste, 103
  melting temperature/level of structuring, manipulation of, 103
  no adverse interactions, 103
  oil structurant, 103
protein hydrogels, 158
protein network, 159
protein–protein and protein–solvent interactions, 155
protein–protein interactions, 157

Raman spectroscopy, 142
raspberry ketone glucoside (RKG), 47
rice bran wax (RBW), 42, 82–84, 219, 232

saturated fatty acids and alcohols, 96
saturated fatty di-acids, 97
SC–alginate (ALG), 177
self-assembled fibrillar networks (SAFiNs), 35, 54, 108, 113
shellac wax (SW), 226
β-sitosterol, 10, 61
sodium caseinate (SC)–LMP, 177
sol and gel transition, 60
solid fat contents (SFC), 73, 231
solvent exchange route, 152
sorbitan monooleate (SMO), 144
sorbitan monostearate (SMS), 144
sorbitan tristearate (STS), 8–9
sorbitol, 46
sphingolipids, 56
  chemical structure of, 55
sphingomyelin, 57
sphingosine, 57
π–π stacking, 5
stearic acid, 29, 32
stratum corneum lipid domains, 61–65
stratum corneum (SC), 55
stress–strain curve, 156

structuring edible oil phases
  fatty acids (FA), 97–98, 98–101
  fatty alcohols, 98–101
  potential oil structurant,
    101–103
    food-grade status, 103
    oil structurant, 103
sugar cane wax (SCX), 85
sunflower (SW), 219
sunflower oil (SFO), 161, 311
sunflower wax (SFW), 42, 82–84, 232

tetrahydrofuran (THF), 152
2-thiobarbituric acid (TBA), 142
three-dimensional crystal
  network, 74
total polar components (TPC), 143
trehalose, 46
triacylglycerides (TAGs), 15, 96, 142
  hydrolysis of, 30

van der Waals attractive
  interactions, 5
vegetable oil structuring
  biobased methods
    low molecular weight
      gelators (LMWGs),
      33–43
    oleogelators with
      crystal particles
      system, 39–43
    oleogelators with
      self-assembly
      mechanism, 35–38

polymeric gelators
  (cellulose derivatives),
  43–44
chemical methods
  fat blending, 30–33
  fractionation, 30–33
  hydrogenation, 28–29
  interesterification,
    29–30
  structuring strategies of, 27
virgin olive oil (VOO), 223
VOO–beeswax oleogel (OBW), 224

water-in-oil and oil-in-water
  emulsions, 103
water-in-oil-in-water (W/O/W),
  double emulsions, 309
wax crystal network microstructure,
  79–82
wax crystal networks
  cooling rate on the properties
    of, 90
  oil binding capacity of, 86–88
  rheological profiling of, 88–90
  shear on properties of, 91–92
wax organogels, panel-defined
  sensory descriptive terms, 222
wheat gluten (WG), 208–210
whey protein isolate (WPI)
  hydrogels, 153, 155
whey protein isolate (WPI)–low
  methoxyl pectin (LMP), 177

zein, 200–205